Corrosion of Ceramic and Composite Materials

CORROSION TECHNOLOGY

Editor
Philip A. Schweitzer, P.E.
Consultant
York, Pennsylvania

1. *Corrosion and Corrosion Protection Handbook: Second Edition, Revised and Expanded*, edited by Philip A. Schweitzer
2. *Corrosion Resistant Coatings Technology*, Ichiro Suzuki
3. *Corrosion Resistance of Elastomers*, Philip A. Schweitzer
4. *Corrosion Resistance Tables: Metals, Nonmetals, Coatings, Mortars, Plastics, Elastomers and Linings, and Fabrics: Third Edition, Revised and Expanded (Parts A and B)*, Philip A. Schweitzer
5. *Corrosion-Resistant Piping Systems*, Philip A. Schweitzer
6. *Corrosion Resistance of Zinc and Zinc Alloys*, Frank C. Porter
7. *Corrosion of Ceramics*, Ronald A. McCauley
8. *Corrosion Mechanisms in Theory and Practice*, edited by P. Marcus and J. Oudar
9. *Corrosion Resistance of Stainless Steels*, C. P. Dillon
10. *Corrosion Resistance Tables: Metals, Nonmetals, Coatings, Mortars, Plastics, Elastomers and Linings, and Fabrics: Fourth Edition, Revised and Expanded (Parts A, B, and C)*, Philip A. Schweitzer
11. *Corrosion Engineering Handbook*, edited by Philip A. Schweitzer
12. *Atmospheric Degradation and Corrosion Control*, Philip A. Schweitzer
13. *Mechanical and Corrosion-Resistant Properties of Plastics and Elastomers*, Philip A. Schweitzer
14. *Environmental Degradation of Metals*, U. K. Chatterjee, S. K. Bose, and S. K. Roy
15. *Environmental Effects on Engineered Materials*, edited by Russell H. Jones
16. *Corrosion-Resistant Linings and Coatings*, Philip A. Schweitzer
17. *Corrosion Mechanisms in Theory and Practice: Second Edition, Revised and Expanded*, edited by Philippe Marcus
18. *Electrochemical Techniques in Corrosion Science and Engineering*, Robert G. Kelly, John R. Scully, David W. Shoesmith, and Rudolph G. Buchheit
19. *Metallic Materials: Physical, Mechanical, and Corrosion Properties*, Philip A. Schweitzer
20. *Encyclopedia of Corrosion Technology: Second Edition, Revised and Expanded*, Philip A. Schweitzer

21. *Corrosion Resistance Tables: Metals, Nonmetals, Coatings, Mortars, Plastics, Elastomers and Linings, and Fabrics: Fifth Edition, Revised and Expanded (Parts A, B, C, and D)*, Philip A. Schweitzer
22. *Corrosion of Ceramic and Composite Materials: Second Edition*, Ronald A. McCauley

ADDITIONAL VOLUMES IN PREPARATION

Corrosion of Ceramic and Composite Materials

Second Edition

Ronald A. McCauley
Rutgers University
Piscataway, New Jersey, U.S.A.

MARCEL DEKKER, INC. NEW YORK • BASEL

The first edition was *Corrosion of Ceramics* (Dekker, 1994).

Library of Congress Cataloging-in-Publication Data
A catalog record for this book is available from the Library of Congress.

ISBN: 0-8247-5366-6

This book is printed on acid-free paper.

Headquarters
Marcel Dekker, Inc., 270 Madison Avenue, New York, NY 10016, U.S.A.
tel: 212-696-9000; fax: 212-685-4540

Distribution and Customer Service
Marcel Dekker, Inc., Cimarron Road, Monticello, New York 12701, U.S.A.
tel: 800-228-1160; fax: 845-796-1772

Eastern Hemisphere Distribution
Marcel Dekker AG, Hutgasse 4, Postfach 812, CH-4001 Basel, Switzerland
tel: 41-61-260-6300; fax: 41-61-260-6333

World Wide Web
http://www.dekker.com

The publisher offers discounts on this book when ordered in bulk quantities. For more information, write to Special Sales/Professional Marketing at the headquarters address above.

Current printing (last digit):

10 9 8 7 6 5 4 3 2 1

PRINTED IN THE UNITED STATES OF AMERICA

To
my father

Harry Sylvester McCauley
1909–1966

Preface to the Second Edition

Although a better understanding of the mechanisms of corrosion has occurred over the past ten years since the publication of the first edition of this book, corrosion still remains a major problem. One area in which some advances have been made is the understanding of the weathering mechanisms of building materials, especially where related to monuments of the past. The weathering of building materials has been added as new sections in Chapters 2 (Fundamentals) and 5 (Corrosion of Specific Crystalline Materials). Although the work on weathering of building materials has been done predominantly by civil engineers, it should be of interest to the ceramic engineer. More information has become available concerning the corrosion of composite materials of all types. The strong interest in composites materials has been enhanced by the aerospace industry and the military. Because of this, a whole new chapter (Chapter 7) has been devoted to composite materials. In addition, new sections have been added on bioceramics (Chapter 5). New literature sources have been added to all chapters where appropriate.

The first edition of this book was written to be used primarily as a reference book. Questions have been added at the end of the most chapters and additional examples have been included along with recommended reading lists so that this second edition may also be used as a textbook for either a senior level undergraduate or a graduate course on corrosion.

Ronald A. McCauley

Preface to the First Edition

One of the most important problems confronting engineers today is the development of materials that are reliable under various environmental conditions. In some cases these conditions are considered extremely hostile—very high temperatures, mechanical loading, and/or aggressive chemical attack. Ambient temperature aqueous attack can also be extremely detrimental, especially over an extended period of time, as in the case of hazardous waste disposal. Engineers and scientists have been combating the attack upon ceramics of molten glass, molten metals and slags, and molten salts for hundreds of years with many improvements. Most of these improvements have occurred through experimentation, eventually finding the material that worked best. Only during the past 25 years has a true understanding of the complexities of corrosion of ceramics begun to develop. Major advances have been made in recent years; however, the details in many cases are still questionable or at least debatable.

The cost to industry due to corrosion is considerable and only a thorough understanding of all the complexities of the process will help to minimize that cost. There will undoubtedly be many applications of ceramics where the ceramic will be consumed during service, but maximizing service life will greatly reduce the overall cost.

While several books, mostly in the form of symposia proceedings, have been published on various aspects of corrosion of crystalline and glassy ceramics, generally on the newer, advanced materials, none has addressed the subject in a comprehensive manner. The most significant works have been reported in the technical literature; however, reading all the published articles is a formidable task. This book is an attempt to discuss all aspects of the corrosion of ceramics, but no attempt has been made to complete an exhaustive literature review. Although not all areas have been described in great detail, a summary of some of the most important work has been given with references for the interested reader.

This book is based upon a combination of lecture notes from the Advanced Refractories course that the author has taught at Rutgers during the past 15 years and the author's industrial and consulting experiences. It is intended predominantly as a reference work for practicing engineers and research scientists but could also be used as a text for a graduate-level course in corrosion of ceramics. Any comments or suggestions about the content of this book will be most welcome.

ACKNOWLEDGEMENTS

The author would like to thank the faculty and students of the Department of Ceramic and Materials Engineering at Rutgers, The State University of New Jersey, for many helpful and thoughtful discussions during the preparation of this book and especially Drs. John Wachtman and M. John Matthewson for reviewing a portion of the manuscript and for their valuable suggestions.

The author would like to extend a very special thank you to Mr. William Englert of PPG Industries, who first introduced the author to the fascinating field of corrosion of ceramics.

Gratitude must also be extended to Mrs. Mary Guerin for her help in preparation of the manuscript and to Paul Mort, Robert Sabia, John Martin, and Ryan McCuiston for their help in preparing the figures.

The author would like to extend a very special thank you to his wife, Eleanora, and his son, Matthew, for their understanding during the many long hours required to complete this task.

Ronald A. McCauley

Contents

Preface to the Second Edition v
Preface to the First Edition vii

1 INTRODUCTION 1

2 FUNDAMENTALS 9

2.1 Introduction 9
2.2 Corrosion by Liquids 11
2.2.1 Introduction 11
2.2.2 Crystalline Materials 15
2.2.3 Glasses 47
2.3 Corrosion by Gas 56
2.3.1 Crystalline Materials 56
2.3.2 Vacuum 63
2.3.3 Glasses 64

2.4 Corrosion by Solid 65
2.5 Surface Effects 67
2.5.1 Surface Charge 67
2.5.2 Porosity and Surface Area 68
2.5.3 Surface Energy 71
2.6 Acid/Base Effects 77
2.7 Thermodynamics 78
2.7.1 Mathematical Representation 79
2.7.2 Graphical Representation 85
2.8 Kinetics 91
2.9 Diffusion 100
2.10 Summary of Important Concepts 105
2.11 Additional Related Reading 107
2.12 Exercises, Questions, and Problems 108
References 109

3 METHODS OF CORROSION ANALYSIS 123

3.1 Introduction 123
3.2 Laboratory Test vs. Field Trials 124
3.3 Sample Selection and Preparation 126
3.4 Selection of Test Conditions 129
3.5 Characterization Methods 130
3.5.1 Microstructure and Phase Analysis 130
3.5.2 Chemical Analysis 134
3.5.3 Physical Property Measurement 136
3.6 Data Reduction 139
3.7 Additional Related Reading 139
3.8 Exercises, Questions, and Problems 140
References 140

4 CORROSION TEST PROCEDURES 143

4.1 Introduction 143
4.2 ASTM Standards 149
4.3 Nonstandard Tests 162
4.4 Additional Related Reading 163
4.5 Exercises, Questions, and Problems 164
References 164

5 CORROSION OF SPECIFIC CRYSTALLINE MATERIALS 165

5.1 Attack by Liquids 165
5.1.1 Attack by Glasses 165
5.1.2 Attack by Aqueous Solutions 177
5.1.3 Attack by Molten Salts 191
5.1.4 Attack by Molten Metals 197
5.2 Attack by Gases 201
5.2.1 Oxides 202
5.2.2 Nitrides and Carbides 209
5.2.3 Borides 231
5.2.4 Silicides 233
5.2.5 Superconductors 234
5.3 Attack by Solids 234
5.3.1 Silica 234
5.3.2 Magnesia 235
5.3.3 Superconductors 235
5.3.4 Attack by Metals 236
5.4 Additional Related Reading 237
5.5 Exercises, Questions, and Problems 238
References 238

6 CORROSION OF SPECIFIC GLASSY MATERIALS 257

6.1 Introduction 257
6.2 Silicate Glasses 258
6.3 Borosilicate Glasses 271
6.4 Lead-Containing Glasses 275
6.5 Phosphorus-Containing Glasses 276
6.6 Fluoride Glasses 278
6.7 Chalcogenide–Halide Glasses 280
6.8 Additional Related Reading 281
6.9 Exercises, Questions, and Problems 281
References 282

7 CORROSION OF COMPOSITE MATERIALS 291

7.1 Introduction 291
7.2 Reinforcement 295

7.2.1 Fibers 295
7.2.2 Fiber Coatings or Interphases 298
7.2.3 Particulates 299
7.3 Ceramic Matrix Composites 300
7.3.1 Oxide–Matrix Composites 301
7.3.2 Nonoxide–Matrix Composites 309
7.4 Metal Matrix Composites 316
7.5 Polymer Matrix Composites 318
7.6 Additional Related Reading 320
7.7 Exercises, Questions, and Problems 321
References 322

8 PROPERTIES AND CORROSION 333

8.1 Introduction 333
8.2 Mechanisms 337
8.2.1 Crystalline Materials 337
8.2.2 Glassy Materials 339
8.3 Degradation of Specific Materials 345
8.3.1 Degradation by Oxidation 345
8.3.2 Degradation by Moisture 352
8.3.3 Degradation by Other Atmospheres 352
8.3.4 Degradation by Molten Salts 354
8.3.5 Degradation by Molten Metals 358
8.3.6 Degradation by Aqueous Solutions 358
8.4 Additional Related Reading 364
8.5 Exercises, Questions, and Problems 364
References 365

9 METHODS TO MINIMIZE CORROSION 373

9.1 Introduction 373
9.2 Crystalline Materials—Oxides 374
9.2.1 Property Optimization 374
9.2.2 External Methods of Improvement 379
9.3 Crystalline Materials—Nonoxides 382
9.3.1 Property Improvement 382
9.3.2 External Methods of Improvement 384

9.4 Glassy Materials 385
9.4.1 Property Optimization 385
9.4.2 External Methods of Improvement 385
References 387

Glossary *389*

Epilog *393*

Index *397*

1

Introduction

> If we begin with certainties, we shall end in doubt; but if we begin with doubt, and are patient in them, we shall end in certainties.
>
> BACON

Most engineers at one time or another will be confronted with corrosion whether it will be their sole endeavor or whether it will be a minor unexpected nuisance. The actual study of corrosion, its causes, effects, and means of elimination is not as common in the field of ceramics as it is in the field of metallurgy. Although many engineers study the corrosion of ceramics all their lives, they normally do not consider themselves as corrosion engineers, but as ceramic engineers or process engineers or possibly some other type of engineer. There are no corrosion engineering courses offered in the several undergraduate ceramic engineering curricula in the United States. Even at the graduate level, there are no courses dedicated to corrosion, although several contain a large amount of information related to corrosion, such as those related to high-temperature materials or thermodynamics, etc. There is definitely no such thing as a bachelor's degree in Corrosion Engineering of Ceramic Materials. The American Ceramic Society does not have a specific division devoted to corrosion like some other major societies. For example, The Electrochemical Society has a Corrosion Division. They also publish a Corrosion Monograph Series.

Throughout the history of the ceramic industries, various material types or compositions have been used because of some particular advantageous, intrinsic property. High strength, low electrical conductivity, or some other property may be the primary concern for a particular application. However, excellent resistance to attack by the environment always plays a role and may, in some cases, be the prime reason for the selection of a particular material. This is especially true for those materials selected for furnace construction in the metal and glass industries.

Almost all environments are corrosive to some extent. For practical applications, it comes down to a matter of kinetics—how long will a material last in a particular environment? In some cases, corrosion may be beneficial, such as in the preparation of samples by etching for microscopic evaluation, in chemical polishing to obtain a flat, smooth surface, or in bioactive materials where reaction with bone or complete disappearance is required. The selective leaching of the sodium- and boron-rich phase in phase-separated borosilicate glass to produce a high silica content glass (called Vycor™*) is an excellent example of how corrosion can be put to a beneficial use. Other examples include dissolution and reprecipitation in

*Corning, Inc., Corning, New York.

liquid phase sintering (also crystal growth studies) and the dissolution of various raw materials in molten glass in the manufacture of glass products.

The proper selection of materials and good design practices can greatly reduce the cost caused by corrosion. To make the proper selection, engineers must be knowledgeable in the fields of thermodynamics, physical chemistry, electrochemistry, and even meteorology. In addition, engineers must be familiar with the corrosion testing of materials, the nature of corrosive environments, the manufacture and availability of materials, and have a good sense of the economics of the whole process. There is a growing need in many ceramic applications to be able to predict the service life based upon laboratory tests. The limiting factors in making such predictions are more often than not due to a lack of a thorough knowledge of the industrial operating conditions rather than to devising the proper laboratory test. A thorough knowledge of the microstructure and phase assemblage of the material, however, is critical to an understanding of the corrosion that may take place. The National Science Foundation has provided funding to several universities* to develop a Digital Library of Ceramic Microstructures (DLCM), which, when completed, will be a tremendous aid to the engineer involved in corrosion studies. This library will contain, in addition to selected micrographs, chemistry, phases, and some properties. Access to the database will be via the Internet. Although a material may be listed in some handbook as having excellent resistance to some particular environment, it is important to know the form of the material. Were the data listed for a single-crystal, a powder, or a dense (or porous) sintered component? Were there any

*University of Dayton Research Institute has the primary role. Georgia Institute of Technology, North Carolina A&T State University, University of Missouri-Rolla, along with Mechanical Test Instrumentation & Control are supplying the information.

secondary phases present or was it a pure material? The form and processing of a material will affect its corrosion.

The cost of corrosion to ceramics in the United States is enormous; however, in many cases, this corrosion is looked upon as a necessary expense in the production of some product. For example, the corrosion of the refractories in a glass melting furnace is an expected phenomenon. These furnaces are shut down periodically to replace the worn-out materials. Not only is the cost of the refractories involved, but also the cost of the tear out, the cost of the reconstruction, and the cost of any lost production is involved. The total cost of such a repair can amount to as much as \$10 million or more for a single furnace. Fig. 1.1 shows an average estimate of the percentages for labor and materials for a typical furnace repair. Business interruption costs have not been included since these will vary considerably depending upon the product being produced and the size of the furnace. The downtime for repairs also varies considerably depending upon the extent of the repair, but varies between 1 and 3 months. The cost due to business interruption can amount to as much as \$1 million per month. There are, however, a few things that add a credit to the costs of a repair,

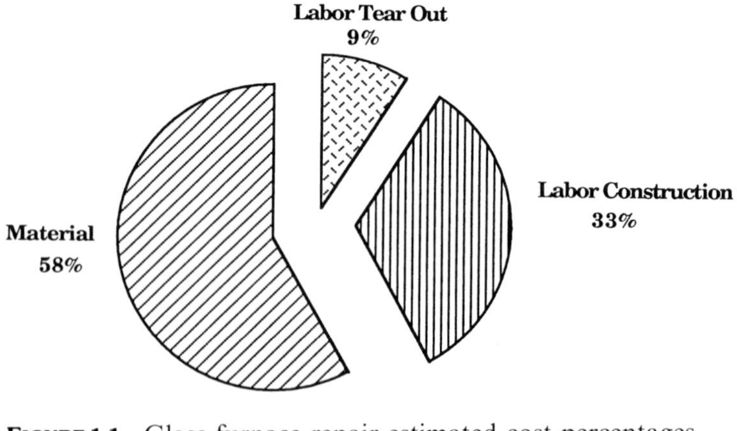

FIGURE 1.1 Glass furnace repair estimated cost percentages.

such as the fuel and raw materials saved during the shutdown. Thus it should be obvious that the equation to determine the exact costs of a particular repair is quite complex. The total cost of such repairs can never be eliminated; however, it can be greatly reduced by the proper selection of refractories and the proper operation of the furnace. There are times when the corrosion of the refractories goes unnoticed and failure occurs prematurely. Since furnaces are insured against premature failure, very large insurance claims have been filed. In addition to the costs related to the refractories, construction, and lost production are the costs related to additional cleanup due to the failure and the costs of insurance adjusters and lawyers. The cost of such a failure can exceed $20 million.

Environmental problems can also add to the total cost of corrosion. For example, a trend toward the use of nonchrome-containing refractories for furnace construction has been ongoing for about the past 20 years. Used refractories have, in the past, been disposed of by burying them in landfills. Chrome-containing refractories have the potential of contaminating groundwaters with hexavalent chrome, a carcinogen. If chrome-containing refractories are used, upon disposal, they must be hauled to toxic waste dumps with an added cost of disposal. To eliminate this problem, some industries have been leaning toward the use of other materials for the construction of their furnaces. In some cases, the replacement material does not last as long as the chrome-containing material, thus shortening the time between repairs and adding to the cost.

The products of corrosion may enter the product being manufactured and lower the quality of the product or decrease the yields. Although this is a cost due to corrosion, it is one that is extremely difficult to quantify. Although no accurate numbers are available for the annual cost to industry for the corrosion of ceramic materials, an estimate of $2 billion does not seem unreasonable. Only through the intelligent selection of ceramic materials can the cost of corrosion be minimized. This intelligent selection of materials can be obtained only through a thorough understanding of all the complexities of ceramic materials and the effect that the environment has upon them.

The corrosion of ceramic materials has not been as well categorized as it has been for metals. Similar terms do, however, appear in the literature. The more common types referred to in the literature are *diffusion corrosion,* which is very similar to concentration cell corrosion in metals, *galvanic cell corrosion*, *grain boundary corrosion*, and *stress corrosion*. A more common trend in ceramic materials is to group corrosion under a more general mechanism, such as *dissolution corrosion* (i.e., corrosion of a solid by a liquid). In this type of corrosion, diffusion, galvanic cell, grain boundary, and stress corrosion may all be present.

There are also many words used to describe corrosion, and if one is looking for information on the subject of corrosion of a particular material, a search including many of the following words should be performed: *dissolution*, *oxidation*, *reduction*, *degradation*, *deterioration*, *instability*, *decomposition*, *consumption*, and *erosion*. Although erosion is technically not the same as corrosion, being a physical effect rather than a chemical one, erosion, in many cases, provides a means for continued corrosion.

Corrosion of ceramic materials and its relationship to various property degradation does not receive as wide a recognition as it probably should at the various technical and professional society meetings. For example, only about 2.5% of the approximately 1500 presentations at the 2002 Annual American Ceramic Society meeting were devoted to topics related to corrosion. As can be seen from Table 1.1, the percentage of presentations on corrosion has been between 4.5% and 2.5% during the decade of the 1990s. The data are truly not sufficient to conclude that a trend exists. Although the main topic of interest changes somewhat from year-to-year, hazardous waste materials appear to remain a major area of interest. Selecting a major topic is complicated by the fact that areas often overlap, for example, general glass durability vs. hazardous waste glass leaching. At meetings of a less general

TABLE 1.1 Presentations on Corrosion at American Ceramic Society Annual Meetings

Year	Total papers[a]	Percentage on corrosion (%)[b]	Main topic[c]
1990	1425	4.5	HWM[d]
1992	1450	3	Refractories/composites
1994	1545	3.5	Composites/basic science
1996	1578	2.5	Refractories
1998	1591	2.5	Refractories/HWM
2000	1455	3.75	Refractories/HWM
2002	1514	2.5	Nonoxide oxidation/HWM

[a] Includes posters.
[b] Only presentations where corrosion was main topic.
[c] Does not include hydration of cements and fabrication reactions.
[d] HWM = hazardous waste materials.

nature, such as the Unified International Technical Conference on Refractories (UNITECR) in 1989, about 12.5% of the approximately 150 papers were on topics related to corrosion of refractories. This is a much better situation than the general field of ceramics, but historically, the corrosion of refractories has received more attention than the corrosion of ceramics. At the First Ceramic Science Technology Congress in 1989, an international symposium was held entitled, *Corrosion and Corrosive Degradation of Ceramics*. Of the 26 papers presented in the symposium, more than 1/2 were devoted to silicon carbide and silicon nitride, indicative of the importance that is placed upon the corrosion resistance of these advanced ceramic materials. This symposium, however, was only a small portion of the parent congress where more than 625 papers were presented. At the 2003 Annual Conference Exposition on Advanced Ceramics and Composites, about 3.75% of the almost 500 presentations were on corrosion, a situation not much different than the Annual Ceramic Society meetings.

1.1 EXERCISES, QUESTIONS, AND PROBLEMS

1. List the different fields that an engineer might want to study to be proficient in evaluating corrosion problems.
2. List all the various courses taught in a standard ceramic engineering curriculum that might contain information related to corrosion.
3. Discuss whether it is more important to know whether a material will corrode or how long it will last.
4. Discuss several applications where the corrosion is a beneficial factor rather than a problem. Find several applications not listed in the text.
5. Develop an equation that represents all the various factors involved in determining the total cost of a particular corrosion problem. Pick a particular product and try to find as much information as possible concerning the actual costs.
6. Discuss the relationship of erosion to corrosion and how it can be a major actor in the overall process of corrosion.

2

Fundamentals

Everything should be made as simple as possible, but not simpler.

ALBERT EINSTEIN

2.1 INTRODUCTION

Corrosion of ceramics can take place by any one or a combination of mechanisms. Various models have been proposed to describe these mechanisms, several of which will be discussed below. In general, the environment will attack a ceramic, forming a reaction product. The reaction product may be either solid, liquid, gas, or any combination of these. This reaction product may remain attached to the ceramic or it may be fugitive, in the case when gaseous species make up the reaction product, or it may be a combination of both. When the reaction product remains as a solid, quite often it forms a protective layer toward further corrosion. At other times, for example, if the reaction product is a combination of solid and liquid, this reaction layer may be removed through the process of erosion. Thus to analyze corrosion, one must have some idea of the type of processes that are in operation. When the reaction product remains as an intact interfacial layer, analysis is relatively easy. When gaseous species are formed, the consumption of the ceramic manifests itself as a weight loss. An understanding of the mechanism, however, requires analysis of the evolved gases. Many times, the interface formed is very porous and/or friable requiring special care in preparing samples for analysis. Because of the various processes that may take place during corrosion, there is no one general model that can explain all cases of corrosion. In addition, a single ceramic material will react differently to different environments and thus there is no single explanation for the corrosion of a particular material for all environments. It is also true that the manufacturing history of a ceramic material will affect its performance. This may manifest itself, for example, as a low corrosion-resistant grain boundary phase or a pore size distribution that greatly increases the exposed surface area to corrosion. Thus it should be obvious that a simple all-encompassing general theory of corrosion of ceramics does not exist and, because of the nature of corrosion and ceramics, will most likely never exist. There does, however, appear to be a common thread connecting all the various studies that have been reported. That is, *corrosion is dependent upon the structural characteristics of the material. The more compact or tightly bonded materials corrode less whether they are glasses or crystalline materials*. Thus it appears that if a general theory is to be developed, a comprehensive investigation of single crystals and some structurally well-characterized glasses should be investigated.

Corrosion, being an interfacial process, requires a thorough understanding of the surface structure of the material being corroded. Thus the study of single crystals is the best method to determine the fundamentals of corrosion mechanisms. It is not always possible, however, to obtain single crystals of sufficient size for appropriate measurements. Although the crystal surface characteristics determine short-term corrosion behavior, they may not be as important for long-term corrosion. Single crystals do lend themselves to evaluation of the effects that various dopants and defects (e.g., dislocations) have upon dissolution kinetics.

When attempting to understand the corrosion of a ceramic, it is a good idea to remember some of the fundamental concepts of chemistry that are too often forgotten. The following are just a few concepts that go a long way in helping one to understand corrosion:

1. A ceramic with acidic character tends to be attacked by an environment with a basic character and vice versa.
2. The vapor pressure of covalent materials is generally greater than that of ionic materials and therefore tend to vaporize or sublime more quickly.
3. Ionic materials tend to be soluble in polar solvents*(e.g., salt in water) and covalent materials tend to be soluble in nonpolar solvents (e.g., SiC in hexane).
4. The solubility of solids in liquids generally increases with increasing temperature.

2.2 CORROSION BY LIQUIDS

2.2.1 Introduction

The solubility of materials in liquids can be obtained from phase diagrams, which give the saturation composition at a given temperature. Unfortunately, for many practical systems, phase diagrams are either very complex or nonexistent. Many data are available, however, for two- and three-component

---

*Polar solvents have a high dielectric constant, whereas nonpolar solvents have a low dielectric constant. At 20°, the dielectric constant of water is 80 and that of hexane is 1.874.

systems, and these should be consulted before attempting to evaluate the corrosion of a specific material* [2.1]. The corrosion of a single pure compound by a liquid can be evaluated by use of the Gibbs Phase Rule. For example, the system of a binary oxide $A_xB_yO_z$ corroded by a liquid M_aO_b contains three components, where a solid and liquid are in equilibrium at some fixed temperature and pressure. This system has only 1 degree of freedom. Thus if the concentration of one dissolved component is changed, the concentrations of the others must also change. A good discussion of the use of phase diagrams in dissolution studies is that by Cooper [2.2].

The corrosion of a solid crystalline material by a liquid can occur through the formation of an interface or reaction product formed between the solid crystalline material and the solvent. This reaction product, being less soluble than the bulk solid, may or may not form an attached surface layer. This type of mechanism has been called *indirect dissolution*, *incongruent dissolution*, or *heterogeneous dissolution* by various investigators. There are many examples of this reported in the literature. In another form, the solid crystalline material dissolves directly into the liquid either by dissociation or by reaction with the solvent. This type of mechanism is called *direct dissolution*, *congruent dissolution*, or *homogeneous dissolution*. The term *selective dissolution* is also found in the literature, but is used to imply that only a portion of the species in the solid are dissolved whether or not an interface is formed. The saturation solution concentrations of the crystalline species in the liquid along with the diffusion coefficients of the species involved all determine whether one mechanism will exist or the other. The most abundant species and their concentrations in the liquid must be known for one to determine the degree of saturation. This, in turn, will determine whether or not the solid will dissolve. The corrosion rate-limiting step in the indirect type may be the

---

*A software package called *Thermo-Calc* is available for calculating phase diagrams from the Dept. of Materials Science & Engineering, KTH, S-100 44 Stockholm, Sweden.

chemical reaction that forms the interfacial layer, diffusion through this interfacial layer, or diffusion through the solvent. When one is involved in the study of single crystals, dissolution of the various crystallographic planes may be different. If the dissolution rate is the same for all planes, it is called *isotropic dissolution*. If the dissolution rate varies among the various planes, it is called *anisotropic dissolution*. It is easy to understand why dissolution may be different for different planes due to the differences in density of atoms, charges, and/or bonding; however, isotropic dissolution is not as easy to understand, even for a single cation–anion combination. In the case of a cubic crystal such as NaCl where the crystal faces exposed to dissolution are all of the same family, then isotropic dissolution is easy to understand.

Examination of the appropriate phase diagram will aid one in determining whether or not a particular combination of liquid and solid will form an interface. An example is given in Fig. 2.1 that shows a portion of the $CaO–Al_2O_3–SiO_2$ diagram. If the 1265°C eutectic composition were selected as the liquid and melted in a crucible of Al_2O_3 at 1500°C, the dissolution would be indirect and an interface of $CaAl_{12}O_{19}$ would form. As the melt temperature increases, the dissolution type changes to direct above 1700°C, with the eutectic melt being in equilibrium with the alumina. A slightly different situation exists if mullite were selected as the crucible material. At 1400°C, the dissolution is indirect with the interface being anorthite. As the temperature increases, the dissolution remains indirect but the interface changes to alumina above 1500°C. As can be seen from these examples, the interface is determined by the intersection of the isotherm of interest and the construction line joining the liquid and solid compositions.

The wetting characteristics between a ceramic and a liquid are very important in determining the degree of contact that is developed. Although the investigation of Kramer and Osborne [2.3] was performed in an attempt to understand better the parameters involved in the processing of glass-ceramic-to-metal seals, their data exhibited the importance of atmospheric effects upon the degree of contact between a ceramic and a

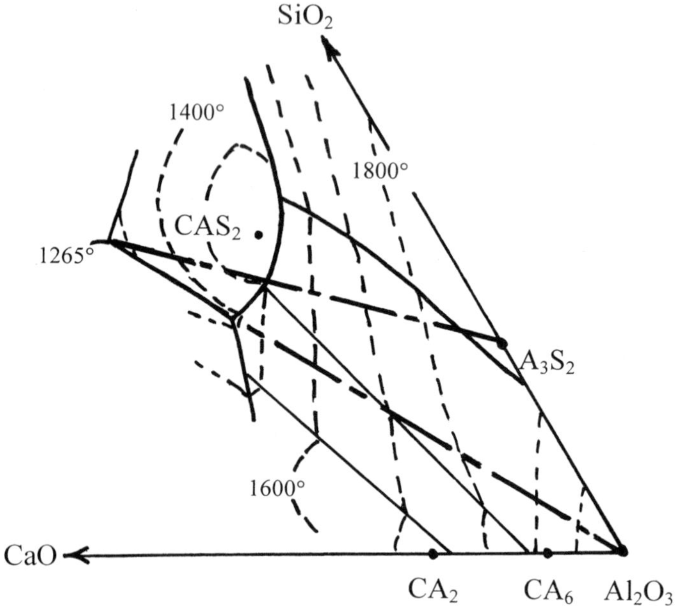

FIGURE 2.1 A portion of the $CaO–Al_2O_3–SiO_2$ phase diagram [2.1]. The heavy lines between 1265°C eutectic and either alumina or mullite are construction lines. The lighter dashed lines are isotherms. The solid lines are liquidus field boundary curves.

liquid. The better the contact, the more aggressive the attack can be. Kramer and Osborne studied the atmospheric effects upon the wetting of a glass-ceramic by Ni-based alloys. They found that helium lowered the contact angle more than argon or an argon/hydrogen mixture. They also found that contact angles were inversely dependent upon the atmospheric dew point.

Noyes and Whitney [2.4], in their classic work of the dissolution of lead chloride in boric acid and water, speculated that the rate of corrosion of a solute by a solvent was controlled by the diffusion rate of atoms away from the solute surface. Nernst [2.5] postulated that a thin layer of solvent adjacent to the solute became rapidly saturated and remained saturated during the dissolution process and that beyond a certain distance, the concentration was that of the bulk solution. The following equation, now called the Noyes–Nernst equation, represents the flux density across the solute interface:

$$VdC_{\infty}/dt = jA = (D/\delta^*)(C_{sat} - C_{\infty})A \tag{2.1}$$

where:

V = volume of solution
C_{∞} = concentration in the bulk
C_{sat} = saturation concentration
A = area of interface
D = diffusion coefficient
δ^* = boundary layer thickness
t = time

By including the surface reaction rate constant, *K*, Berthoud [2.6] derived the following equation:

$$j = \frac{K}{\{1 + (K\delta^*/D)\}}(C_{sat} - C_{\infty}) \tag{2.2}$$

which indicated that the driving force for dissolution was a combination of both the interface chemical reaction and the interdiffusion of the products and reactants. The derivation from first principles of the empirical constant, δ^*, came after the development of boundary layer theory by Prandtl [2.7] and Levich [2.8]. The most important consequence of these theories for the experimentalist was that the effective boundary layer thickness, δ^*, of a rotating disk was independent of its radius and proportional to the square root of the angular velocity [2.9].

2.2.2 Crystalline Materials

Attack by Molten Glasses

The use of a single diffusion coefficient, as was done in Eqs. (2.1) and (2.2), even in multicomponent systems was verified by Cooper and Kingery [2.10]. They, along with Samaddar et al. [2.11] and Oishi et al. [2.12], described in detail the theory of corrosion by liquids in ceramic systems (i.e., alumina, mullite, fused silica, and anorthite in Al–Ca–silicate liquid). Diffusion through the boundary layer was determined to be the rate-limiting step during dissolution. The composition of the boundary layer may vary depending upon whether diffusion is more or less rapid than the boundary reaction. The basic equation describing the rate of solution under free convection with density being the driving force is:

$$j = \frac{-dR}{dt} = 0.505\left(\frac{g\Delta\rho}{v_i x}\right)^{1/4} D_i^{3/4} C^* \exp\left(\frac{\delta^*}{R + \delta^*/4}\right) \tag{2.3}$$

where:

g = acceleration due to gravity
$\Delta\rho$ = $(\rho_i - \rho_\infty)/\rho_\infty$ (ρ_i = saturated liquid density and ρ_∞ = original)
v = kinematic viscosity
x = distance from surface of liquid
D_i = interface diffusion coefficient
C^* = a concentration parameter
δ^* = effective boundary layer thickness
R = solute radius

The exponential term was introduced as a correction for cylindrical surfaces. Since experimental tests often involve cylindrical specimens, these equations have been developed for that geometry. In practical applications, the condition relating to the corrosion of slabs is most predominant. However, if the sample diameter is large compared to the boundary layer thickness, the two geometries give almost identical results.

After a short induction period (which is of no consequence in practical applications) in which molecular diffusion predominates, the rate of corrosion becomes nearly independent of time. As a surface corrodes, the interface, if denser than the corroding medium, will be eroded away due to free convection caused by density variation. Use of this equation implies that one has at his disposal data relating to the variation of density and viscosity with temperature. In cases where these data are not available, the investigator will need to determine them prior to any calculation of corrosion rates.

Hrma [2.13] has used the work of Cooper and Kingery to discuss further the rates of corrosion of refractories in contact with glass. The following equation given by Hrma describes the corrosion under the condition of free convection due to density difference:

$$j_c = k\Delta c\left(\frac{D^3\Delta\rho g}{vL}\right)^{1/4} \tag{2.4}$$

where:

j_c = rate of corrosion
c = solubility of material in liquid
D = coefficient of binary diffusion
g = acceleration due to gravity
v = kinematic viscosity
L = distance from surface of liquid
ρ = relative variation of density
k = constant = 0.482

This is essentially the same equation as that of Cooper and Kingery, without the exponential term.

Many corrosive environments associated with ceramic materials involve diffusion into the corroding medium, and thus increased velocity of the medium increases corrosion. Thus, if transport in the liquid were important, the corrosion rate must be evaluated under forced convection conditions. In such cases, the rate depended upon the velocity of forced convection:

$$j = 0.61D^{*2/3}v^{*-1/6}\omega^{1/2}C^{*} \tag{2.5}$$

The terms D^* and v^* were introduced since diffusivity and viscosity may be composition-dependent. The important point of this equation was that the rate of corrosion depended upon the square root of the angular velocity ω.

In the majority of practical cases, the solubility of the material in the liquid and the density of the liquid change much more slowly than the viscosity of the liquid. Under isothermal conditions, the viscosity change is due to compositional changes. Thus, the predominant factor in the corrosion of a material by a liquid is the viscosity of the liquid [2.14,2.15]. This, however, does not hold for every case since liquid composition does affect the solubility of the solid [2.16]. These relationships hold quite well for the corrosion of a solid below the liquid surface. At the surface, where three states of matter are present, the corrosion mechanism is different and much more severe.

At the liquid surface, a sharp cut normally develops in the vertical face of the solid material being corroded as shown in Fig. 2.2. This region has been called *flux-line*, *metal-line*, or *glass-line* corrosion (also called the *Marangoni* effect [2.17]). Pons and Parent [2.18] reported that the flux-line corrosion rate was a nonlinear function of the oxygen potential difference between the surface and the interior of a molten sodium silicate. Cooper and Kingery [2.10] reported that flux-line corrosion was the result of natural convection in the liquid caused by changes in surface forces due to an increase in surface energy of the liquid as solid is dissolved. They also reported that if the surface energy of the liquid were independent of the amount of

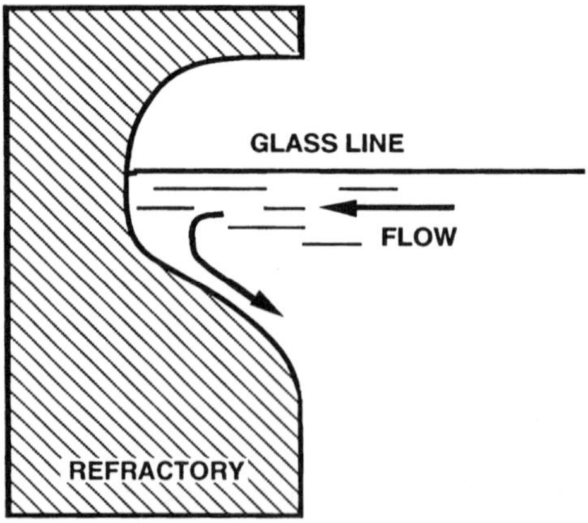

FIGURE 2.2 Corrosion of a vertical face by a liquid.

solid dissolved, no such excessive flux-line corrosion would occur. Hrma [2.13] reported that the additional corrosion at the flux-line depended only upon the variation in surface tension and density, with surface tension being the more important factor. Although this is a well-known phenomenon, no one has investigated it thoroughly to determine a definitive mechanism. In actual practice, quite often, a thermal gradient exists such that the highest temperature exists at the flux line. This temperature difference, however, cannot be the sole driving force for excessive corrosion at the flux line since the same phenomenon is observed in laboratory isothermal studies. This same excessive corrosion occurs at any location where three substantially different materials come in contact with one another. In the above case, it was ceramic, liquid, and atmosphere. It may also occur where two liquids come in contact with a ceramic—a well-known phenomenon in metallurgy. The two liquids in that case are molten metal and an oxide slag.

The temperature dependence of corrosion can be represented by the Arrhenius equation:

$$j = A \exp(E/RT) \tag{2.6}$$

Excellent fit of some experimental data to this equation reported by Samaddar et al. [2.11] has indicated that corrosion corresponds to an activated process. Blau and Smith [2.19] have attempted such an interpretation. However, the fact that variations of liquidus compositions, diffusion coefficients, and liquid structure change with temperature suggests that interpreting corrosion as an activated process may be very misleading and at least ineffective. The Arrhenius dependence should be used only for cases where the liquid is far from being saturated with components from the solid, which, according to Woolley [2.20], is the case for practical glassmaking applications.

The corrosion of a flat vertical slab under a thermal gradient is depicted in Fig. 2.3. As the convective flow of the liquid (caused by either forced convection or density changes) removes some of the reaction product interface, the total thick-

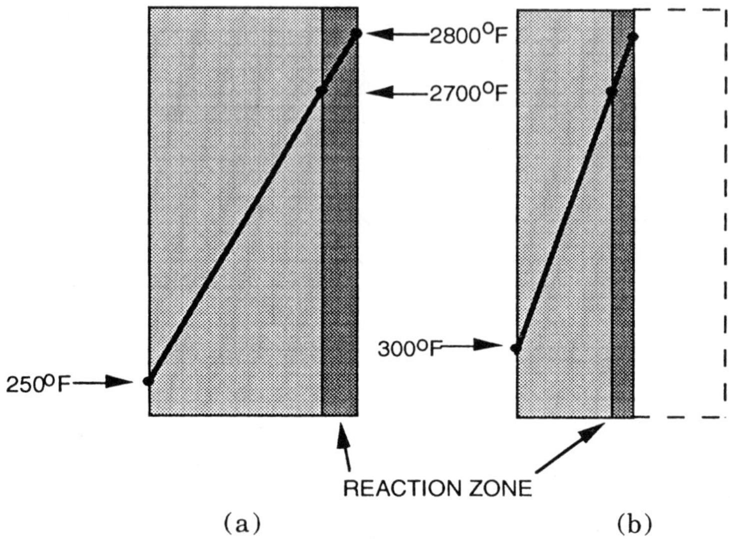

FIGURE 2.3 Effect of thermal gradient upon corrosion interface: (a) short time and (b) extended time.

ness of the slab decreases and the thermal gradient becomes steeper, assuming that the hot face temperature remains constant, which is very close to actual furnace operations. The actual cold face temperature will rise slightly, but the overall result is a steeper thermal gradient. The thermal gradient through a wall as depicted in Fig. 2.3 is more complex than presented here (actually, it is not linear but a complex 3-D shape), but the overall effect is the same. If the reaction product layer can form between certain temperature limits (2800°F and 2700°F in Fig. 2.3), it is obvious that the layer thickness must become smaller as corrosion proceeds. Thus the corrosion rate decreases with time. It is not uncommon for the flux line of the basin wall of a glass furnace to corrode away approximately one-half of its thickness in less than 1 year, while the remaining half may take four or five times as long to exhibit the same amount of corrosion.

A downward-facing horizontal surface also exhibits greater corrosion than does a vertical or upward-facing hori-

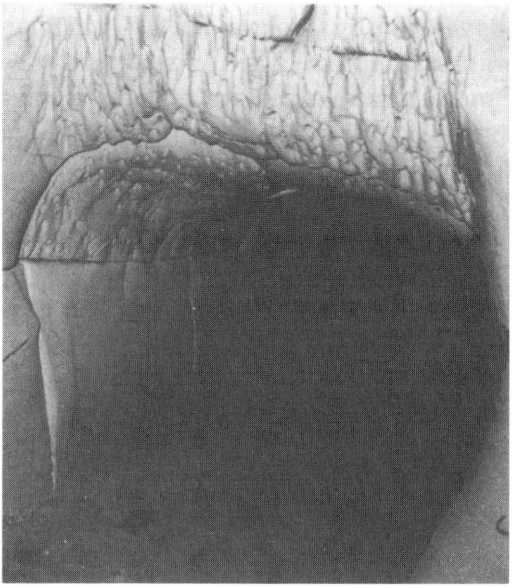

FIGURE 2.4 A fusion cast alumina–zirconia–silica refractory throat of a TV panel glass furnace exhibiting upward drilling of the throat cover. (Courtesy of Corning, Inc.)

zontal surface. A downward-facing surface can exhibit excessive corrosion if bubbles are trapped beneath the horizontal surface. This is known as *upward-drilling* since it results in vertically corroded shafts (see Figs. 2.4 and 2.5). Surface tension changes around the bubble cause circulatory currents in the liquid that cause excessive corrosion very similar to flux-line corrosion. Although no scientific comparisons have been made to geological corrosion, examples of something similar to upward drilling can be found. Fig. 2.5 shows a comparison of upward drilling of an AZS* fusion cast refractory paver from a glass furnace and a dolomite boulder (compare also with

* AZS is the common abbreviation used by the industry to represent refractories composed of alumina, zirconia, and silica. Quite often, they are of the fusion cast variety.

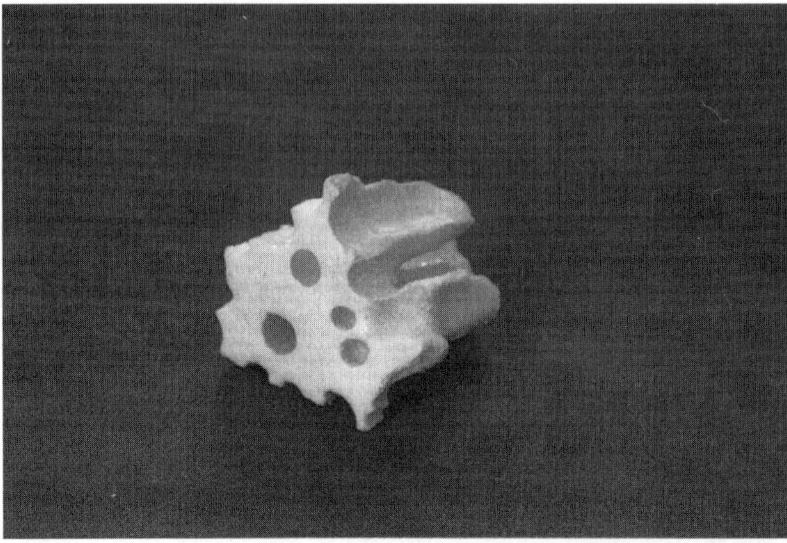

(a)

(b)

FIGURE 2.5 Comparison of upward drilling between (a) an AZS fusion cast glass furnace paver (top surface towards left) and (b) a dolomite boulder.

Fig. 2.4). The surface of the boulder shown in Fig. 2.5 was not horizontal like the paver but sloped approximately 45°.

Attack by Molten Salts

The corrosion of ceramic components in gas turbine engines generally occurs through the action of condensed salts formed from impurities in the fuel and/or combustion air. Similar corrosion mechanisms occur in glass furnace regenerators and on glass furnace crowns. The condensation of molten salts occurs below their dew point and is thus dependent upon the temperature and pressure of operation, along with the concentration of the impurities in the fuel or air (compare discussion under Dew Point Corrosion on page 44). Fox et al. [2.21] listed the dew points for sodium sulfate deposition, a few of which are given in Table 2.1. Not only did a higher pressure raise the dew point for condensation, but it also increased the deposition rate, which generally led to more severe corrosion [2.22]. The effects of molten Na_2SO_4 upon the dissolution of silica and the importance of Na_2O activity and the partial pressure of oxygen is discussed in more detail in Chap. 5, Carbides and Nitrides, page 193.

A model developed by Cook et al. [2.23] in their study of hot corrosion of ceramic (alumina) barrier coatings by sodium, sulfur, and vanadium molten salts gave the rate of solution of

TABLE 2.1 Dew Points (°C) for Na_2SO_4 Condensation

		Sodium (ppm)		
Pressure (atm)	Sulfur (ppm)	0.1	1.0	10
1	500	876[a]	937	991
10	500	969	1045	1111
1	5000	887	961	1025
10	5000	984	1075	1155

[a] Solid, since melting point is 884°C.
Source: Ref. 2.21.

a ceramic when a steady-state condition prevailed for the rate of salt removal equal to the rate of salt deposition. This provided a salt layer of constant thickness. The ceramic solution rate was then dependent upon the rate of salt deposition:

$$d(M_c/A)/dt = \left(\frac{C}{1-C}\right) d(M_s/A)/dt \tag{2.7}$$

where:

M_c = mass of ceramic dissolved
A = surface area
M_s = mass of salt deposited
C = concentration of ceramic in layer

At low deposition rates when salts become saturated, the solubility in the salt becomes important. Use of this model requires the calculation of the gas phase and condensed solution equilibria using a computer program such as that developed by the NASA-Lewis Research Center [2.24]. In addition to the steady-state assumption for salt deposition and removal, other assumptions included the parabolic rate law, known equilibrium solubilities, and congruent dissolution.

Corrosion by molten salts has several beneficial applications. One very important application where dissolution of a ceramic is desired is in the removal of the ceramic cores from metal castings manufactured by the investment casting technique. The solvent used for core removal must be highly reactive to the ceramic at rather low temperatures while not damaging the metal. The ceramic must be stable toward molten metal attack at high temperatures and highly reactive toward solvent attack at low temperatures. In a study of the leaching rates of Al_2O_3, Y_2O_3, La_2O_3, ZrO_2, ThO_2, and MgO by molten Li_3AlF_6, Borom et al. [2.25] found that the corrosion appeared to involve a solid reaction layer and a boundary layer in the liquid. Vigorous solvent circulation was required to overcome the diffusion-controlled process. Thus it appeared that congruent dissolution was required for optimum core removal since incongruent dissolution may form reaction layers that require forced convection for removal.

Electrochemical Corrosion

Very few studies have been reported over the last 30 years; however, much work was performed in the 1950s and 1960s on what has been called *galvanic corrosion* of refractories by glasses. Galvanic corrosion as defined by the physical chemist must occur between two materials in contact with one another and both must be in contact with the same electrolyte. Much of what has been reported should more appropriately be called *electrochemical corrosion*. One of the first reports of the existence of an electrical potential between refractory and glass was that of Le Clerc and Peyches [2.26] in 1953. The setup is schematically represented in Fig. 2.6. In such a case, the molten glass acts as the electrolyte and the platinum wire acts as a reference electrode (i.e., standard oxygen electrode). The use of platinum as a reference electrode requires that the atmosphere

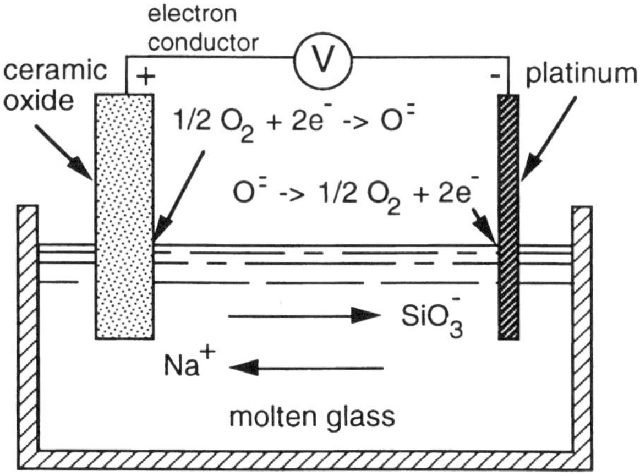

FIGURE 2.6 Electrochemical cell to determine potential difference between a ceramic and a molten glass.

above the melt contain a reasonable oxygen partial pressure since the reaction:

$$O^{2-} \leftrightarrow 1/2\ O_2 + 2e^- \tag{2.8}$$

must be present at the site where the platinum comes in contact with the melt. The overall accuracy of such potential measurements is critically dependent upon obtaining excellent electrical contact among the various components of the galvanic cell. An additional problem that undoubtedly causes variation or drift in the measurements is the formation of a reaction interface layer between the refractory and the molten glass.

Godrin [2.27] has published a review of the literature on electrochemical corrosion of refractories by glasses. It has been shown that a potential difference does exist in such systems; however, no quantitative relationship between corrosion and potential has been reported. Since a potential difference exists in corroding systems, it has been tempting to assume that the potential is at least partly responsible for the corrosion; however, the application of a bias potential has been unsuccessful in eliminating corrosion. Although not totally reliable, Godrin concluded that refractories that had an electrical potential with respect to glass that was positive 0.4 to 0.7 V were fairly resistant to corrosion, that refractories with a potential greater than 1.0 V had rather poor resistance, and that refractories that had a negative potential with respect to glass should not be used.

Pons and Parent [2.18] have concluded that the oxygen ion activity was a very important parameter in corrosion and that its role was determined by the difference in oxygen potential between the molten glass and the refractory oxide. An additional interesting case was that of two different oxide materials (i.e., a multiphase polycrystalline material) in contact with the same glass that had oxygen potentials on either side of that of the glass. In such a case, it was assumed that oxygen migrated from the oxide of higher potential toward that of lower potential. If the conduction mechanism of the two oxides were different (ionic vs. electronic), the situation would become more complex. When the oxygen potentials of the oxides were greater than the glass, oxygen ions were assumed to be transported from the ionic conductive oxide to the electronic conductive one, which may ultimately result in pitting caused by the release of oxygen. If the oxygen potential of the oxides was lower than the glass, alkali ions of the glass would be transported to the electronic conductive oxide with oxygen release at the interface between the two oxides.

Although in theory the application of a bias potential to minimize or eliminate corrosion, which implies that the corrosion process is one that involves charge transfer, should produce noticeable results, a major practical problem has been that of making the electrical connection to the ceramic. The other problems relating to the success of a bias potential in eliminating corrosion are the other factors in corrosion—chemical reaction, diffusion, viscosity, solubility, etc. This topic is one of considerable importance and should receive a lot more attention than it has in recent years. A standard text that discusses electrode effects in liquid electrolytes should be consulted by the interested reader [2.28].

Wall et al. [2.29], in their studies of graphite fiber/BMI (bismalimide) composites in contact with various metals immersed into an electrolyte, developed an electrical equivalent circuit of the electrochemical interface. The *classic electrical double layer* (see discussion on page 30, Attack by Aqueous Media) is established at the surface of the graphite fibers where the electrons at the surface are separated from the ionic charges in solution. This forms a capacitor, called the *double-layer capacitance*, C_{DL}. This capacitance is dependent upon the electrode surface area, which allows one to monitor surface roughening, surface adsorbed species, and ingress of solution between fiber and matrix by changes in capacitance. Charge leakage rate across the interface can be represented by a resistance, R_F, in parallel with C_{DL} that is inversely proportional to the reaction rate. Changes in exposed surface area of the fibers or changes in surface chemistry will affect this rate.

The electrical equivalent circuit is completed by the addition of resistors in series representing the solution resistance, R_S, and the material resistance, R_M (circuit depicted in Fig. 2.7). Imposing a high frequency AC signal to this circuit will allow one to determine the resistance $R_S + R_M$ since C_{DL} will short out R_F. If the solution resistance, R_S, is assumed to remain constant, changes in the sum, $R_S + R_M$, can be related to composite integrity. Wall et al. used electrochemical impedance spectroscopy to evaluate all the circuit elements under steady-state conditions.

Attack by Molten Metals

The potential reaction of molten metals with oxide ceramics can be easily obtained from an Ellingham-type diagram, which is a compilation of the free energies of formation of oxides from their metals and oxygen at different temperatures. The simple redox mechanism is:

$$xM + y/2\ O_2 \rightarrow M_xO_y \tag{2.9}$$

Thus one can easily determine compatibility between a metal and an oxide since any metal will reduce any oxide that has a

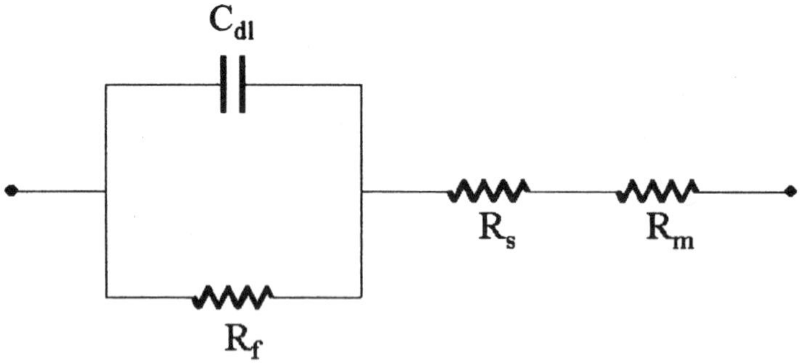

FIGURE 2.7 Equivalent circuit model for electrochemical interface. (From Ref. 2.29.)

less negative free energy of formation for the oxide. The reaction of aluminum metal with silica is a good example:

$$3SiO_2 + 4Al \rightarrow 2Al_2O_3 + 3Si \quad (2.10)$$

Since metals such as aluminum and magnesium have very large negative free energies of oxide formation, determining what to use as a container when melting these metals becomes a serious problem.

Another mechanism for metal attack upon a ceramic is by the formation of a new compound by a reaction of the type:

$$A_xO_y + zM \rightarrow A_{x-w}M_zO_y + wA \quad (2.11)$$

An example of this is the formation of spinel:

$$Al_2O_3 + 3Mg \rightarrow 3MgAl_2O_4 + 2Al \quad (2.12)$$

This reaction yields a lower free energy of reaction ($\Delta G^\circ_{1000} = -52$ kcal/mol) than the simple redox reaction ($\Delta G^\circ_{1000} = -28$ kcal/mol) shown below:

$$Al_2O_3 + 3Mg \rightarrow 3MgO + 2Al \quad (2.13)$$

Thus a compound is formed that would appear to be a good container for molten Al/Mg alloys. However, Lindsay et al. [2.30] reported that reaction (2.13) was preferred to reaction (2.12) due to the magnesium activity being sufficiently high to form MgO and the reaction of Al_2O_3 and MgO to form spinel being kinetically slow.

Another possible reaction of a ceramic with a metal is that of reduction to the metal and solution into the attacking molten metal as shown below:

$$A_xO_y \rightarrow xA + yO \quad (2.14)$$

The metal that forms may be in the gaseous state depending upon the environmental conditions, which may also be true for the oxide that forms in reaction (2.13).

It has been found in the operation of commercial glass (especially lead-containing glasses) furnaces that metals cause a unique corrosion pattern on upward-facing horizontal surfaces by drilling vertical shafts into the bottom paving refractories. This is called *downward drilling* and is very similar to the upward drilling described on page 21. According to Busby [2.31], the excessive corrosion caused by molten metal droplets is due to a surface tension gradient on the surface of the droplet. The corrosion drilling rates are not dependent upon the type of metal, but are dependent upon the quality of the corroded material. Smaller droplets are more corrosive than larger ones.

Attack by Aqueous Media

Probably some of the more significant work being done today concerning the understanding of ceramic–water interfaces is that being done in the area of ab initio calculations [2.32]. Large modern computers have made it possible to obtain accurate calculations that describe the potential surface of silicates through solutions to the Schrodinger equation. The major assumption in these calculations is that the local chemical forces of the first and second nearest neighbors determine the largest portion of the dynamics and energetics of the chemisorption process. Calculations of this type have shown that the adsorption of water onto silica terminal OH groups (silanols) is more stable than onto bridging OH groups. If these silanol groups are removed through heating, leaving a surface of essentially siloxane bonds (Si–O–Si), the surface becomes hydrophobic. In addition to the ab initio calculations, the use of molecular dynamics has allowed the description of the collective atomic motions on mineral surfaces and the surrounding fluids. One result of these studies has been the finding that for at least a few atomic layers, the surface structure of solids can deviate considerably from that of the bulk. In general, all these studies indicate that the dissolution rate is controlled by the chemical species adsorbed onto the surface. In addition, the amount and type of surface species is dependent upon the pH, which also determines the surface charge.

The structure of the water molecule is very important in the understanding of aqueous dissolution. The single oxygen atom has four sp^3 atomic hybrid orbitals (2s, $2p_x$, $2p_y$, and $2p_z$) that are directed toward the corners of a tetrahedron. Each of the hybrid orbitals contains two electrons. Two of these hybrid orbitals react with a 1s hydrogen atomic orbital, leaving the two remaining sp^3 hybrid orbitals uncommitted (called the *lone pairs*). This configuration thus produces an asymmetric electric charge across the water molecule making water a *polar molecule*. The small mass of the hydrogen allows the water molecule to rotate so that the negative charge of the lone pairs can interact with a positively charged ceramic surface or, alternatively, form hydrogen bonds with a negatively charged surface. The dissociation of water into ions of H^+ and OH^- most likely occurs as an excess or deficiency of protons on a water molecule creating H_3O^+ and HO^-, maintaining electrical neutrality. The relative concentration of these ions is the bases for the pH scale of acidity and alkalinity.

A tremendous amount of information is available concerning the leaching or dissolution of minerals, especially silicates, in the soils literature, and anyone interested in corrosion in aqueous systems should avail themselves of that literature. The Jackson weathering sequence discussed by Marshall [2.33] exhibits a trend that may be applicable, to some degree, to the dissolution of silicate ceramics. The greater the degree of bonding of the silica tetrahedra, the more difficult the weathering. Some variation and overlap occurs due to the specific chemistry for an individual mineral, with the minerals containing alkalies and alkaline earths being less stable than those containing alumina and the transition metals. This is supported by Huang [2.34] who, in his studies of olivines, pyroxenes, and amphiboles, reported that the relative stability of these materials appeared, among other things, to be related to the degree of polymerization of the tetrahedra, with more highly polymerized materials being more stable. A related phenomenon reported by Casey and Bunker [2.35] was that minerals with a low density of cross-links tend to dissolve congruently and rapidly, while minerals with a high density of cross-links, such as the tektosilicates, dissolve incongruently producing a leached surface layer. Incongruent dissolution and selective leaching results from the three processes of hydration, hydrolysis, and ion exchange. Hydration is more prevalent in materials with a low degree of covalent character to the cross-links and a structure that allows water penetration into the structure (i.e., those containing pore sizes >2.8 Å).

This relationship between structure and leaching was described by Casey and Bunker in a comparison of quartz containing a small percentage of aluminum, forsterite, and albite. The quartz structure, being a completely linked network of silica tetrahedra with very small pores, exhibits only a near-surface leaching of the aluminum. Forsterite, on the other hand, is a structure with a cross-link density of zero, containing independent silica tetrahedra bonded by magnesium ions. In acid solutions, the silica tetrahedra are converted intact into silicic acid without hydrolysis. If a leached layer were to form, it would be very thin since no bridging oxygen bonds would remain when the magnesium ions were removed. The albite structure contains exchangeable ions (sodium), hydrolyzable silica and alumina tetrahedra, and a cross-link density the same as quartz. One-third of the cross-links in albite are of the Al–O–Si type, a more reactive cross-link than the Si–O–Si bonds. The resultant hydrolysis of these Al–O–Si bonds opens the structure, allowing deep penetration of the solutes and water. As this occurs, structural integrity is maintained by the residual silicate framework, which allows a very thick leached layer to form.

Dissolution studies of smectite minerals by acids were shown by Borchardt [2.36] to take place through the following steps:

1. Exchange of cations with H_3O^+
2. Removal of the octahedral Al, Mg, and Fe
3. Removal of the tetrahedral Si and Al

Humic substances were reported by Schnitzer and Kodama [2.37] to exhibit strong solvent activity toward minerals, with silicate minerals generally being more resistant to attack by humic and fulvic acids than nonsilicate minerals. Fulvic acid and low molecular weight humic acid attack minerals by forming water-soluble complexes with the divalent and trivalent cations via CO_2H and phenolic OH groups. In general, any strong bonding to the metal cations promotes dissolution.

The model of dissolution of minerals is based upon the diffusion of leachable species into a thin film of water ~ 110 μm thick that is stationary. The movement of soil water also receives these soluble materials by diffusion from this thin film. The following equation represents this process:

$$\text{Mineral A} + nH^+ + mH_2O \leftrightarrow \text{Mineral B} + qM^+ \tag{2.15}$$

where M^+ is the soluble species. The equilibrium constant is:

$$k = \left(\frac{[M^+]^q}{[H^+]^n[H_2O]^m}\right) \tag{2.16}$$

From Eq. (2.16), it should be obvious that the dissolution of minerals, and actually any ceramic, is dependent upon the pH of the water. The mineral B may actually not be crystalline but may form a *gel layer* with a variation in composition through its thickness. Mineral B may also be a metastable form that may vary its structure and composition depending upon the test conditions as reported by Jennings [2.38] for the action of water. In dilute aqueous systems at 1 atm, the activity of H_2O can be assumed to be unity. At higher pressures, the activity of water is approximately proportional to the pressure. Thus the dissolution of mineral A is directly affected by the pH of the system. In complex minerals where several ions may be released into solution, the theory of Donnan discussed by Marshall [2.39] predicted that the activity ratio of monovalent cations to the square root of divalent cations in solution should tend to be constant.

Elmer [2.40] has shown how the water activity of the leachant decreases with increasing acid strength. The ion exchange reaction between the various leachable ions in the ceramic with hydronium ions would be expected to increase with acid strength. He showed that the reaction rate of cordierite (and a borosilicate glass) decreased after reaching a maximum at some intermediate acid strength. This he attributed to the reduction in the water activity with increasing acid strength.

When a silicate is leached by an aqueous solution, an ion is removed from a site within the crystal structure and is placed into the aqueous phase. Most of the transition metal ions and many other ions occur in sixfold coordination in the crystalline structure and also in solution as hexahydrated ions. Since the crystal field stabilization energies of the transition metal ions in oxide crystal structures and in aqueous solutions are about the same, whether or not leaching occurs depends upon the ease with which the ions are removed from the crystal structure.

The mechanism reported by Burns [2.41] requires a water molecule to approach the metal ion along a vacant t_{2g} orbital (see Burns also for a discussion of orbital theory) forming a sevenfold coordinated intermediate state, which is the rate determining step. This intermediate state spontaneously disproportionates into a metal-hydroxysilicate and a hydroxysilicate residue. Continued repetition of the process ultimately produces a metal hydroxide or hydrated oxide and a hydroxysilicate residue.

Ions with d^3, d^8, and low-spin d^6 configurations are the most resistant to leaching since these configurations contain electrons in the low-energy t_{2g} orbitals, producing a larger energy barrier for the formation of the intermediate state. In those ions with at least one empty t_{2g} orbital (i.e., d^1 and d^2), the energy barrier is much lower, and thus these ions exhibit less resistance to leaching. In those ions with more than three d electrons, additional energy is required to pair electrons in the t_{2g} orbital thus creating a vacant t_{2g} orbital. Thus ions with d^4, d^5, d^6 hi-spin, and d^7 hi-spin should exhibit intermediate leaching characteristics. Data reported by Hawkins and Roy [2.42] are in very good agreement with the predicted results.

To understand the interaction of a material with an electrolyte, one must have a good understanding of the electrical double-layer characteristics of the immersed material. The structure of this electrical double layer is dependent upon the decay of the potential from that at the solid surface to the zero potential of the bulk electrolyte. A more detailed discussion of the electrical double-layer concept can be found in the book by Shaw [2.43]. The methods that are used to study the interaction of ceramics with electrolytes are generally that of pH changes, such as potentiometric titration. One result of these studies is the determination of the pH at which a net zero surface charge exists [called the *zero point of charge* (ZPC)], which may or may not correspond to a zero zeta potential. The condition of zero zeta potential is called the *isoelectric point* (IEP). For those materials that exhibit some solubility, it is more appropriate to use the IEP value since this relates to an equal number of positively and negatively charged dissolution species present at the solid surface. Because of this, the pH of the IEP also represents the pH of minimum solubility. Above and below the pH of the IEP, the dissolution increases due to the presence of species that weaken the metal–oxygen bonds. At low pH, dissolution studies of alkaline earth oxides and transition metal oxides have indicated that dissolution is controlled by the strength of the metal–oxygen bond [2.44]. Thus dissolution rates can be predicted from the oxide free energy of formation. The driving force for dissolution is the difference in potentials between the solid surface and the Stern plane or outer Holmholtz plane (i.e., the closest distance of approach of hydrated ions). According to Parks [2.45], the probable IEP of a material falls into a range of values depending upon the cation oxidation state as shown in Table 2.2. Parks also lists the IEP of many materials.

In addition to the soils literature, a large amount of work has been reported on the dissolution of oxides that form as protective or semiprotective coatings on metals. Diggle [2.46] has reported a good review, although dated, of this literature up to about 1971. Diggle has divided the dissolution of these oxide coatings into two major groupings: those cases where the rate determining step involves electronic charge transfer, which are

TABLE 2.2 Probable IEP Values

Oxide type	IEP pH range
M_2O_5, MO_3	<0.5
MO_2	0–7.5
M_2O_3	6.5–10.4
MO	8.5–12.5
M_2O	>11.5

Source: Ref. 2.45.

called *electrochemical dissolution*, and those where no charge transfer is involved in the rate determining step, which are called *chemical dissolution*. Electronic conductivity, which is related to the oxide structure and bonding, is more important the greater the covalent character of the oxide. In chemical dissolution, crystallographic and metal–oxygen bond strengths play a very important role, whereas in electrochemical dissolution, the electronic structure is of prime importance.

Dissolution of solids in solutions is sometimes dependent upon surface-controlled reactions at the solid/solution interface. The exchange rate between the solid and the solution ligands decreases as the cation charge increases, forming stronger bonds between the cation and the ligands. An example reported by Bright and Readey [2.47] of this dependence is the comparison between Ti^{4+} and Mg^{2+} where the dissolution of MgO is much faster than that of TiO_2.

Bioceramics. When one thinks of ceramic materials that are used in the human body, the first that comes to mind are dental applications—false teeth, caps, and crowns. The use of ceramics in the body, however, is becoming more widespread than just dental applications. Some of these now include bone and joint replacements, heart valves, eye lenses, and therapeutic glasses for treatment of tumors. The materials that are most often used include hydroxyapatite, alumina, zirconia, bioactive phosphate glasses, and rare earth aluminosilicate glasses. The crystalline materials can be either single crystals or polycrystalline. Some applications require the bioceramic to react with and bond to the bone. An interesting review of the history, although dated, of bioceramics can be found in the article by Hulbert et al. [2.48].

Four types of interface reaction can occur between a ceramic implant and human tissue. These are formation of a fibrous capsule around the implant, formation of an interfacial bond between the implant and tissue (called a *bioactive* interface), and dissolution of the implant that may be inert or toxic [2.49]. In the first type of interface reaction, the implant (essentially inert) does not form a bond with the bone. In the second type, *bioactive*, the implant forms a bond with the bone through a chemical reaction at the interface. Bioactive implants characteristically form a layer of hydroxyl-carbonate apatite on their surface along with collagen fibrils that bind the ceramic implant to the organic tissues. This reaction forms an interface that is essentially identical to the naturally occurring interfaces between bone and ligaments or tendons. If the reaction at the interface were purely mechanical, where the bone grows into a porous implant, it would not be termed *bioactive*. In the last type of reaction, dissolution, the implant is replaced by bone. Very small variations in chemistry can change a material from being inert to one that is bioactive or resorbable. Although alumina and zirconia tend to be inert, they do form thin layers of fibrous tissue at the interface. The thickness of the fibrous layer is dependent upon the chemical, physical, and mechanical properties of both the implant and the tissue. These various types of implants along with their different reactions were developed for specific applications in the human body. Although at first thought the applications of ceramic implants into the human body appear to be very different than the conventional applications of ceramics (e.g., in furnaces), there are many similarities. For example, mechanical loading at joints in the human body and their movement can be related to convective flow of metal at the bottom of a blast furnace—both environments exhibit a mechanical load and an erosive effect.

These two environmental factors produce similar effects upon the interface reactions that take place. One must know the environment and the ceramic thoroughly to understand and explain any reactions or corrosion that takes place.

Although not an area that one would normally classify as bioceramic, the whole area of the effects of mineral dusts upon the human body are closely related to and actually very similar to bioceramics. Anyone interested in this topic should consult the review by Guthrie and Mossman [2.50]. In general, mineral dusts that are inhaled may affect the tissues of the lungs. Since lung fluids are undersaturated with respect to most chemical species contained in minerals, these minerals will dissolve in the lung. The health aspect of all this is a kinetics problem—how long will it take to dissolve completely? Minerals that dissolve within a matter of days are thought to have no health effect, although this may not be true. Many studies have been done to determine whether the health problem lies with the chemistry of the minerals or with their morphology. Stanton [2.51] and his colleagues have proposed a hypothesis that mesothelioma (i.e., ability to form tumors) is caused by the morphology of ceramic fibers. The general result was that decreasing fiber diameter and increasing fiber length were related to increased activity.

Weathering. The weathering of ceramics is generally restricted to those materials used in the construction industry, mostly structural clay products, sandstone, limestone, marble, and glass. Sec. 2.3.3 and also Chapter 6 discuss all the various corrosion aspects of glassy materials including weathering; therefore there is no need to repeat them here. A large amount of literature is available on the subject of weathering of rocks and minerals; however, this vast source is generally not tapped when concerned with the weathering of ceramics, although many similarities exist.

Historically, it has been a general practice to bury the used refractories from a furnace rebuild in landfills. When chromium-containing basic refractories are buried, a serious potential problem may arise. This problem relates to what happens to the chromium. Most references about basic refractories consider the chromium to exist as Cr^{3+} and reside on the octahedral sites of the mineral phases present. In the fusion cast type of chromium-containing refractories, the chromium is located either in the spinel or the sesqueoxide phase. In any of the various unused refractories, the chromium is thus generally accepted to be in the trivalent state. It is the hexavalent state of chromium that has been identified as a carcinogen.* Thus it is a question of whether or not the chromium will be oxidized while in service or disposed in a landfill and whether or not this oxidized ion will be leached into the surrounding groundwaters. At present, there still has not been a definitive study performed to determine accurately the fate of the chromium in these materials.

Today, the vast majority of corrosion problems related to weathering are those of historic buildings, monuments, and landmarks. The civil engineering community has made great strides in recent years in understanding the mechanisms of weathering of ancient building materials. One of the more common building materials was a rock called *sandstone* (mainly quartz grains bonded by lime), which is a rather porous, weak material. Other materials used were *limestone* (calcium carbonate rock, often from marine sources) and *marble* (crystalline metamorphic limestone). All these materials, being basic in nature, are attacked by acids. The porosity (discussed below in more detail) provides a route through capillary action, which is large for pore diameters less than 1 μm, for water, moisture, etc. to enter the sandstone. While inside, some dissolution can take place depending on the chemical makeup of the liquid. The most predominant species are CO_2 and SO_2, which, when mixed with water, form weak

*Hexavalent chrome was confirmed to be a carcinogen by R.L. Joiner, J.D. Rench, M.A. Zanetos, and S.E. Brauning published in a report of The Refractories Institute, Feb. 1983.

acids. After dissolution, the liquid migrates to the surface and evaporates. When evaporation takes place at the surface, various salts (e.g., gypsum) are left behind either on the surface or within the pores that can disrupt the mechanical integrity of the sandstone depending on their expansion and growth characteristics upon precipitation from the liquid. When salts are precipitated and left behind on the surface of the material, the term *efforescence* is used. When precipitation occurs within the pores, the terms *subflorescence* or *crytoflorescence* are used. It should be remembered that with many salt solutions, considerable supersaturation may take place before precipitation occurs, and for mixed salts, the individual solubilities may be increased due to various interactions. Supersaturation may occur by decreasing temperature, increasing concentration, or a combination of both decreasing temperature and increasing concentration (the more common situation). Supersaturated solutions generally produce larger crystals that, in turn, produce greater pressures for disruption. For one-directional crystal growth, a pressure within small pores is developed that can be represented by the following equation [2.52]:

$$P = (RT/V_s)\ln(C/C_s) \tag{2.17}$$

where:

P = pressure produced by crystal growth
R = gas constant
T = temperature
V_s = solid salt molecular volume
C = solute concentration
C_s = solute saturation concentration

If hydrated salts occur during the growth process, volume changes will occur if humidity changes take place causing the salts to adjust their water of hydration. These volume changes will cause pressure changes that have been represented by the following equation [2.53]:

$$P = (nRT/V)\ln(P_o/P_h) \tag{2.18}$$

where:

P = hydration pressure
n = moles of water gained during hydration
R = gas constant
T = temperature
V = change in volume (volume of hydrate minus volume of original salt)
P_o = partial pressure of water vapor
P_h = vapor pressure of hydrated salt

An added aggravation can result from the freezing of water with its associated expansion that can lead to cracking of the building material. In addition, many marble monuments develop a golden patina on the surface due to the selective dissolution of the calcium contained in marbles with a small content of iron [2.54]. This causes an enrichment of iron compounds at the surface.

The porosity of limestone can increase if temperature fluctuations (on the order of $-15°C$ to $+35°C$) are present [2.55]. It has been reported that some marbles have exhibited as much as 50% increase in porosity when subjected to temperature fluctuations. The porosity increase is a function of the differential thermal expansion of the calcite crystals present in limestone and marbles. Calcite actually expands along the *c*-axis while contracting in the perpendicular direction ($\alpha_c = +26.5 \times 10^{-6}/°C$ vs. $\alpha_a = -5.7 \times 10^{-6}/°C$). This expansion/contraction of the calcite crystals causes them to separate forming small pores. Thus the circulation of corrosive liquids increases causing increased deterioration of the limestones.

In all the processes discussed above, it should be recognized that water is an essential ingredient. If there were no water present, there would be no dissolution or chemical reactions, no transport of corrosive species, or no disruptive forces. The chemical makeup of the water depends upon its origin: rainwater, groundwater, or condensation (similar to rain). Condensation is generally thought of as dew, but one must not forget that frost and fog are also forms of condensation that may lead to corrosion problems. The corrosion that results from contact with fog can be more serious than that from rainwater since, as reported by Hoffmann [2.56], pollutants are more concentrated and the pH is generally much lower. Although somewhat dated, values for the pH of fog in the Los Angeles basin were reported to range from 2 to 4. The reason for these differences from rainwater was attributed to the droplet size being much smaller for fog (2–50 μm for fog and 0.1–3 mm for rain). Higher concentrations of pollutants have also been reported for dew by Mulawa et al. [2.57]. In addition to the higher pollutant concentrations in dew, Mulawa et al. reported that the acidity on downward-facing surfaces was higher (i.e., lower pH) than on upward-facing surfaces. The chemical makeup of all these sources has changed over the years, and thus the exact mechanisms of weathering including the kinetics have also changed. The sad thing is that the deterioration is becoming more rapid for some very famous landmarks.

Rainwater chemistry is affected by local pollutants (see Table 2.3) which can have their source many hundreds of miles away and also change over time depending upon the socioeconomic progress of the region. Weather patterns are also of major importance since the wind direction plays a role in where various pollutants end up. The acidity of rainwater is governed mostly by the content of sulfate, nitrate, and ammonium ions. The acidity is generally represented by the pH, which has been reported to be less than 5 over almost the entire eastern United States [2.58]. Neal et al. [2.59] reported that the leaching rate of

TABLE 2.3 Common Atmospheric Pollutants

CO_2 (combustion processes)
Cl (mostly marine atmosphere but also coal burning)
SO_2/SO_3 (combustion processes, especially those of coal)
NH_3 (waste treatment)
NO_x (combustion processes)

limestone doubled for each pH = 1 decrease. Thus one can see that to be able to understand the problems of weathering of ancient buildings, one must, in addition to being an engineer, also be a historian and meteorologist.

Groundwater-related weathering implies that one knows a little geology since dissolved species play a role in attacking the sandstone. Local water tables and flooding history are important aspects of the problem. To complicate the whole process, a combination of the above mechanisms can take place.

The dissolution of limestones and marbles by atmospheric corrosion takes place by one or a combination of the following reactions [2.54]:

$$H_2O_{(l)} + CO_{2(g)} \rightarrow H_2^{2+}CO_3^{2-}{}_{(aq)} \tag{2.19}$$

$$CaCO_{3(s)} + H_2^{2+}CO_3^{2-}{}_{(aq)} \rightarrow Ca^{2+}(HCO_3)_2^{2-}{}_{(aq)} \tag{2.20}$$

Since all atmospheres contain carbon dioxide, these reactions will take place in the presence of rainwater. The temperature and the CO_2 partial pressure in the solution were reported by Reddy et al. [2.60] also to affect the weathering rate. If any sulfur is present in the atmosphere (generally from industrialization that also increases the carbon dioxide concentration), the following reactions take place:

$$SO_{2(g)} + {}^1/_2\ O_{2(g)} \rightarrow SO_{3(g)} \tag{2.21}$$

$$SO_{3(g)} + 2H_2O_{(l)} \rightarrow H_2^{2+}SO_4^{2-}{}_{(aq)} \tag{2.22}$$

$$CaCO_{3(s)} + H_2^{2+}SO_4^{2-}{}_{(aq)} \rightarrow Ca^{2+}SO_4^{2-}{}_{(aq)} + H_2^{2+}CO_3^{2-}{}_{(aq)} \tag{2.23}$$

These reactions are more severe than those shown in Eqs. (2.19) and (2.20) above and the subsequent dissolution is more rapid.

$$CaCO_{3(s)} + SO_{3(g)} + 2H_2O_{(l)} \rightarrow CaSO_4 \cdot 2H_2O_{(s)} + CO_2 \tag{2.24}$$

This final reaction is one that leaves a white scum on the surface of the limestone or marble through the evaporation of the water and formation of gypsum. Vanadium oxides (a pollutant contributed by internal combustion engines) can act as catalysts for reaction (2.21), along with colloidal particulates suspended in the atmosphere and photochemical reactions [2.54]. Due to the variations of the catalysts upon the oxidation of SO_2, the solid that forms is generally not completely gypsum but contains as much as 15% $CaSO_3 \cdot 2H_2O$.

Skoulikidis [2.54] has reported that the sulfation of marble monuments is through galvanic corrosion, where the anode is the environment (SO_2, air, and water vapor), the cathode is marble ($CaCO_3$), and the electrolyte is the gypsum ($CaSO_4 \cdot 2H_2O$) formed during reaction. He also reported that the formation of gypsum was linear up to a thickness of about 300 Å and was thereafter parabolic. This was attributed to the pores of the original film being filled with new material blocking the diffusion of SO_2, O_2, and H_2O to the marble/gypsum interface. Subsequent formation of gypsum (greater than 300 Å) was then due to diffusion of Ca^{2+} and CO_3^{2-} to the surface forming more gypsum at the gypsum/environment interface.

Dew Point Corrosion. Corrosion by condensation, which can be another form of rainwater, is actually called *dew point corrosion*. But again, an understanding of meteorology is helpful since one must know what species are contained within the atmosphere and at what temperatures they will condense.

Dew point corrosion is the deterioration of a solid ceramic material caused by the condensation of a corrosive liquid from a saturated gas when the temperature is lowered below the point (the dew point) where the liquid will condense. Although a considerable amount of dew point corrosion is from manmade industrial gases, a portion of dew point corrosion could actually be a subgroup of *atmospheric corrosion*, which is the degradation of materials by natural environments. *Atmospheric corrosion* is a term used by the metallurgist, whereas an equivalent term used by ceramists is *weathering*.

Some industrial gases may contain many species that will condense at different temperatures and thus exhibit more than one dew point. Since the dew point is the temperature at which evaporation and condensation rates are equal for a saturated vapor, it should be obvious that the dew point will vary depending upon the concentration (equal to the vapor pressure) of the species in the vapor. Stated in another way, the dew point is the highest temperature at which a condensate can exist in equilibrium with the vapor from which it condensed. Many of the condensable species condense in combination with water vapor forming acid condensates. For example, for a specific concentration of SO_3, the dew point temperature will increase as the water vapor concentration increases (see Fig. 2.8). The actual corrosion rate reaches a maximum about 30°C below the dew point [2.61].

In a reaction similar to that shown in Eq. (2.22), atmospheric SO_2 as it diffuses upward may react with hydroxide radicals produced by sunlight forming H_2SO_4 [2.62]. Just as with rainwater, these species may cause dissolution of the sandstone and subsequent mechanical problems.

Some common constituents of industrial flue gases and their dew points are given in Table 2.4.

The relationship between excess air and SO_3/SO_2 levels has been used by glass furnace operators for many years to adjust the fining* operation. Above about 10% excess air, there is little additional conversion of SO_2 to SO_3.

Determination of the dew point in gaseous environments dates back to 1929 [2.63]. Commercial equipment is available from several suppliers.

Cussler and Featherstone [2.64] reported the action of acids upon porous ionic solids and concluded that within the porous solid, dissolution would take place only if the valence of the solid cation were between 0 and 1 and that material would precipitate if the valence were greater than 1. They verified this conclusion with experiments on $Ca(OH)_2$ that showed $Ca(OH)_2$

---

*The process of fining is the removal of small bubbles and general homogenization. The furnace atmospheric SO_3/SO_2 level is related to the residual sulfate level that remains within the glass and whether or not it exists as bubbles.

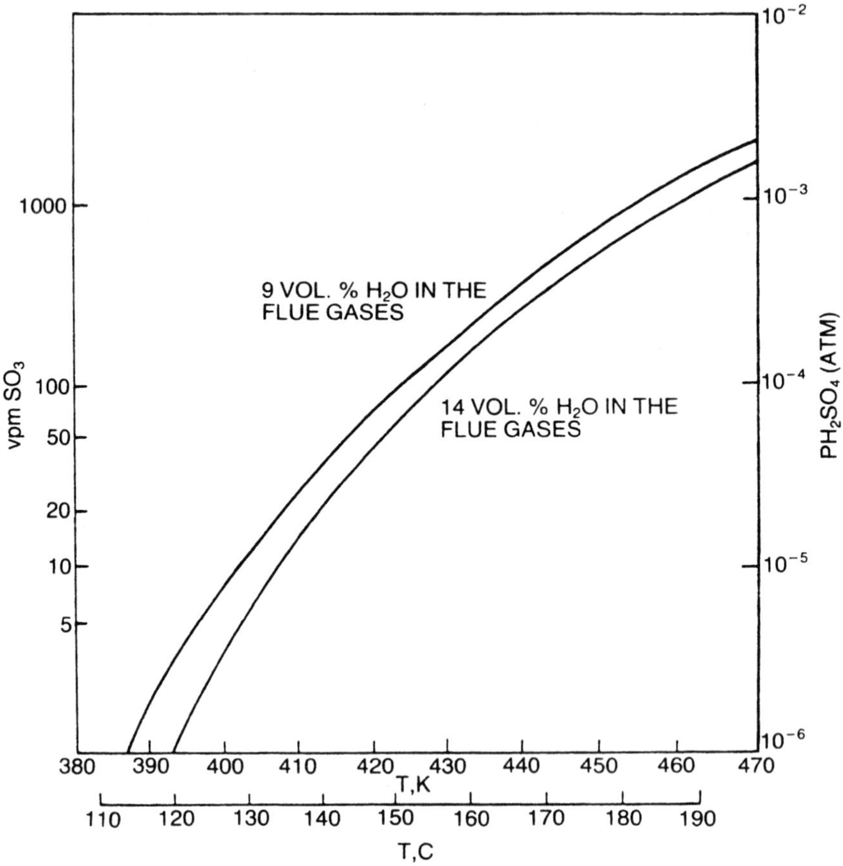

FIGURE 2.8 Sulfuric acid dew point curves. (Reprinted with permission of The Institute of Corrosion, United Kingdom. From Ref. 2.61.)

TABLE 2.4 Dew Points of Common Constituents of Industrial Flue Gases

Species	Dew point (°C)
SO_3/H_2O	100–140
HCl	40–50

precipitated within the pores of the $Ca(OH)_2$ being dissolved. The assumptions of Cussler and Featherstone were that all reactions in the solid were much faster than diffusion so that the reactions reached equilibrium, the diffusion coefficients of all species were equal, and the porous solid was present in excess. Although these assumptions may yield reasonable first approximations for simple systems, they generally do not hold true, especially for the more complex type systems often encountered.

Another effect of water has been reported in the literature in which the reaction with water resulted in the transformation of a metastable phase to the more stable form. This has been reported by Yoshimura et al. [2.65] for partially stabilized zirconia (PSZ) where the reaction with yttria causes the transformation of the metastable tetragonal zirconia to the stable monoclinic form. Similarly, the adsorption of water onto the surface of zirconia has been reported by Sato et al. [2.66] to cause this transformation. Yoshimura et al. concluded that if the reactivity of Y_2O_3 in YSZ was the same as in Y-PSZ, the transformation would not be caused by strain release but by the formation of nucleating defects caused by the chemisorption of water that forms stress concentration sites.

One of the more practical problems associated with service life of ceramics is the often observed degradation of mechanical properties attributed to attack by atmospheric water vapor. This is commonly called *stress corrosion*, is time-dependent, and is capable of decreasing both Young's modulus and fracture strength [2.67]. For more information concerning property degradation caused by corrosion, see Chap. 8.

2.2.3 Glasses

Bulk Glasses

Probably the most abundant examples of glass corrosion are those caused by a liquid. Release of toxic species (such as PbO or radioactive waste) from various glass compositions has received worldwide interest during the past 20–30 years. Although glass is assumed by many to be inert to most liquids, it does slowly dissolve. In many cases, however, the species released are not harmful.

The corrosion resistance of glasses is predominately a function of structure, which is determined by the composition. Although some have related glass durability to the number of nonbridging oxygens, a function of composition, White [2.68] has suggested that glass durability is more closely related to the presence of specific depolymerized units. He arrived at this conclusion through the correlation of vibration spectra with the effective charge on bridging and nonbridging oxygens. In a study of the leaching behavior of some oxynitride glasses, Wald et al. [2.69] reported that the nitrogen-containing glasses exhibited a greater durability (i.e., silicon release) by at least a factor of 2 than either fused silica or quartz tested under identical conditions at 200°C in deionized water for 28 days. This they attributed to the increased amount of cross-linking of the silica network and the resultant reduction in hydrolysis.

Glasses can be soluble under a wide range of pH values from acids to bases, including water. Water-soluble sodium silicates form the basis of the soluble silicate industry that supplies products for the manufacture of cements, adhesives, cleansers, and flocculants. At the other extreme are glasses designed for maximum resistance to corrosion.

The mechanism of silicate glass corrosion by water involves competition between ion exchange and matrix dissolution [2.70] that are affected by glass composition and the possible formation of a protective interfacial layer. The characteristics of this interfacial layer control subsequent dissolution. Dealkalization of this layer, which generally causes further matrix dealkalization and dissolution, is dependent upon the ease of alkali diffusion through this layer, the physical properties of the layer (i.e., porosity, thickness, etc.), and the pH of the solution. The increase in pH of the solution caused by dealkalization causes increased silica dissolution. High initial reaction rates are quite often observed and are generally caused by an excessively large exposed surface area due to microcracks or generally rough surfaces. This excessive surface area can be eliminated by proper cleaning procedures.

Jantzen [2.71] has used a thermodynamic approach to the corrosion of glasses, especially applied to nuclear waste glass leachability. The earlier work of Newton and Paul [2.72] on a wide variety of glasses was expanded and then combined with that of Pourbaix [2.73] and Garrels and Christ [2.74] to describe the effects of natural aqueous environments. Using thermodynamic hydration equations, Newton and Paul predicted glass durability from composition. Jantzen showed that the kinetic contribution was primarily a function of the test conditions (SA/V ratio,* time, and temperature). The major assumptions in Jantzen's approach were that the total free energy of hydration of the glass was the sum of the free energies of hydration of the components and that the glass structure was a primary function of glass composition. The activity–pH diagrams of Pourbaix provided the needed correlation between free energy of hydration and ion concentration in solution. Thus Jantzen was able to determine glass durability from glass composition by use of a pH-adjusted free energy of hydration term for several hundred compositions of nuclear waste glasses, manmade glasses, and natural glasses. The more negative the pH-adjusted free energy of hydration term, the less durable the glass.

Species may be leached from a glass as a result of ion exchange with protons from solution, or silica may be leached as the siloxane bonds of the matrix are attacked by hydroxyl ions from the solution. The former mechanism is predominant at low pH, whereas the latter is predominant at high pH. Hench and Clark [2.75] categorized leached glass surfaces into five groups. These groupings are listed in Table 2.5. In Types I, II,

*SA/V ratio is the ratio of the surface area of the sample to volume of the corroding liquid.

TABLE 2.5 Leached Glass Surface Types

Type I	Thin surface hydrated layer; <5 Å thick; high durability
Type II	Surface layer depleted in alkali; medium durability
Type III	Silica-rich layer adjacent to bulk and cation-rich (leached from bulk) adjacent to solution; medium durability
Type IV	Silica-rich nonprotective layer; low durability
Type V	No layer formation; lowest durability

Source: Ref. 2.75.

and III, the surface reaction layer that forms has a low solubility if composed of metal hydrates or hydrated silicates. Quite often, these layers are protective, essentially stopping further attack.

Many consider the pH of the solvent to be the most important parameter that affects glass durability [2.76]. At pH <5, ion exchange is the predominant mechanism, and at pH >9, matrix dissolution is predominant. Between pH 5 and 9, the corrosion is a minimum. This is represented schematically in Fig. 2.9. Thus dissolution is rapid when the metal–oxygen bond is extensively coordinated to hydrogen or hydroxyl ions and is a minimum under neutral conditions.

Since the driving force for dissolution of silicate glasses is the hydrogen ion activity, the loss of hydrogen ions during dissolution causes a continuous decrease in the dissolution rate. The change of the pH of the solution can drift into the basic region causing breakdown of the silica matrix with subsequent increasing dissolution rates. To determine accurate rate constants, the experimenter must hold the pH constant by use of buffer solutions or an automatic titration system. Silica solubility increases significantly above about pH = 9.

In systems containing ions of variable valence, one must consider the redox potential, Eh, of the system since the solubility of ions is dependent upon their valence state. A general rule of thumb is that higher valence states are more soluble.

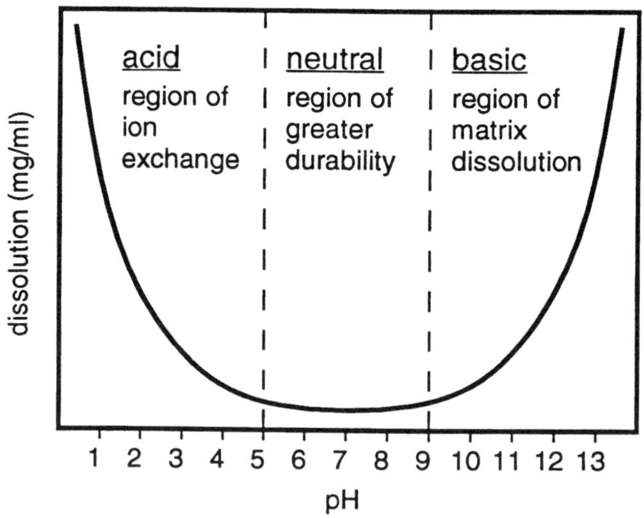

FIGURE 2.9 Effect of pH upon glass dissolution.

Hogenson and Healy [2.77] developed the following equation:

$$W = a\phi^{b_2}\exp(-b_1/T) \tag{2.25}$$

where:

W = weight loss
a = experimentally determined coefficient
b_1 = experimentally determined coefficient
b_2 = experimentally determined coefficient
ϕ = time
T = temperature

for describing the effects of time and temperature upon the acid (10% HCl) corrosion of silicate glasses. This equation, since it relates total multicomponent weight loss to time and temperature assuming a uniform surface corrosion, does not take into account the mechanism of dissolution, but instead determines the total overall corrosion. This is probably sufficient for practical problems but does not allow one to study mechanisms.

Budd [2.78] has described the corrosion of glass by either an electrophilic or a nucleophilic mechanism, or both. The surface of the glass has electron-rich and electron-deficient regions exposed. Various agents attack these regions at different rates. Exposed negatively charged nonbridging oxygens are attacked by H^+ (or H_3O^+), whereas exposed network silicon atoms are attacked by O^2, OH^-, and F^-.

Budd and Frackiewicz [2.79] found that by crushing glass under various solutions, an equilibrium pH value was reached after sufficient surface area was exposed. The value of this equilibrium pH was a function of the glass composition, and it was suggested that it was related to the oxygen ion activity of the glass. When foreign ions were present, the amount of surface required to reach an equilibrium pH was greater.

The rate of hydrolysis of a glass surface is one of the major factors that delineates the field of commercial glasses. The rate of hydrolysis is of great importance because it determines the service life of a glass with respect to weathering or corrosion and also because it influences the mechanical properties. Glass fracture is aided by hydrolysis. The rate of hydrolysis of alkali-silicate glasses of the same molar ratios proceeds in the order Rb > Cs > K > Na > Li.

The mechanism of corrosion of fluorozirconate glasses is substantially different from that of silicate-based glasses [2.80]. The fluorozirconate glass corrodes by matrix dissolution, with the components going into solution as fluorides, without first hydrolyzing as in the silicates. These glasses are also characterized by the formation of a nonprotective porous hydrated interfacial layer. Compounds highly insoluble in water remain in the porous layer. The formation of a hydroxylated zirconia fluoride complex in solution causes the pH of the solution to decrease considerably increasing the solubility of zirconia fluoride, thus increasing the overall dissolution rate by orders of magnitude.

The properties of the leached layers that build up can dramatically affect the dissolution rate since the silanol groups present can polymerize, various solutes and colloids present can react with the leached layer, and stress buildup can cause cracking and spalling. The characteristics of the leaching solution are very important, especially in long-term test, where the solution may become saturated and various crystalline phases may precipitate altering the concentration of leached species and the pH of the solution. The evaluation of glasses for hazardous waste disposal, where dissolution is over a very long time, requires careful examination of the solution characteristics.

Fiber Glass

A discussion of glass would not be complete if some mention of glass fibers were not made. The corrosion of fibers is inherently greater than bulk glass simply because of the larger surface-to-volume ratio. Since one of the major applications of fibers is as a reinforcement to some other material, the main property of interest is that of strength. Thus, any corrosion reactions that would lower the strength are of interest. This effect is important both when the fiber is being manufactured and after it has been embedded in another material. For example, the strength of E-glass (borosilicate) fibers in dry and humid environments was studied by Thomas [2.81], with the observation that humid environments lower strength. The mechanisms of environmentally enhanced stress corrosion of glass fiber are discussed in more detail in Chap. 8, page 360, Glassy Materials.

Wojnarovits [2.82] reported that multicomponent glass fibers exhibited a variation in dissolution in acid and alkaline environments due to the existence of a layered structure, each having a different dissolution rate, with the core generally having the highest rate. Single component fibers (i.e., silica) did not show this layering effect and thus no variation in dissolution rate.

Bioactive Glass

Bioactive glasses were first discovered by Hench in 1969. The special chemistry of these glasses allowed them to bond to living bone. These $Na_2O-CaO-P_2O_5-SiO_2$ glasses have been trademarked as Bioglass® and marketed under several other names depending upon the application. The beneficial effect of these glasses is their controlled release of soluble silicon and calcium ions. In this way, the glass acts as a substrate for the growth of new cells. Newer forms of these glasses have been prepared via sol–gel routes that contain numerous very fine interconnected pores. Dissolution kinetics are a function of the following variables [2.83]:

1. Composition
2. Particle size
3. Pore size distribution, average size, and volume percentage
4. Surface area
5. Thermal stabilization temperature
6. Chemical stabilization temperature

The alumina content of bioactive glasses is very important in controlling the durability of the glass surface. The bioactivity, although dependent upon the bulk composition of the glass, decreases beyond acceptable levels once the alumina content rises above 1.0–1.5 wt.% [2.49]. This same phenomenon is present for glass compositions containing cations such as Ta_2O_5 except higher levels are tolerable (1.5–3.0 wt.%).

Rare earth aluminosilicate (REAS) glasses have been developed for applications as delivery agents for radiation in the treatment of various cancerous tumors [2.84]. In these cases, the glass must be sufficiently durable to allow the release of beta-radiation over a specified period of time (about 2 weeks) while being lodged within the malignant tumor. Once the radiation treatment has been completed, then the REAS can be resorbed into the body. It is important that these glasses not dissolve while being radioactive, which would release radioactive species into the other parts of the body damaging healthy tissue. These glasses are generally incorporated into the body as microspheres about 30 μm in diameter. A ^{90}Y-containing radiotherapeutic REAS is sold under the trade name Thera-

Sphere™.* White and Day [2.84] reported no detectable weight loss of a 1×1×0.2 cm glass sample before 6 weeks in 100 mL of distilled water (pH=7) or saline (pH=7.4) at 37°C, 50°C, or 70°C. Dissolution rates of =3 × 10^{-9} g/cm^2·min were determined after 6 weeks. In a comparison study of fused silica, a Corning glass (CGW-1723™), and yttria aluminosilicate (YAS), Oda and Yoshio [2.85] showed that YAS was significantly more durable than fused silica in saturated steam at 300°C and 8.6 MPa. The dissolution mechanism is very important for applications in the human body; however, it is very difficult to determine whether these glasses exhibit congruent or incongruent dissolution. Surface analyses of microspheres and bulk glasses indicated that the mechanism was congruent [2.84]. Using inductively coupled plasma and atomic adsorption spectroscopy, it has been determined that the yttrium release from YAS microspheres in distilled water or saline at 37°C or 50°C was below detectable limits [2.86].

More recently, Conzone et al. [2.87] have reported the development of borate glasses for use in treatment of rheumatoid arthritis since these glasses are potentially more reactive with physiological liquids. Borate glasses containing only alkali ions dissolved uniformly (i.e., congruently) in simulated physiological liquids at temperatures ranging from 22°C to 75°C. When the borate glasses contained other cations (such as Ca, Mg, Fe, Dy, Ho, Sm, and Y) in amounts ranging from 2 to 30 wt.%, dissolution was nonuniform (i.e., incongruent) with the formation of new compounds. Day [2.88] gave an example of Dy_2O_3-containing borate solid glass microspheres that reacted to form hollow spheres, shells of concentric layers, or microspheres filled with homogeneous gel-like material depending upon the Dy_2O_3 content. The dissolution mechanism involved the selective leaching of lithium and boron allowing the rare earth (i.e., Dy) to react and form an

---

*TheraSphere™ is manufactured by MDS Nordion located in Ottawa, Ontario, Canada.

insoluble phosphate.* When calcium-containing borate glasses were reacted, a semicrystalline or gel calcium phosphate formed that had a composition very similar to hydroxyapatite. Although early work by Hench et al. has indicated the need for the formation of a silica gel surface layer for silicate glasses to be bioactive, the work of Day et al. has indicated that a silica gel is not always necessary for bioactivity.

In addition to the beneficial bioactive glasses discussed above, there is the extremely important area of hazardous health effects from glasses. One such case is that of inhalation of glass fibers. The dissolution of these fibers is very critical in determining their health risk. Bauer [2.89] reported the work of Eastes and Hadley that glass fibers greater than 20 μm, if inhaled, have been correlated to respiratory disease in laboratory animals. The dissolution was dependent upon the fiber surface chemistry and physical nature. The continuous movement of fluids in the human lung can increase the dissolution rate and also transport the dissolved species to other parts of the body via the blood stream. Aluminosilicate fibers were the most durable, while the dissolution rate of borosilicate fibers (e.g., home insulation) was 1000 times greater. The biopersistence of 1-μm diameter fibers varied from several days to as long as 14 years depending upon their chemistry. Annealing fibers at temperatures below the transition temperature decreased the dissolution rate in simulated extracellular fluid (pH = 7.4) by 2 to 3 times. The fact that they have not shown any major adverse reaction in human lungs was attributed by Bauer to the high dissolution rate of glass fibers.

2.3 CORROSION BY GAS

2.3.1 Crystalline Materials

The corrosion of a polycrystalline ceramic by vapor attack can be very serious, much more so than attack by either liquids or

*The phosphorus is from a phosphate-buffered saline simulated physiological liquid.

solids. One of the most important material properties related to vapor attack is that of porosity or permeability. If the vapor can penetrate the material, the surface area exposed to attack is greatly increased and corrosion proceeds rapidly. *It is the total surface area exposed to attack that is important.* Thus not only is the volume of porosity important, but the pore size distribution is also important. See Chap. 3, page 137, Porosity-Surface Area, for a discussion on porosity determination.

Vapor attack can proceed by producing a reaction product that may be either solid, liquid, or gas, as in the equation:

$$A_s + B_g \rightarrow C_{s,l,g} \tag{2.26}$$

As an example, the attack of SiO_2 by Na_2O vapors can produce a liquid sodium silicate.

In another type of vapor attack, which is really a combined sequential effect of vapor and liquid attack, the vapor may penetrate a material under thermal gradient to a lower temperature, condense, and then dissolve material by liquid solution. The liquid solution can then penetrate further along temperature gradients until it freezes. If the thermal gradient of the material is changed, it is possible for the solid reaction products to melt, causing excessive corrosion and spalling at the point of melting.

The driving force for ionic diffusion through a surface reaction layer and for continued growth is thermal energy. If sufficient thermal energy is not provided, layer growth falls off rapidly. Across very thin (<5 nm) films at low temperatures, strong electric fields may exist that act to pull cations through the film, much like that which occurs in the room-temperature oxidation of metals [2.90]. The growth of the reaction layer generally can be represented by one of the following equations for thin films:

$$y = K_1 \log t \text{ (logarithmic)}, \tag{2.27}$$

$$1/y = K_2 - K_3 \log t \text{ (inverse log), and} \tag{2.28}$$

$$y = K_4(1 - \exp[-K_5 t])\text{(asymptotic)} \tag{2.29}$$

and for thick films:

$$y^2 = K_6 t \text{ (parabolic) and} \tag{2.30}$$

$$y = K_7 \text{ (rectilinear)} \tag{2.31}$$

where:

y = film thickness
t = time
K_i = rate constant

Oxidation processes are generally more complex than the simple mechanism of a single species diffusing through an oxide layer. Preferential diffusion along grain boundaries can alter the oxide layer growth substantially. Grain boundary diffusion is a lower energy process than bulk diffusion and thus will be more important at lower temperatures. Quite often, a higher reaction rate will be observed at lower temperatures than expected if one were to extrapolate from high-temperature reaction rates. Thus the microstructure of the layer, especially grain size, is particularly important. In addition, fully stoichiometric reaction layers provide more resistance to diffusion than anion- and/or cation-deficient layers, which provide easy paths for diffusion.

Readey [2.91] has listed the possible steps that might be rate-controlling in the kinetics of gas–solid reactions. These are given below:

1. Diffusion of the gas to the solid
2. Adsorption of the gas molecule onto the solid surface
3. Surface diffusion of the adsorbed gas
4. Decomposition of reactants at surface-specific sites
5. Reaction at the surface
6. Removal of products from reaction site
7. Surface diffusion of products
8. Desorption of gas molecules from the surface
9. Diffusion away from solid

Any one of these may control the rate of corrosion.

Much attention has been given recently to the oxidation of nonoxide ceramics, especially silicon carbide and nitride. In general, the stability of nonoxides toward oxidation is related to the relative free energy of formation between the oxide and nonoxide phases. When studying the oxidation of nitrides, one must not overlook the possibility of the formation of an oxynitride, either as the final product or as an intermediate. The stability of the oxide vs. the nitride, for example, can be represented by the following equation:

$$2M_xN_y + O_2 \leftrightarrow 2M_xO + yN_2 \tag{2.32}$$

As the difference in free energy of formation between the oxide and the nitride becomes more negative, the greater is the tendency for the reaction to proceed toward the right. Expressing the free energy change of the reaction in terms of the partial pressures of oxygen and nitrogen, one obtains:

$$\Delta G^\circ = -RT \ln \frac{(pN_2)^y}{pO_2} \tag{2.33}$$

One can then calculate the partial pressure ratio required for the oxide or nitride to remain stable at any temperature of interest. For example, the oxidation of silicon nitride to silica at 1800 K yields a partial pressure ratio of nitrogen to oxygen of about 10^7. Thus very high nitrogen pressures are required to stabilize the nitride. Anytime the permeability of the product gas through the reaction layer is less than that of the reactant gas, the product gas pressure can build at the interface to very high levels with the result being bubbles and/or cracks in the reaction interface layer. This subsequently leads to continued reaction.

The reduction of oxide ceramics at various partial pressures of oxygen may also be of interest and can be obtained from the examination of Ellingham plots of $\Delta G^\circ = -RT \ln pO_2$ vs. temperature (see Fig. 2.14 in Sec. 2.7.2). If one is interested in the reduction of a binary compound, such as mullite, the presence of a second more stable oxide that forms the compound increases the stability of the less stable oxide by decreasing RT ln pO_2. Although increasing the stability of the less stable oxide, the magnitude of this change is not large enough to increase the stability of the more stable oxide. Thus the free energy of formation of mullite will be between that of silica and alumina but closer to that of silica.

The reduction of binary compounds can take place by one of the constituent oxides being reduced with decreasing oxygen partial pressure:

$$4\text{LaCo}^{3+}\text{O}_3 \rightarrow \text{La}_4\text{Co}_2^{3+}\text{Co}^{2+}\text{O}_{10} + \text{CoO} + 1/2\ \text{O}_2 \qquad (2.34)$$

a reaction that is very common when transition metals are present. These reactions become very important when applications of double oxides (or multicomponent oxides) require placement in an environment containing an oxygen potential gradient. In more general terms, this is true for any gaseous potential gradient if the gas phase is one of the constituents of the solid.

As reported by Yokokawa et al. [2.92], a double oxide may decompose kinetically even if the oxygen potential gradient is within the stability region of the double oxide. This kinetic decomposition is due to cation diffusivity differences along the oxygen potential gradient.

Another factor that might enhance the reduction of an oxide is the formation of a more stable lower oxide and the vaporization of the reaction products. An example of this is the reduction of silica by hydrogen at elevated temperature to the monoxide, which is highly volatile above 300°C.

A loss of weight by oxidation to a higher oxide that is volatile can also occur. A good example of this is the assumed vaporization of Cr_2O_3 that actually occurs through oxidation to CrO_3 gas by the following equation:

$$\text{Cr}_2\text{O}_3 + 3/2\ \text{O}_2 \leftrightarrow 2\text{CrO}_3(\text{g}) \qquad (2.35)$$

This reaction is one that is not easily proven experimentally since CrO_3 upon deposition/condensation dissociates to Cr_2O_3 and O_2. CrO_3 gas; however, it has been identified by mass spectrometry [2.93]. Diffusion of CrO_3 gas through a stagnant gaseous boundary layer was determined to be rate-controlling as opposed to the surface reaction for the reaction above [2.94].

A gas that is often encountered in practical applications is water vapor. An increase in corrosion rates when moisture is present has been reported by many investigators. This is apparently related to the ease with which gaseous hydroxide species can form.

A possible rate-controlling step in vapor attack is the rate of arrival of a gaseous reactant and also possibly the rate of removal of a gaseous product. One should realize that many intermediate steps (i.e., diffusion through a gaseous boundary layer) are possible in the overall reaction, and any one of these may also be rate-controlling. It is obvious that a reaction cannot proceed any faster than the rate at which reactants are added, but it may proceed much more slowly. The maximum rate of arrival of a gas can be calculated from the Hertz–Langmuir equation:

$$Z = \frac{P}{(2\pi MRT)^{1/2}} \tag{2.36}$$

where:

- Z = moles of gas that arrive at surface in unit time and over unit area
- P = partial pressure of reactant gas
- M = molecular weight of gas
- R = gas constant
- T = absolute temperature

Using P and M of the product gas, the rate of removal of gas product can be calculated using the same equation. To determine if service life was acceptable, these rates may be all that would be needed. Actual observed rates of removal may not agree with those calculated if some surface reaction must take place to produce the species that vaporizes. The actual difference between observed and calculated rates depends on the activation energy of the surface reaction. If the gaseous reactant was at a lower temperature than the solid material, an additional factor of heat transfer to the gas must also be considered and may limit the overall reaction.

According to Readey [2.91] in the corrosion of spheres, the rate of corrosion is proportional to the square root of the gas velocity. If the gas vapor pressure and velocity were held constant, the corrosion rate then would be proportional to the square root of the temperature. At low gas vapor pressures, transport of the gas to the surface controls the corrosion rate. At high vapor pressures, the reaction at the surface is controlling. The gaseous reaction products many times cause formation of pits and/or intergranular cracking. This can be very important for materials containing second phases (e.g., composites) that produce gaseous reaction products.

Pilling and Bedworth [2.95] have reported the importance of knowing the relative volumes occupied by the reaction products and reactants. Knowing these volumes can aid in determining the mechanism of the reaction. When the corrosion of a solid by a gas produces another solid, the reaction proceeds only by diffusion of a reactant through the boundary layer when the volume of the solid reactant is less than the volume of the solid reaction product. In such a case, the reaction rate decreases with time. If the volume of the reactant is greater than the product, the reaction rate is usually linear with time. These rates are only guidelines since other factors can keep a tight layer from forming (i.e., thermal expansion mismatch).

When a surface layer is formed by the reaction through which a gas must diffuse for the reaction to continue, the reaction can generally be represented by the parabolic rate law, which is discussed in more detail in Sec. 2.8. Jorgensen et al. [2.96] have shown that the theory put forth by Engell and Hauffe [2.97] that described the formation of a thin oxide film on metals was applicable to the oxidation of nonoxide ceramics. In this case, the rate constant being dependent upon oxygen partial pressure had the form:

$$\mathrm{k} = \mathrm{A} \ln pO_2 + \mathrm{B} \tag{2.37}$$

where A and B are constants. The driving force for diffusion was reported to be mainly an electric field across the thin film (100–200 nm thick) in addition to the concentration gradient.

2.3.2 Vacuum

It is generally believed that all materials vaporize; however, several modes of vaporization are possible. Some materials will vaporize congruently to a gas of the same composition as the solid, which is also called *sublimation*. Others will vaporize incongruently to a gas and a different condensed phase. It is also possible for more than one stable gas molecule to form. Decomposition to the elements may also occur, which is called *direct vaporization*. In multicomponent materials where the various components exhibit greatly different heats of vaporization, *selective vaporization* may occur.

The deterioration of ceramics in a vacuum in many cases is the equilibration of the material with a low partial pressure of oxygen. In such a case, a lower oxide of the metal may form along with some oxygen represented by the following equation:

$$MO_{2(s)} \rightarrow MO_{(g)} + O_{2(g)} \tag{2.38}$$

Sublimation of solid spheres controlled by gaseous diffusion through a boundary layer was first suggested by Langmuir [2.98] in 1918. The reduction in size was given by the equation:

$$r_o^2 - r^2 = \frac{KDV_oP}{RT}t \tag{2.39}$$

where:

r_o = initial radius
r = radius at time t
K = geometrical constant (~2)
D = diffusion coefficient of gas through boundary layer
V_o = molar volume of evaporating species
P = equilibrium partial pressure of gas
R = gas constant

T = temperature
t = time

2.3.3 Glasses

The corrosion of glasses by atmospheric conditions, referred to as *weathering*, is essentially attack by water vapor. Weathering occurs by one of two mechanisms. In both types, condensation occurs on the glass surface; however, in one type, it evaporates, whereas in the other, it collects to the point where it flows from the surface, carrying any reaction products with it. The latter type is very similar to corrosion by aqueous solutions. The former type is characterized by the formation of soda-rich films, according to Tichane and Carrier [2.99]. This soda-rich film has been shown to react with atmospheric gases such as CO_2 to form Na_2CO_3, according to the work of Simpson [2.100] and Tichane [2.101].

The electronics industry is one area where vapor attack of glasses may be of importance. Sealing glasses and glass envelopes have been developed that resist attack by alkali vapors and mercury vapors. In their study of some CaO- and Al_2O_3-containing glasses, Burggraaf and van Velzen [2.102] reported that alkali vapor attack increased greatly above a temperature that coincided approximately with the transformation range* (T_g) of the glass, indicating that one should use a glass with the highest possible T_g.

In the manufacture of flat glass by the Pilkington or PPG processes, glass is floated onto a bed of molten tin in a chamber containing a reducing atmosphere ($N_2 + \sim 10\% H_2$). The hydrogen present in the atmosphere above the glass can act upon the top surface of the glass causing reduction of the most reducible species present. All commercial flat glass contains some iron

* The transformation range of a glass is the range of temperatures where the glass transforms from a viscous liquid to an elastic solid upon cooling. The actual temperature of this range depends upon the cooling rate.

and that present near the top surface is predominantly in the reduced ferrous state. This is generally not a problem; however, those glasses containing NiO can exhibit small metallic droplets on the top surface that are cause for rejection. Based upon Fig. 2.14, this should not occur if the pO_2 is maintained greater than 10^{-9} atm, assuming a maximum temperature no greater than 1100°C. Johnston and Chelko [2.103] proposed the mechanism of reduction of ions in glass by hydrogen diffusion through the glass to the reducible ions that act as immobile traps reacting with the hydrogen and stopping further diffusion.

2.4 CORROSION BY SOLID

Many applications of materials involve two dissimilar solid materials in contact. Corrosion can occur if these materials react with one another. Common types of reactions involve the formation of a third phase at the boundary, which can be a solid, a liquid, or a gas. In some cases, the boundary phase may be a solid solution of the original two phases. Again, phase diagrams will give an indication of the type of reaction and the temperature where it occurs.

When the reaction that takes place is one of diffusion as a movement of atoms within a chemically uniform material, it is called *self-diffusion*. When a permanent displacement of chemical species occurs, causing local composition change, it is called *interdiffusion* or *chemical diffusion*. The driving force for chemical diffusion is a chemical potential gradient (i.e., concentration gradient). When two dissimilar materials are in contact, chemical diffusion of the two materials in opposite directions forms an interface reaction layer. Once this layer has been formed, additional reaction can take place only by the diffusion of chemical species through this layer.

Solid–solid reactions are predominantly reactions involving diffusion. Diffusion reactions are really a special case of the general theory of kinetics (discussed in Sec. 2.8) since the diffusion coefficient, D, is a measure of the diffusion reaction rate. Thus diffusion can be represented by an equation of the Arrhenius form:

$$D = D_o \exp(-Q/RT) \tag{2.40}$$

where:

D = diffusion coefficient
D_o = constant
Q = activation energy
R = gas constant
T = absolute temperature

The larger the value of Q, the activation energy, the more strongly the diffusion coefficient depends upon temperature.

The diffusion in polycrystalline materials can be divided into *bulk diffusion*, *grain boundary diffusion*, and *surface diffusion*. Diffusion along grain boundaries is greater than bulk diffusion because of the greater degree of disorder along grain boundaries. Similarly, surface diffusion is greater than bulk diffusion. When grain boundary diffusion predominates, the log concentration decreases linearly with the distance from the surface. When bulk diffusion predominates, however, the log concentration of the diffusion species decreases with the square of the distance from the surface. Thus by determining the concentration gradient from the surface (at constant surface concentration), one can determine which type of diffusion predominates.

Since grain boundary diffusion is greater than bulk diffusion, it would be expected that the activation energy for boundary diffusion would be lower than that for bulk diffusion. *The boundary diffusion is more important at lower temperatures, and bulk diffusion is more important at high temperatures.*

Chemical reactions wholly within the solid state are less abundant than those which involve a gas or liquid, owing predominately to the limitation of reaction rates imposed by slower material transport. The solid–solid contact of two different bulk materials also imposes a limitation on the intimacy of contact—much less than that between a solid and a liquid or gas.

Applications of ceramic materials commonly involve thermal gradients. Under such conditions, it is possible for one component of a multicomponent material to diffuse selectively along the thermal gradient. This phenomenon is called *thermal diffusion* or the *Sorét effect*. This diffusion along thermal gradients is not well understood, especially for ceramic materials. See Sec. 2.9 for a discussion of diffusion.

2.5 SURFACE EFFECTS

2.5.1 Surface Charge

Gibson and LaFemina [2.104] offered an excellent discussion of the various aspects of mineral surfaces and how these affect dissolution. Surfaces that have the same atomic structure (i.e., symmetry) as the bulk are termed *relaxed*. Those that are different are termed *reconstructed*. There is an excess electronic charge density associated with the broken or dangling bonds at the surface that is not present with bonds within the bulk. Different crystal faces exhibit different numbers of dangling bonds for the ions. If electrons can transfer between dangling bonds of anions and cations, then a situation arises where one ion has completely filled bonds and the other has completely empty bonds. When this occurs, the surface is charge-neutral. This will occur for surfaces that have a stoichiometric ratio of anions to cations. If the contribution of electrons from the different ions causes an excess charge density, then the surface becomes charged. The atoms on the surface will move to minimize the excess charge density associated with the dangling bonds and thus cause a localized strain. Dissolution of mineral surfaces depends upon the surface structure or arrangement of the atoms on the surface [2.105]. Gibson and LaFemina reported that the exact chemical species forming the surface is of secondary importance and that it is the atomic connectivity that dominates surface relaxations. This is important since *one need not have data on a specific material, chemically, but only on one of identical structure to estimate its dissolution characteristics.*

2.5.2 Porosity and Surface Area

The corrosion of ceramics (i.e., weight gain/loss) is proportional to the porosity; the more porous the sample, the more corrosion that is exhibited. This is in reality related to the surface area exposed to corrosion. The fact that one material may yield a better corrosion resistance than another does not necessarily make it the better material, if the two materials have different porosities. This is very important, for example, when comparing different sintering aids for silicon nitride and their effects upon oxidation. The more oxidation-resistant material may not be due to the chemical species of the sintering aid used, but, in actuality, may be due to the fact that one particular sintering aid yields a denser sintered ceramic. *One must remember that it is not the total porosity that is important, but the surface area of the total porosity*, thus making the pore size distribution an important parameter to determine.

The porosity of a ceramic can affect the overall corrosion only if the attacking medium can penetrate the porosity. Washburn [2.106] derived the following equation to determine the pore size distribution by mercury intrusion:

$$P = \frac{-2\gamma_{lv}\cos\phi}{r} \tag{2.41}$$

where P is the pressure required to force liquid into a cylindrical pore of radius r, γ is the surface tension of the liquid, and ϕ is the contact angle between the liquid and the ceramic. Although some have applied this equation to liquids other than mercury, the results are generally inaccurate due to the wetting of the solid by the liquid. Several assumptions were made by Washburn; the applied force required to force a nonwetting liquid into the pore is equal to the opposing capillary force, the void space is one of nonintersecting cylindrical pores, and that the pores exist in a graded array with the largest ones toward the outside of the ceramic as shown in Fig. 2.10. A quick glance at Fig. 2.10 should convince anyone that Washburn's assumptions are far from reality.

One of the more controversial aspects of this technique is the discrepancy between intrusion and extrusion data, which has been explained by contact angle hysteresis by Smithwick and Fuller [2.107]. Conner et al. [2.108] have shown the sensitivity of this technique to pore morphology. Moscou and Lub [2.109] reported that the hysteresis stems from a combination of both contact angle differences for intrusion and extrusion and pore morphology.

Lapidus et al. [2.110] and Conner and Lane [2.111] have compared computer simulations of mercury flow through a pore space assumed to be a pore-throat network to actual porosimetry data and found that the throats determine the intrusion behavior and the pores determine extrusion behavior. The reader is referred to any of several review papers for more detailed information [109,112,113].

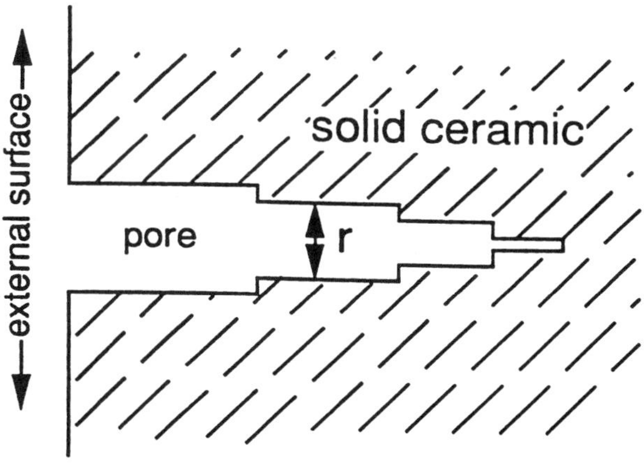

Figure 2.10 Nonintersecting cylindrical pores in a graded array becoming larger as the surface is approached, as assumed by Washburn.

One effect that is directly related to the pore size distribution is a phenomenon called *thermal transpiration*. This is the transport of gases through a ceramic caused by a thermal gradient. The relationship between pressure and temperature is given by:

$$P_1/P_2 = \sqrt{(T_1/T_2)} \tag{2.42}$$

where the subscript 1 denotes the hot face. If the gas pressure is essentially the same on both sides, gases will migrate up the thermal gradient in an attempt to make the pressure on the hot face higher. The rate of migration is inversely proportional to the square root of the molecular weight of the gas. Pore size will affect the migration since very fine pores create too great a resistance to flow and very large pores allow ordinary flow due to pressure differences. Thus at some intermediate pore size, transpiration will occur. In ceramics with a large pore size distribution, ordinary flow tends to equalize the pressures, minimizing flow by transpiration. There are no known reports in the literature indicating that thermal transpiration influences corrosion of ceramics; however, it may suggest a means to minimize the effects from corrosive ordinary flow. *If sufficient flow of the transpiring gas is present, dilution of the corrosive gas at the hot face may lower the corrosion rate to an acceptable level.*

The manufacturers of flat glass by one of the float processes* are well aware of the problems that thermal transpiration may cause. Although not a corrosion process, defective glass has been produced by gases transpiring up through the tin bath bottom blocks, rising through the tin, and then causing an indent in the bottom surface of the glass. In some cases, the gas pressure has been sufficient to puncture completely through

*Two somewhat different processes are currently being used today to manufacture flat glass by floating molten glass onto molten tin.

the glass ribbon. To eliminate this problem, bath bottom blocks are manufactured to a specific pore size distribution.

A purely surface area effect, which is very important in the corrosion of asbestos or chrysotile fibrils, is that related to ledge effects. As one can see from Fig. 2.11, ledges can greatly increase the exposed surface area. This is extremely important in the dissolution of spiral fibrils and their related health effects.

Similar structural effects can be present due to dislocations and other defects (see Fig. 2.11a). Chrysotile is a two-layer sheet silicate with a dimensional misfit between the octahedral and tetrahedral layers. This causes the sheet to curl forming spiral fibrils. This property causes some confusion since chrysotile is a sheet silicate, not a chain silicate, although both have properties related to fibrous materials.

Surface areas determined from sample geometry are generally many times smaller than that determined from BET measurements. This difference can be attributed to the presence of microscopic surface features. Thus one must be careful how surface areas are determined and how these data are related to the subsequent dissolution data.

2.5.3 Surface Energy

The surface energy of a material is the ratio of the potential energy difference obtained when moving an atom from the bulk to the surface to the area of the surface. A term that is closely related is the surface tension, which is the force required to move the atom to the surface divided by its diameter. Since liquids cannot maintain a shear stress, the surface energy and surface tension of liquids are equivalent. This is not the case for solids where the surface energy is generally greater than the surface tension. In general, the symbol used to represent surface tension is σ, whereas the one used to represent surface energy is γ.

One area where the surface energies play a very important role is in the movement of liquids into capillaries. For ceramics,

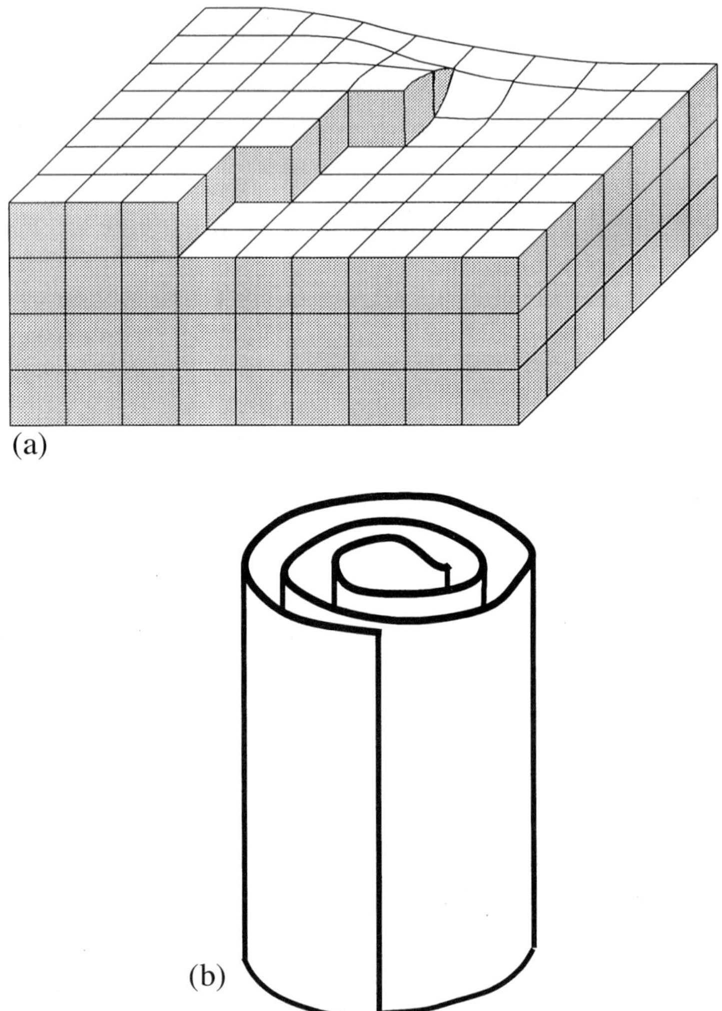

FIGURE 2.11 Surface area increase (a) due to dislocations and defects and (b) due to ledge effect in fibrils.

capillaries can be considered to be very small connected pores. Equation (2.43) represents the equilibrium pressure difference at the interface (i.e., liquid–vapor) due to surface energy:

$$\Delta P = 2\gamma(\cos\phi)/R \tag{2.43}$$

where:

ΔP = pressure drop across interface
γ = surface energy of liquid
ϕ = contact angle of meniscus at wall of capillary
R = radius of capillary

In actual materials, the porosity is not a cylindrical cavity and therefore one must use an effective radius that represents the weighted average of the contributing porosity. In addition, Eq. (2.43) is valid for only an empty capillary. As soon as any liquid penetrates the capillary, the driving force, ΔP, decreases.

The relationship of the surface energies among the solid–vapor interface, solid–liquid interface, and the liquid–vapor interface is given by:

$$\cos\phi = \frac{\gamma_{sv} - \gamma_{sl}}{\gamma_{lv}} \tag{2.44}$$

When the contact angle, ϕ, is less than 90°, capillary attraction will allow the liquid to fill the pores displacing the gas within without any applied force. When the contact angle is greater than 90°, an applied force, P, is required to force the liquid into the pores. The pressure exerted upon a ceramic in service will depend upon the height and density of the liquid. When this pressure is greater than P, the liquid will enter the pores that have a radius greater than R.

Carrying this one step further, the penetration of liquids between like grains of a ceramic can be predicted from the interfacial surface energies of the liquid–solid and solid–solid interfaces according to Smith [2.114] since if:

$$\gamma_{ss} \geq 2\gamma_{sl} \tag{2.45}$$

complete wetting will occur. If

$$\gamma_{ss} \leq 2\gamma_{sl} \tag{2.46}$$

solid–solid contact is present and the liquid will occur in discrete pockets. A balance of forces exists when:

$$\gamma_{ss} = 2\gamma_{sl}\cos\left(\frac{\phi}{2}\right) \tag{2.47}$$

where ϕ is the dihedral angle between several grains and the liquid. Thus Eq. (2.45) is valid when $\phi > 60°$ and Eq. (2.46) is valid when $\phi < 60°$. For this reason, 60° has been called the *critical dihedral angle* that separates the conditions of complete wetting and nonpenetration of the second phase between grains of the major phase. Although some data on dihedral angles exist as discussed later, very little actual data have been reported. The general factors that cause variation in the dihedral angle, however, are often mentioned (see discussion below).

The balance of forces (see Fig. 2.12) holds well for grains that tend to be rounded. If marked crystallographic faces exist, Eq. (2.47) is no longer valid. Surface forces are then no longer tangential and isotropic, which was assumed in the derivation of Eq. (2.47). However, if

$$\gamma_{ss} > \sqrt{3}\,\gamma_{sl} \tag{2.48}$$

the liquid occurs only at three grain intersections or triple points. Thus one would desire that γ_{ss} be $<2\gamma_{sl}$ and at least $<\sqrt{3}\gamma_{sl}$ to minimize liquid penetration into the ceramic. This balance of forces is affected, however, by many things, one important factor being the temperature. Composition and grain size will also affect the overall balance of forces as discussed below.

Due to the random orientation of the three-grain junctions in polished sections, the determination of ϕ varies between 0° and 180°, even when it is constant throughout the structure. In

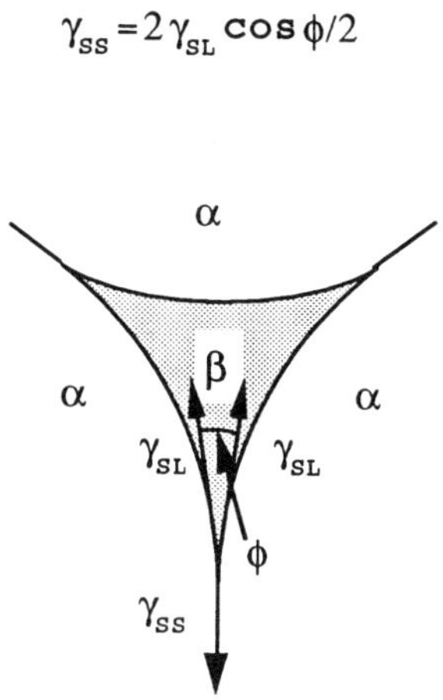

FIGURE 2.12 Balance of surface energy forces between a major and a secondary grain boundary phase (α = solid grains; β = liquid grain boundary phase; γ_{ss} = surface energy between two solid grains; γ_{sl} = surface energy between solid and liquid; and ϕ = dihedral angle).

this case, the median of a large number of determinations is taken as the dihedral angle value.*

White [2.115], in his studies of refractory systems, has shown that as the temperature increased, the dihedral angle decreased. He has also shown the effects of composition upon the dihedral angle in 85% MgO–15% Ca–Mg–silicate liquids at 1550°C in air. These effects are shown in Table 2.6. White reported that as the concentration of solid in the saturated liquid increased, the dihedral angle decreased, which is the same as the effect of temperature. Since the curvature of the grains must decrease as the dihedral angle increases, larger grains will produce a smaller dihedral angle. In addition, White

*The evaluation of a 3-D parameter by use of a 2-D array will always lead to a variation in values. Obtaining a representative value requires the determination of a large number of individual dihedral angles. The final average value is never the same as the actual value, but something less.

TABLE 2.6 Effects of Composition upon the Dihedral Angle

Substitution for MgO[a]	Amount (%)	Temperature (°C)	Dihedral Angle (°)
Cr_2O_3	0–10	1550	25–45
Fe_2O_3	0–5	1550	25–20
Al_2O_3	0–5	1550	25–20
TiO_2	0–2	1550	25–15
Cr_2O_3	5	1550–1725	40–30

[a] Substitution for MgO in an 85% MgO–15% CMS composition.
Source: Ref. 2.115.

showed that the dihedral angle between like grains was smaller than that between unlike grains, indicating that the penetration of liquid between unlike grains should be less than between like grains.

The nature of the bonding type of the solid being attacked compared to that of the attacking medium often can give an indication as to the extent of wetting that may take place. For example, transition metal borides, carbides, and nitrides, which contain some metallic bond character, are wet much better by molten metals than are oxides, which have ionic bond character [2.116]. Various impurities, especially oxygen, dissolved in the molten metal can have a significant effect upon the interfacial surface energies. For example, Messier [2.117] reported that silicon wet silicon nitride at 1500°C in vacuum but did not spread due to oxygen contamination. In most cases, it is the nature of the grain boundary or secondary phases that is the controlling factor.

Puyane and Trojer [2.118] examined the possibility of altering the wettability of alumina by using additives to their glass composition. They found that V_2O_5 and CeO_2 additions changed the surface tension of the glass in opposite directions, with V_2O_5 decreasing it and CeO_2 increasing it. They concluded that the *glass characteristics were more important than the solid parameters in corrosion*.

2.6 ACID/BASE EFFECTS

The chemical species present in the liquid will determine whether it is of an acidic or basic character. *Ceramics with an acid/base character similar to the liquid will tend to resist corrosion the best.* In some cases, the secondary phases of a ceramic may be of a slightly different acid/base character than the major component, and thus whether the major phase or the secondary bonding phase corrodes first will depend upon the acid/base character of the environment.

Several acid–base reaction theories have been proposed. The Brönsted and Lowry theory may be sufficient to explain those reactions in aqueous media where the acid/base character of a surface is determined by its zero point of charge (zpc) or the pH where the immersed surface has a zero net surface charge. In nonaqueous media, the Lewis theory is probably more appropriate when acids are defined as those species that accept a pair of electrons thus forming a covalent bond with the donor, and bases are defined as those species that donate a pair of electrons thus forming a covalent bond with the acid. Ionization may follow formation of the covalent bonds. Those species that can both accept or donate electrons depending upon the character of its partner are called *amphoteric*. Thus a particular species may act as an acid toward one partner but as a base toward another. Oxidizing agents are similar to acids since they tend to accept electrons; however, they keep the electrons to themselves rather than share them.

Carre et al. [2.119] have devised a simple approach to calculations of the zpc from ionization potentials of the metallic elements contained in pure oxides. Those values differ very little from those determined by Parks [2.45]. They used an additive method to calculate the zpc of multicomponent glasses. *The importance of the zpc in corrosion is that it is the pH of maximum durability*. The approach of Carre et al. is fundamentally very similar to that of Lewis since oxide acidity depends upon the electron affinity of the metal, whereas O^{2-} anions act as the basic component.

According to Carre et al., abrading or grinding the surface of various glasses increases the zpc (e.g., soda-lime glass zpc increased from about 8.0 to 12.0) supposedly by increasing the alkalinity at the surface. Acid washing produces just the opposite effect, decreasing zpc caused by leaching the alkali from the surface.

2.7 THERMODYNAMICS

The driving force for corrosion is the reduction in free energy of the system. The reaction path is unimportant in thermodynamics, only the initial and final states are of concern. In practice, intermediate or metastable phases are often found when equilibrium does not exist and/or the reaction kinetics are very slow. In general, a reaction may occur if the free energy of the reaction is negative. Although the sign of the enthalpy (or heat) of reaction may be negative, it is not sufficient to determine if the reaction will proceed. The spontaneity of a reaction depends upon more than just the heat of reaction. There are many endothermic reactions that are spontaneous. To predict stability, therefore, one must consider the entropy. Spontaneous, irreversible processes are ones where the entropy of the universe increases. Reversible processes, on the other hand, are those where the entropy of the universe does not change. At low temperatures, exothermic reactions are likely to be spontaneous because any decrease in entropy of the mixture is more than balanced by a large increase in the entropy of the thermal surroundings. At high temperatures, dissociative reactions are likely to be spontaneous, despite generally being endothermic, because any decrease in the thermal entropy of the surroundings is more than balanced by an increase in the entropy of the reacting mixture.

In the selection of materials, an engineer wishes to select those materials that are thermodynamically stable in the environment of service. Since this is a very difficult task, knowledge of thermodynamics and kinetics is required so that materials can be selected that have slow reaction rates and/or harmless reactions. Thermodynamics provides a means for the engineer to understand and predict the chemical reactions that take place. The reader is referred to any of the numerous books on thermodynamics for a more detailed discussion of the topic [2.120–2.122].

2.7.1 Mathematical Representation

The enthalpy and entropy are related through the free energy. The change in free energy of an isothermal reaction at constant pressure is given by:

$$\Delta G = \Delta H - T\,\Delta S \tag{2.49}$$

where:

G = Gibbs free energy
H = enthalpy or heat of formation
T = absolute temperature
S = entropy of reaction

The change in free energy of an isothermal reaction at constant volume is given by:

$$\Delta F = \Delta E - T\,\Delta S \tag{2.50}$$

where:

F = Helmholtz free energy
E = internal energy

From Eqs. (2.49) and (2.50), it is obvious that the importance of the entropy term increases with temperature. *The reactions of concern involving ceramic materials are predominately those at temperatures where the entropy term may have considerable effect on the reactions.* In particular, species with high entropy values have a greater effect at higher temperatures.

Gibbs free energy is a more useful term in the case of solids since the external pressure of a system is much easier to control than the volume. The change in free energy is easy to calculate at any temperature if the enthalpy and entropy are known.

Evaluation of Eq. (2.49) will determine whether or not a reaction is spontaneous. *If the reaction is spontaneous, the change in free energy is negative, whereas if the reaction is in equilibrium, the free energy change is equal to zero.*

The free energy change for a particular reaction can be calculated easily from tabulated data, such as the JANAF Tables [2.123], by subtracting the free energy of formation of the reactants from the free energy of formation of the products. An example of the comparison of free energy of reaction and the enthalpy of reaction at several temperatures is given below for the reaction of alumina and silica to form mullite:

$$3Al_2O_3 + 2SiO_2 \rightarrow Al_6Si_2O_{13} \tag{2.51}$$

Using the following equations to calculate the enthalpy and free energy change from enthalpy and free energy of formation data given in the JANAF tables, assuming unit activity for all reactants and products, one can easily determine if the formation of mullite is a spontaneous reaction at the temperature in question:

$$\Delta G_r = \sum \Delta G_f \text{ (products)} - \sum \Delta G_f \text{ (reactants)} \tag{2.52}$$

$$\Delta H_r = \sum \Delta H_f \text{ (products)} - \sum \Delta H_f \text{ (reactants)} \tag{2.53}$$

Using the values from Table 2.7, one then calculates:

$$\begin{aligned}
\Delta H_r &= (-6846.78) - \{3(-1688.91) + 2(-899.808)\} \\
\Delta H_r &= +19.587 \text{ kJ/mol} \\
\Delta G_r &= (-5028.75) - \{3(-1229.39) + 2(-661.482)\} \\
\Delta G_r &= -17.609 \text{ kJ/mol}
\end{aligned}$$

It can be seen that although the enthalpy of reaction is positive, the free energy of reaction is negative and the reaction is spontaneous at 1400 K and mullite is the stable phase, allowing one to predict that alumina will react with silica at that temperature.

Tabulations of the standard free energy, ΔG°, at 1 bar and 298 K, as a function of temperature are available for the more common reactions [2.123,2.124]. For less-common reactions,

TABLE 2.7 Enthalpy and Free Energy of Formation at 1400 K

Material	ΔH_f (kJ/mol)	ΔG_f (kJ/mol)
Mullite	−6846.78	−5028.75
Alumina	−1688.91	−1229.39
Cristobalite	−899.81	−661.48

one must calculate the free energy of reaction by using values of ΔH°, ΔS°, and heat capacity data. Heat capacities can be experimentally determined by differential scanning calorimetry up to about 1000 K [2.125] as can heats of reaction. The change in entropy cannot be obtained directly from thermal measurements. If one must do his own calculations, various computer programs are also available to aid the investigator [2.24,2.126]. The data of these tables are always in different stages of the confirmation process and can thus vary widely in accuracy. Therefore it is in the best interest of the user to check the source of the data.

The real problem with predicting whether a reaction may take place or not is in selecting the proper reaction to evaluate. Care must be taken not to overlook some possible reactions.

Other forms of the free energy equation can be useful when evaluating corrosion by specific mechanisms. If the reaction is one of electrochemical nature, the free energy change for the reaction can be calculated using:

$$\Delta G = -nFE \tag{2.54}$$

where:

n = number of electrons involved
F = Faraday constant (96,490 C/mol)
E = standard cell potential (volts)

Tabulations of standard half-cell potentials (standard emf series) are available and are more commonly called *redox potentials* [2.127]. The use of the emf series for studies in aqueous solutions has been established for a long time and has now been extended to nonaqueous electrolytes such as molten salt mixtures. According to Brenner [2.128], who reported average errors of 32% between calorimetric and emf measurements, the use of Eq. (2.54) is not accurate and it should be modified as required for each galvanic cell evaluated.

Although industrial process gas streams are generally not in thermodynamic equilibrium, their compositions are shifting toward equilibrium at the high temperatures normally encountered. Using equilibrated gas mixtures for laboratory studies then is a basis for predicting corrosion but is not necessarily accurate. Which reaction products form at solid/gas interfaces can be predicted from free energy calculations using the following equation:

$$\Delta G^{\circ} = -RT \ln\left\{\frac{(p_c)^u (p_d)^w}{(p_a)^x (p_b)^y}\right\} \tag{2.55}$$

where p = partial pressure of each component of the reaction

$$xA + yB = uC + wD \tag{2.56}$$

The bracketed expression inside the logarithm in Eq. (2.55) is the equilibrium constant for the reaction, thus:

$$\Delta G^{\circ} = -RT \ln k_p \text{ (the well–known Nernst equation)} \tag{2.57}$$

When pure solids are involved in reactions with one or more nonideal gaseous species, it is more relevant to work with activities rather than compositions or pressures. Therefore the equilibrium constant can be expressed in terms of activities:

$$k = \frac{(a_c)^u (a_d)^w}{(a_a)^x (a_b)^y} \tag{2.58}$$

where the subscripts a and b denote reactants and c and d denote the products. The activity is the product of an activity coefficient and the concentration for a solute that does not dissociate. The solute activity coefficient is taken as approaching unity at infinite dilution. If the solute were an electrolyte that is completely dissociated in solution, the expression for the activity would be more complicated. A few assumptions that are made in the use of Eqs. (2.55) and (2.58) are that the gases behave as ideal gas mixtures, that the activity of pure solids is equal to 1, and the gas mixture is in equilibrium. In those cases where the ideal gas law is not obeyed, the fugacity is used in place of the activity to maintain generality. The assumption that the gases are ideal is not bad since one is generally concerned with low pressures. The assumption of unity for the activity of solids is true as long as only simple compounds are involved with no crystalline solution. The assumption of equilibrium is reasonable near surfaces since hot surfaces catalyze reactions.

If one is interested in the dissociation pressure of an oxide, Eq. (2.57) can be used where the equilibrium constant is replaced with the partial pressure of oxygen (pO_2) since, for ideal gas behavior, the activity is approximately equal to the partial pressure. If the oxide dissociates into its elements, the measured vapor pressure is equal to the calculated dissociation pressure. If the oxide dissociates into a lower oxide of the metal forming a stable gas molecule, the vapor pressure measured is greater than the calculated dissociation pressure. A compilation of dissociation pressures was given by Livey and Murray [2.129]. At moderate to high temperatures and atmospheric pressure, however, the fugacity and partial pressure are almost equal. Thus for most ceramic systems, the partial pressure of the gas is used, assuming ideality.

An example where a pure solid reacts to form another pure solid and a gas is that of calcite forming lime and carbon dioxide. The equilibrium constant is then independent of the amount of solid as long as it is present at equilibrium.

$$CaCO_3 \rightarrow CaO + CO_2 \tag{2.59}$$

$$k = \frac{(a_{CaO})(a_{CO_2})}{a_{CaCO_3}} \tag{2.60}$$

rearranging:

$$a_{CO_2} = \frac{(a_{CaCO_3})k}{a_{CaO}} = k \tag{2.61}$$

or:

$$pCO_2 = k_p \,(\text{equilibrium reaction constant at constant pressure}) \tag{2.62}$$

At constant temperature, if the partial pressure of CO_2 over $CaCO_3$ is maintained at a value less than k_p, all the $CaCO_3$ is converted to CaO. If the partial pressure of CO_2 is maintained greater than k_p, then all the CaO will react to form $CaCO_3$. This type of equilibrium, involving pure solids, is different from other chemical equilibria that would progress to a new equilibrium position and not progress to completion.

An example, similar to the above description for Eq. (2.57), for a reaction when both the reactants and products are all solid phases was given by Luthra [2.130] for the reaction of an alumina matrix with SiC reinforcement fibers. The following equation depicts this reaction:

$$2Al_2O_3 + 3SiC \leftrightarrow 3SiO_2 + Al_4C_3 \tag{2.63}$$

where the silica activity is dependent upon the alumina activity, assuming the activities of both SiC and Al_4C_3 are unity. This is given by:

$$a_{SiO_2} = \left[k(a_{Al_2O_3})^2\right]^{1/3} \tag{2.64}$$

If the silica activity in the matrix is greater than the equilibrium silica activity, no reaction will occur between the matrix and the fiber. Since the activities of both silica and alumina are very small, minor additions of silica to the alumina matrix will prevent matrix/fiber reaction. Thus the use of small mullite additions prevents this reaction.

Since the corrosion of ceramics in service may never reach an equilibrium state, thermodynamic calculations cannot be strictly applied because these calculations are for systems in equilibrium. Many reactions, however, closely approach equilibrium, and thus the condition of equilibrium should be considered only as a limitation, not as a barrier to interpretation of the data.

2.7.2 Graphical Representation

The thermodynamics of reactions between ceramics and their environments can be best represented by one of several different types of stability diagrams. Graphs provide the same information as the mathematical equations; however, they can display unexpected relationships that provide new insight into solving a problem. Various types of graphical representations emphasize different aspects of the information and thus are well suited only to a specific problem. Fig. 2.13 is a schematic representation for each of the various types of diagrams that one may find in the literature. Probably the most common type of graphical representation of thermodynamic data is the *equilibrium phase diagram* [2.1]. These are based upon the Gibbs Phase Rule, which relates the physical state of a mixture with the number of substances or components that make up the mixture and with the environmental conditions of temperature and/or pressure. The region above the *solidus* is of greatest importance in most corrosion studies. The *liquidus* lines or the boundary curves between the region of 100% liquid and the region of liquid plus solid determine the amount of solid that can be dissolved into the liquid (i.e., saturation composition) at any temperature. For this reason, these curves are also called *solubility or saturation curves*. Thus, these curves give the mole fraction (or weight fraction) at saturation as a function of temperature. To obtain concentrations, one must also know the density of the compositions in question.

Another type of diagram is a graphical representation of the standard free energy of formation of the product between a metal and 1 mol of oxygen as a function of temperature at a constant total pressure. These are called *Ellingham diagrams* [2.131]. Richardson and Jeffes [2.132] added an oxygen

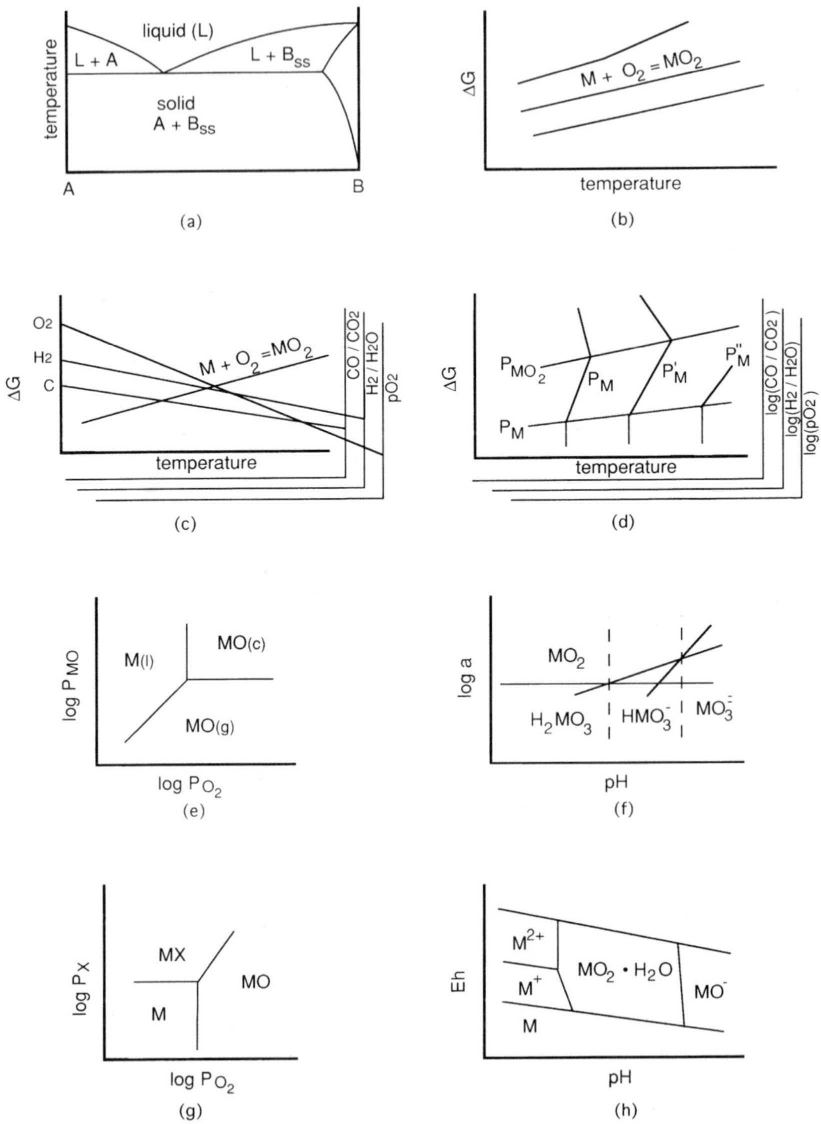

FIGURE 2.13 Representation of thermodynamic data: a) phase equilibrium diagram, b) Ellingham diagram, c) Darken and Gurry modified Ellingham diagram, d) Lou et al. modified Ellingham diagram, e) volatility diagram, f) stability diagram, g) phase stability or Kellogg diagram, h) Pourbaix diagram.

nomograph scale to the Ellingham diagram so that one could also determine the reaction for a certain partial pressure of oxygen in addition to the temperature. Since CO/CO_2 and H_2/H_2O ratios are often used in practice to obtain various partial pressures of oxygen (especially the very low values), Darken and Gurry [2.133] added nomograph scales for these ratios. These diagrams now can be found in many places containing various numbers of oxidation/reduction reactions and have been referred to as *Ellingham*, *Ellingham–Richardson*, *Darken* and *Gurry*, or *modified Ellingham diagrams*. On these plots (Fig. 2.14), the intercept at $T = 0$ K is equal to ΔH° and the slope is equal to $-\Delta S^\circ$.

To use the diagram shown in Fig. 2.14, one needs only to connect the point representing zero free energy at the absolute zero of temperature (e.g., the point labeled O to the left of the diagram) and the point of intersection of the reaction and temperature in question. As an example, for alumina at 1400°C, this line intersects the pO_2 scale at about 10^{-24} atm, the equilibrium partial pressure of oxygen for the oxidation of aluminum metal to alumina. Any pressure lower than this will cause alumina to be reduced to the metal. *This leads to the general tendency for oxides to be reduced at higher temperatures at constant oxygen partial pressures. One should also be aware that any metal will reduce any oxide above it in this diagram.*

One should remember that all condensed phases of the reactions plotted in Fig. 2.14 are assumed to be pure phases and therefore at unit activity. Deviations from unit activity are encountered in most practical reactions. The correction that is applied is proportional to the activities of the products to that of the reactants by use of Eqs. (2.55) and (2.58). As an example for the manufacture of glass containing nickel, the NiO activity is less than unity due to its solution in the glass. The correction term would then be negative and the free energy plot would be rotated clockwise. This change in slope can considerably affect the equilibrium partial pressure of oxygen required to maintain the nickel in the oxidized state. In this case, the lower activity is

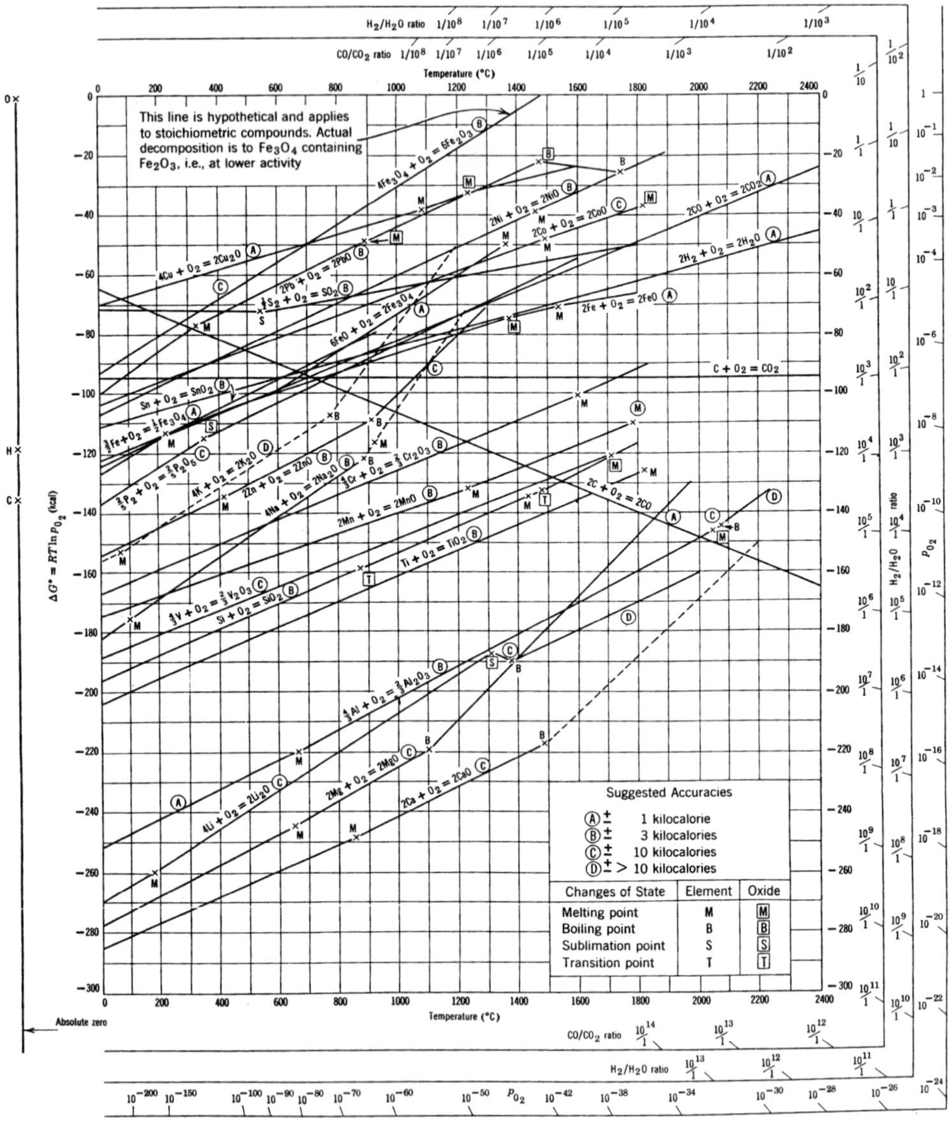

FIGURE 2.14 The standard free energy of formation of many metal oxides as a function of temperature. (From Ref. 2.133, reproduced with permission of McGraw-Hill Companies.)

beneficial since the nickel will remain in the oxidized state at lower partial pressures of oxygen at any given temperature. Many reactions that do or do not occur based upon examination of Fig. 2.14 can be explained by nonunit activities.

Since greater values of negative ΔG° indicate greater stability of an oxide with respect to its elements, Ellingham diagrams are excellent for determining the relative stability of oxides in contact with metals; however, they contain no information about the various vapor species that may form. Lou et al. [2.134] have described a modified Ellingham diagram containing vapor pressure information. They have combined the information of *volatility diagrams* (isothermal plots of partial pressure relationships between two gaseous species in equilibrium with the condensed phases) with that of Ellingham-type information to derive a diagram for the free energy changes vs. temperature at various vapor pressures for individual oxides. The example for aluminum is shown in Fig. 2.15. This diagram is a plot of pO_2 (actually, $RT \ln pO_2$) and temperature for various $pAlO_x$ values. Line 6 is the boundary for the transition from Al solid or liquid to Al_2O_3 solid or liquid; line 7 is the boundary for transition of the principal vapors from Al to AlO_2. The vapor pressure of Al over solid Al_2O_3 is shown as a series of lines sloping toward the right in the center portion of the diagram. The upper dashed line is the isomolar line that defines the maximum pAl over Al_2O_3 in a nonreactive system (i.e., vacuum or inert gas). The lower dashed line is constructed from isobaric points that represent the maximum Al vapor pressure allowed for any hydrogen pressure at a particular temperature (based upon the reaction $Al_2O_3 + 3H_2 \rightarrow 2Al_{(g)} + 3H_2O_{(g)}$). For example, at 1800°C, the maximum predicted vapor pressure of Al over solid Al_2O_3 would be 10^{-3} Pa and the maximum pO_2 would be $10^{-3.3}$ Pa.

The free energy is also related to the dissociation pressure of the product; thus other types of graphical representations are also available in the literature. These are generally isothermal plots of the gaseous partial pressures in equilibrium with the condensed phases and have been called *volatility diagrams*,

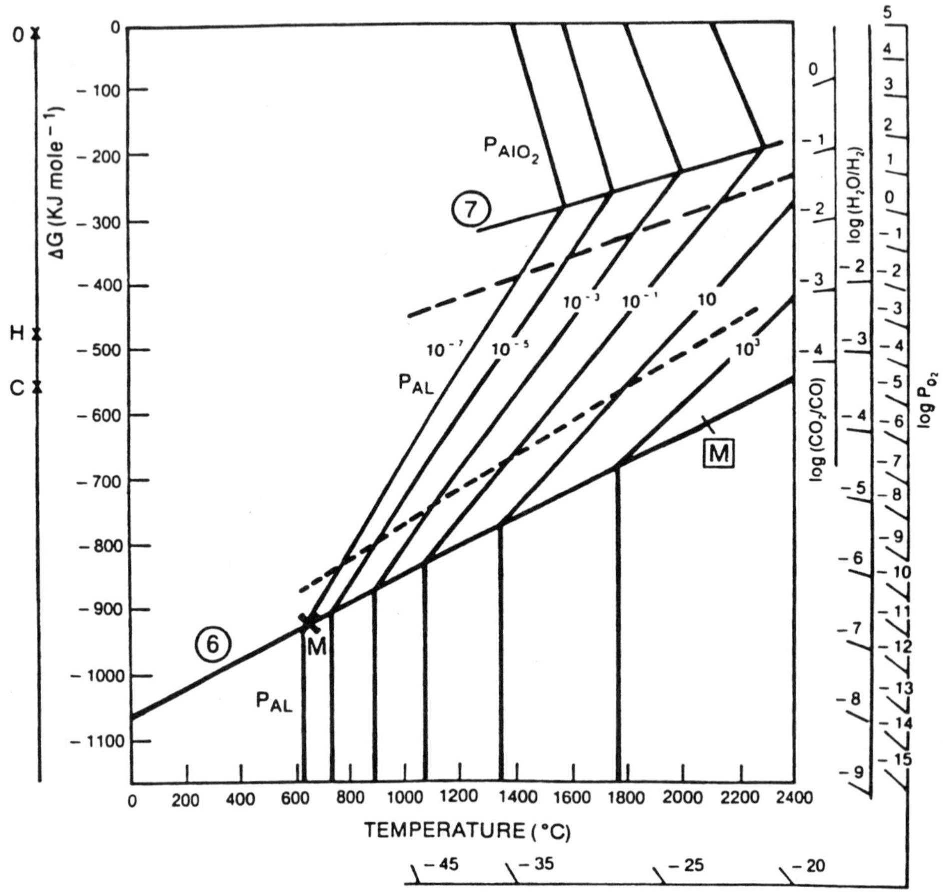

FIGURE 2.15 Ellingham type diagram for the Al–O system. (From Ref. 2.134, reprinted with permission of The American Ceramic Society, www.ceramics.org.)

volatility maps, or *phase stability diagrams* [2.132,2.133]. A similar type of diagram can be obtained when two oxidants are present (i.e., O_2 and N_2) as long as all possible condensed phases are known. Diagrams for systems such as metal–oxygen–carbon are available [2.135]. An assumption that is usually made that is not always true is that the condensed phases are at unit activity. *Unit activity should be applied only to species in the pure state*. When more than one gaseous species is involved in the reaction, *volatility diagrams* are more appropriate.

Many cases of corrosion of ceramic materials take place in an aqueous media (e.g., weathering of window glass). In these cases, the pH of the system becomes important. Pourbaix [2.73] first suggested the use of redox potential (E) vs. pH plots to predict direction of reaction and the phases present. These plots, now called *Pourbaix diagrams*, are graphical representations of thermodynamic and electrochemical equilibria in aqueous systems. Fig. 2.16 is a *Pourbaix diagram* of the system aluminum–water at 25°C. The two dashed lines labeled a and b in Fig. 2.16 enclose the region where water is stable. At any potential and pH above the top line (b), water decomposes evolving oxygen. At any potential and pH below the lower line (a), water decomposes evolving hydrogen. These diagrams delineate three major regions of interest. The first is the region where no reaction occurs to the metal (i.e., the region of immunity), generally the lower portion of the diagram. The second is the region of corrosion where the metal reacts to form an ion, generally the upper left region of the diagram. This second region is the one of most interest to the ceramist studying corrosion. The third is the region of passivity where the metal reacts to form an insoluble species that may be protective (generally an oxide), generally the upper right portion of the diagram. Garrels and Christ [2.74] have extensively developed Pourbaix's concept for use in describing the action of water upon soils. These diagrams, related to soil–water systems, have been called *Garrels and Christ diagrams*. In aqueous dissolution studies, it is also convenient to plot the pH of the solution vs. the logarithm of the concentration of the species dissolved (*solubility diagrams*).

2.8 KINETICS

It is normally expected that materials will corrode, and thus it is important to know the kinetics of the reaction so that predictions of service life can be made. *Thus the most important*

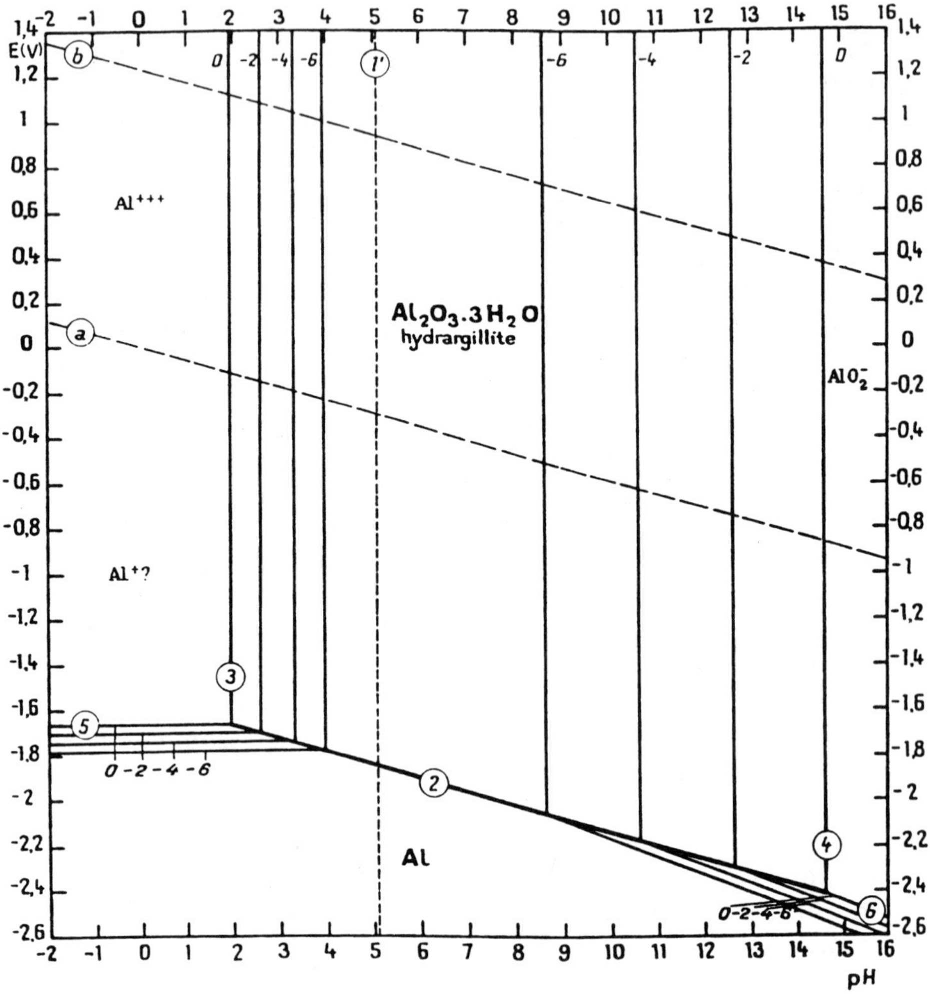

FIGURE 2.16 Potential-pH equilibrium diagram for the system alumina–water at 25°C. (From Ref. 2.73, reprinted with permission of NACE International and CEBELCOR.)

parameter of corrosion from the engineering viewpoint is the reaction rate. Systems can often exist for extended periods of time in a state that is not the equilibrium state or the state of lowest free energy. These states are called *metastable states* and may occur for many reasons. One case is where a surface reaction forms a diffusion barrier that blocks or drastically diminishes further reaction. In another more important case, for the reaction to proceed to the lowest free energy state, it must first pass through an intermediate state where the energy is higher than either the initial or final states. The energy required to overcome this barrier is called the *activation energy* and the net energy released is the *heat of reaction*. This is depicted in Fig. 2.17 where the movement of an atom from an initial metastable state (a) to the final stable state (c) requires passage through the higher energy unstable state (b). The reaction is exothermic in going from (a) to (c) and endothermic in the reverse direction. The activation energy for the reverse direction obviously must be greater than for the forward direction. The speed of the reaction is dependent upon the total number of atoms in the metastable state, the vibration frequency of the atoms, and the probability that an atom during vibration will have the necessary energy to overcome the

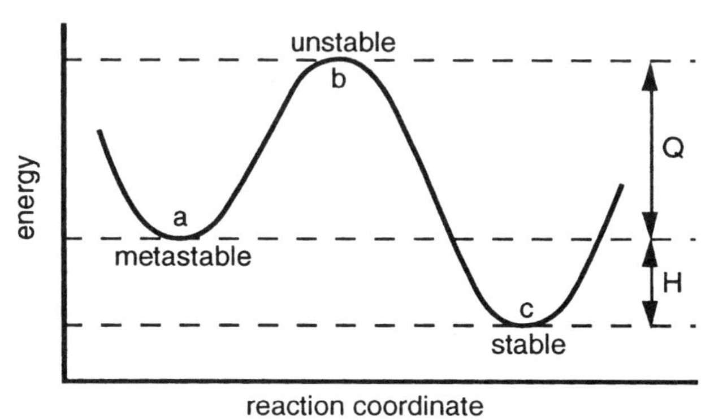

FIGURE 2.17 Energy barrier diagram (Q = activation energy and H = heat of reaction).

barrier. If sufficient energy is not acquired to overcome the activation energy barrier, the system will remain indefinitely in the metastable state. The number of atoms that pass over the barrier is then the rate of the reaction and is given by:

$$\text{Reaction rate} = Ae^{-Q/RT} \tag{2.65}$$

where A is a constant containing the frequency term and Q is the activation energy. Expressing this equation in logarithmic form, one obtains:

$$\ln(\text{rate}) = \ln A - (Q/R)/T \tag{2.66}$$

A plot of ln(rate) vs. reciprocal temperature yields Q/R as the slope and the intercept at $1/T = 0$ yields A.

The effect of temperature upon the reaction rate can be seen by the following example. Suppose that $Q = 45$ kcal/mol, a number not unreasonable for many ceramic reactions, and that R = 2.0 cal/mol K. Calculation of the exponential term yields a rate that is approximately 10^{23} times as fast at 1000 K as it is at 300 K. Thus if a reaction takes 1 sec at 1000 K, it takes on the order of 10^{12} years at room temperature. This is the basis of quenching and allows one to examine reactions at room temperature that have occurred at high temperature.

Quite often, a plot of the logarithm of the corrosion rate vs. the inverse temperature yields a straight line, indicating that corrosion is an activated process. Attempting to correlate various ceramic material properties to these activation energies, however, can be very misleading. *Generally, the range of activation energies experimentally observed for different materials is very large and any interpretation is difficult since diffusion coefficients depend upon the composition and structure of the material through which diffusion occurs*. Since the interface composition generally changes with temperature, the driving force for diffusion also changes with temperature, neither of which has any relationship to an activated process.

When a substance increases the reaction rate but is not itself consumed in the reaction, it is called a *catalyst*. Catalysts operate by many different mechanisms, but all essentially go through a cycle where they are used and then regenerated.

When a catalyst occurs in solution as a molecule or ion, it operates through a process called *homogeneous catalysis*; when the reaction occurs on a surface, the process is called *heterogeneous catalysis*. The reaction path provided by the catalyst is one of lower activation energy and/or higher frequency factors. If the products of the reaction act as a catalyst, the reaction is said to be *autocatalytic*.

Reaction rates for condensed-phase processes normally involve the transport of products away from the boundary. Thus the rate of the overall process is determined by the rate of each individual step and on the reaction rate constant and concentration of reactants for that step. The reaction with the lowest rate determines the overall rate of the corrosion process. Some of the more important factors that may influence the rate of reaction are diffusion rates, viscosity, particle size, heat transfer, and the degree of contact or mixing.

The stoichiometric chemical equation of the overall process does not reveal the mechanism of the reaction. To determine the overall reaction rate, one must determine all the intermediate steps of the process. Prediction or identification of the reaction mechanism is never certain since other mechanisms could account for the experimental data.

The rate of the reaction expressed as the rate of change of concentration, dc/dt, depends upon the concentration of the reactants. Rates may also depend upon the concentrations of other substances not involved in the stoichiometric equation. The rate equation as a function of concentration of each substance that affects the rate is called the *rate law* for the reaction. When the rate equation contains powers of the concentration, the *order* of the reaction equals the exponent. Rate laws can be determined only experimentally and cannot be predicted from the chemical equation.

The first-order rate equation is given by:

$$dc/dt = -kc^n \tag{2.67}$$

where:

k = rate constant

c = concentration of reacting species
n = reaction order = 1 for first order
t = time

If log c is plotted against time, a straight line is obtained for a first-order reaction. If the reaction is one of the first order, it will take twice as long for three-fourths to react as it will for one-half to react. A discussion of the order of reactions and the various equations can be found in any book on kinetics [2.136].

Integration of Eq. (2.67) between concentration limits of c_1 and c_2 at time limits of t_1 and t_2 yields:

$$k = \frac{1}{(t_2 - t_1)} \ln (c_1/c_2) \tag{2.68}$$

Thus it should be apparent from this equation that to determine k, it is necessary to evaluate only the ratio of the concentrations at the two times. This can make analysis easier since one can substitute any measurable property that is proportional to the concentration. Changes in properties such as volume, partial pressure of gases, light absorption, and electrical conductivity are often used.

Equation (2.67) is often written in a form relating the fraction of product formed to the reaction time:

$$d\alpha/dt = k(1 - \alpha)^n \tag{2.69}$$

where α is now the total amount of product formed. According to Sharp et al. [2.137], when n = 1/2 or 2/3, the equations represent phase-boundary-controlled reactions for circular disks (or cylinders) or spheres, respectively.

Diffusion-controlled reactions have been represented by various functions of the amount of product formed given by the general equation:

$$F(\alpha) = kt \tag{2.70}$$

The frequently cited article by Sharp et al. gives numerical data that allows one to evaluate $F(\alpha)$ from experimental data for the commonly used equations, which are given in Table 2.8. Since

TABLE 2.8 Kinetic Equations

Diffusion-controlled	
$\alpha^2 = (k/x^2)t$	One-dimensional; $2x$ = reaction layer thickness
$(1-\alpha)\ln(1-\alpha)+\alpha = (k/r^2)t$	Two-dimensional; r = radius of cylinder
$[1-(1-\alpha)^{1/3}]^2 = (k/r^2)t$	Three-dimensional; r = radius of sphere Commonly called the Jander equation
$(1-2\alpha/3)-(1-\alpha)^{2/3} = (k/r^2)t$	Three-dimensional; r = radius of sphere
Phase-boundary controlled	
$[1-(1-\alpha)^{1/2}] = (u/r)t$	Circular disk or cylinder of radius = r and with u = velocity of interface; assuming an instantaneous nucleation
$[-\ln(1-\alpha)]^{1/2} = kt$	Avrami–Erofe'ev random nucleation equation for the disk/cylinder case
$[1-(1-\alpha)^{1/3}] = (u/r)t$	Sphere of radius = r and u = velocity of interface; assuming an instantaneous nucleation
$[-\ln(1-\alpha)]^{1/3} = kt$	Avrami–Erofe'ev random nucleation equation for the sphere case

Source: Ref. 2.137.

these equations have been derived for specific geometric monosized shapes, which are seldom present in actual cases, Sharp et al. concluded that considerable experimental accuracy was required to distinguish among the various possibilities.

It should be realized that the solutions to the kinetic equations discussed by Sharp et al. are only approximate. Frade and Cable [2.138] pointed out that the deviation observed between experimental data and theoretical models is often due only to the approximations that were made in the original theoretical analysis. Frade and Cable reexamined the basic theoretical model for the kinetics of solid state reactions by considering spherical particles, moving reaction boundaries, and changes in volume. *The discrepancies between the experimental data and the theoretical models are often due to nonspherical particles, a range in sizes, poor contact between reactants, formation of multiple products, and the dependency of the diffusion coefficient upon composition*. The commonly used Jander equation was originally derived for reactions between flat slabs and is therefore inappropriate for use with spherical particles, although the Jander model fits reasonably well for low values of conversion.

Many reactions are not simply zero-, first-, second-, or third-order reactions since they proceed by a multistep mechanism. Multistep reactions may, however, behave as zero-, first-, etc. order reactions. Some of the complexities that may be encountered are parallel steps, consecutive steps, and reversible steps or may even be other types of steps. Many times, a complex reaction may appear to be zero-, first-, second-, or third-order only because the rate-limiting step is of that order and all other steps are very fast.

Nonisothermal thermogravimetry (TG) has been used by many investigators over the past 30 years to study the kinetics of decomposition reactions. The amount of data that can be collected by dynamic methods is considerably more than by isothermal methods, which has led investigators to rely more heavily upon the dynamic method. The convenience today is so great with modern computerized thermal analyzers that one need only scan a sample at several different heating rates, and then push the appropriate buttons to obtain the kinetic data! One must be extremely careful in collecting kinetic data in this fashion. A thorough understanding of the various effects that the sample characteristics, machine operation, etc. have upon the kinetics is important along with all the various assumptions that may have been made by the software programmer to use that data to calculate the kinetic parameters. Too often, these *psuedokinetic* data are published in the technical literature and can be misleading to the unwary reader.

Although many advantages exist for the use of nonisothermal studies over isothermal studies, the main disadvantage is that the reaction mechanism usually cannot be determined, which leads to uncertainties in the activation energy, order of reaction, and frequency factor. There must be at least two dozen different methods and variations reported in the literature to calculate kinetic parameters from dynamic thermogravimetric studies. The most widely used is that of Freeman and Carroll [2.139]. Sestak [2.140] performed a comparison of five methods and found a variation of approximately 10% in the calculated values of the activation energy. *Arnold et al. [2.141] concluded that dynamic thermogravimetric studies provide insufficient data for calculation of reaction kinetics, that the data are influenced by the experimental procedures, and that the results are uncertain.*

Differential thermal analysis (DTA) and differential scanning calorimetry (DSC) have also been used to study reaction kinetics. The equation to evaluate the rate depends on the mechanism, and thus the mechanism must be known before these methods can be used. The rate of heat generation must also be proportional to the rate of reaction for these methods to be valid. A recent review of the use of DTA to determine reaction kinetics has been given by Sestak [2.142].

Probably the most important parameter that is uncertain in nonisothermal studies is the temperature of the sample. *The enthalpy of the reaction is often sufficient to raise or lower the sample temperature by as much as 1000°C.* This fact is overlooked or unavailable if one uses nonisothermal thermogravimetry, which is most often the case. DTA or DSC may be more appropriate than TG since these techniques either determine the sample temperature or maintain the sample at a constant temperature relative to a reference material. Generally, the temperature range studied in thermal analysis to evaluate the kinetics of a reaction is on the order of 100–150°C. This range covers only about 10–20% of the total reaction and leads to excessive scatter in the calculated values of the activation energy and the preexponential term of the Arrhenius equation.

For corrosion rates to be useful to practicing engineers, it is best that they be expressed in a useful manner. In most cases, the engineer is involved with the amount of material corroded away during a specified time period or the depth of penetration per unit time. In the literature, corrosion rates are often given as the mass of material reacted per unit area for a unit time. These can easily be converted to the depth of penetration per unit time by dividing by the density of the material as shown below:

$$P = \frac{M}{\rho A t} \tag{2.71}$$

where:

P = depth of penetration
M = mass loss
ρ = density
A = area of exposure
t = time of exposure

In using the above equation to calculate corrosion rates from laboratory experiments, one must be very conscious of the total surface area exposed to corrosion. This will include a determination of the open porosity of the specimen. Many investigators have attempted to compare corrosion resistance of various materials incorrectly by omitting the porosity of their samples. Omitting the porosity, although not giving a true representation of the material's corrosion, will give a reasonable idea of the corrosion of the as-manufactured material.

2.9 DIFFUSION

When the transport of ions or molecules occurs in the absence of bulk flow, it is called *diffusion*. Substances will spontaneously diffuse toward the region of lower chemical potential. This transport or flux of matter is represented by Fick's first law and is proportional to the concentration gradient. This is represented by:

$$J_{ix} = -D\left(\frac{\partial c_i}{\partial x}\right) \tag{2.72}$$

where:

J_{ix} = flux of component i in the x direction
D = diffusion coefficient
c_i = concentration of component i

The flow of material is thus proportional to the concentration gradient and is directed from the region of high concentration to one of low concentration.

Fick's second law describes the nonstationary state of flow where the concentration of a fixed region varies with time:

$$\frac{\partial c}{\partial t} = \frac{\partial}{\partial x}\left(D \frac{\partial c}{\partial x}\right) \tag{2.73}$$

Since diffusion is directional, one must be aware of anisotropic effects. The rate of diffusion may be very different in different crystalline directions. *In isometric crystals, the diffusion coefficient is isotropic, as it is in polycrystalline materials as long as no preferred orientation exists.* The second-order tensor defined by the equations for the flux, J, in each of the x, y, and z directions, contains a set of nine diffusion coefficients designated D_{ij}. Due to the effects of the various symmetry operations in the tetragonal, hexagonal, orthorhombic, and cubic crystal classes, only a few of these D_{ij} have nonzero values. All the off-diagonal D_{ij} $(i=j)$ are equal to zero. Thus only the three diagonal values are of any consequence; however, symmetry again causes some of these to be equivalent. In the remaining two crystal classes, the number of independent coefficients increases; however, the total number is decreased somewhat since $D_{ij}=D_{ji}$. The possible nonzero diffusion coefficients for each of the crystal classes are shown in Table 2.9.

A solution of Eq. (2.73) for nonsteady-state diffusion in a semi-infinite medium (D is independent of concentration) is:

$$C(x, t) = \frac{C_o}{2}\left\{1 + \mathrm{erf}\left(\frac{x}{2\sqrt{(Dt)}}\right)\right\} \tag{2.74}$$

where:

$C(x,t)$ = concentration after time t
C_o = initial concentration in the medium

TABLE 2.9 Effect of Symmetry Upon the Second-Rank Tensor Diffusion Coefficients

Crystal class	Number of independent coefficients	Nonzero coefficients
Cubic	1	$D_{11} = D_{22} = D_{33}$
Tetragonal and hexagonal	2	$D_{11} = D_{22} \neq D_{33}$
Orthorhombic	3	$D_{11} \neq D_{22} \neq D_{33}$
Monoclinic	4	$D_{11} \neq D_{22} \neq D_{33} \neq D_{31} = D_{13}$
Triclinic	6	$D_{11} \neq D_{22} \neq D_{33} \neq D_{21} = D_{12} \neq$ $D_{31} = D_{13} \neq D_{23} = D_{32}$

Solutions to Eq. (2.74) depend upon the boundary conditions that one selects in the evaluation. More than one set of boundary conditions have been selected by various investigators, and thus several solutions to the equation exist in the literature that may provide some confusion to the uninitiated. In the above case [Eq. (2.74)], which is appropriate for the diffusion between two solids, the boundary conditions were selected such that as time passes, the diffusing species are depleted on one side of the boundary and increased on the other. This will yield a constant midpoint concentration at the boundary of $C_o/2$. In the case of corrosion of a solid by a liquid, one assumes that the concentration of diffusing species from the liquid into the solid remains constant at the boundary (C_s) at a value equal to that in the bulk. The solution to Eq. (2.73) is then:

$$C(x,t) = C_s\left\{1 - \mathrm{erf}\left(\frac{x}{2\sqrt{(Dt)}}\right)\right\} \tag{2.75}$$

where C_s is the concentration at the surface. One should note that the sign within the brackets changes when the boundary conditions are changed. $\sqrt{(Dt)}$ is a measure of the order of magnitude of the distance that an average atom will travel and thus approximates the distance over which the concentration will change during diffusion. The use of error functions (erf) in evaluating diffusion is relatively easy by use of published tables [2.143] for various values of erf(z).

Most of the solutions to Fick's equations assume that D is constant; however, *in most real cases, the diffusion coefficient can vary with time, temperature [see Eq. (2.40)], composition, or position along the sample, or any combination of these*. If these are included in the equation, the mathematics become very difficult if not impossible; thus the equations used to describe diffusion generally assume constant D. See Table 2.10 for some typical values of diffusion coefficients.

TABLE 2.10 Diffusion Coefficients for Some Typical Ceramics

Diffusing ion	System	D_o (cm^2/sec)	Q (kcal/mol)	Comment	Ref.
$O^=$	Al_2O_3	1.9×10^3	152	Single crystal, >1600°C	2.146
$O^=$	Al_2O_3	2.0×10^{-1}	110	Polycryst, >1450°C	2.146
$O^=$	Al_2O_3	6.3×10^{-8}	57.6	Polycryst, <1600°C	2.146
$O^=$	MgO	2.5×10^{-6}	62.4	1300–1750°C	2.146
$O^=$	SiO_2	1.5×10^{-2}	71.2	Vitreous, 925–1225°C	2.147
$O^=$	ZrO_2(Ca)	1.0×10^{-2}	28.1	15 mol% CaO, 700–1100°C	2.148
Al^{3+}	Al_2O_3	2.8×10	114	Polycryst, 1670–1905°C	2.149
Ca^{2+}	ZrO_2(Ca)	3.65	109	16 mol% CaO	2.150
Mg^{2+}	MgO	2.3×10^{-1}	78.7	1400–1600°C	2.151
Mg^{2+}	$MgAl_2O_4$	2.0×10^2	78		2.152
Zr^{4+}	ZrO_2(Ca)	1.97	109	16 mol% CaO	2.150
$Pb^?$	$PbSiO_3$	1.0×10^{-4}	24.8	Glass, 300–600°C	2.153

Several mechanisms for diffusion have been hypothesized and investigated. One of the more important in ceramic materials is diffusion by vacancy movement in nonstoichiometric materials. Another mechanism involves diffusion by movement from one interstitial site to another. The ease with which this mechanism can occur, however, is not as great as that by vacancy movement. Other mechanisms that provide high-diffusivity paths include diffusion aided by dislocations, free surfaces, or grain boundaries.

Permeability constants as a function of temperature give an indication of the ease of diffusion of a species through a material. Silica has the lowest permeability to oxygen. This has been attributed to the difference in mechanism of transport among silica and most other materials. Transport in silica is by molecular species, whereas in other materials, it is by ionic species [2.144]. *For this reason, silica-forming reactions are the most desirable for protection against oxygen diffusion* [2.145].

Since many applications of ceramics involve thermal gradients, some mention of thermal diffusion should be made. Based upon studies in liquids, this has been called the *Sorét effect*. To evaluate the effect using Fick's first law, an additional term must be added to Eq. (2.72) that involves the temperature gradient. The flux is then given by:

$$J_i = -D\frac{\partial c_i}{\partial x} - \beta_i\frac{dT}{dx} \tag{2.76}$$

where β_i is a constant independent of the thermal gradient for component i and may be positive or negative depending upon whether diffusion is down or up the thermal gradient, respectively. This constant is proportional to D and is given by:

$$\beta_i = \frac{D_i Q_i^* c_i}{RT^2} \tag{2.77}$$

where Q_i^* is an empirical parameter that describes the sign and magnitude of the thermal diffusion effect. It has also been called the *heat of transport*. One interesting phenomenon that comes from an analysis of thermal diffusion is that *a diffusion flux will set up a thermal gradient in an isothermal system.*

When an elastic stress gradient is present along with a concentration gradient, a potential term must be included in the equation for total flux, just as was necessary for the thermal gradient. Thus the total flux of atoms in a particular direction is increased (or decreased) over that due only to concentration differences. This effect is called *stress-assisted diffusion.*

Diffusion is probably the most important rate-controlling step when one is evaluating the kinetics of a reaction by thermal analysis. Diffusion in the gas phase is about 10^4 times greater than that in the liquid phase. For a more complete description of diffusion, the reader is referred to any one of the texts on diffusion [2.154,2.155].

2.10 SUMMARY OF IMPORTANT CONCEPTS

1. It is the total surface area exposed to attack that is important.
2. Grain boundary diffusion is more important at low temperatures, and bulk diffusion is more important at high temperatures.
3. One need not have data on a specific material, chemically, but only on one of identical structure to estimate its dissolution characteristics.
4. One must remember that it is not the total porosity that is important, but the surface area of the total porosity.
5. If sufficient flow of a transpiring gas along a thermal gradient is present, dilution of the corrosive gas at the hot face may lower the corrosion rate to an acceptable level.
6. The characteristics of the corroding glass are more important than the solid parameters in corrosion.
7. Ceramics with an acid/base character similar to the liquid will tend to resist corrosion the best.
8. The importance of the zero point of charge (zpc) in corrosion is that it is the pH of maximum durability.

9. The spontaneity of a reaction depends upon more than just the heat of reaction. To predict stability, one must consider also the entropy.
10. If the reaction is spontaneous, the change in free energy is negative, whereas if the reaction is in equilibrium, the free energy change is equal to zero.
11. The real problem with predicting whether a reaction may take place or not is in selecting the proper reaction to evaluate. Care must be taken not to overlook some possible reactions.
12. Since the corrosion of ceramics in service may never reach an equilibrium state, thermodynamic calculations cannot be strictly applied because these calculations are for systems in equilibrium. Many reactions, however, closely approach equilibrium, and thus the condition of equilibrium should be considered only as a limitation, not as a barrier to interpretation of the data.
13. There is a general tendency for oxides to be reduced at higher temperatures at constant oxygen partial pressures. One should be aware that any metal will reduce any oxide above it in the Ellingham diagram.
14. Unit activity should be applied only to species in the pure state.
15. The most important parameter of corrosion from the engineering viewpoint is the reaction rate.
16. Diffusion coefficients depend upon the composition and structure of the material through which diffusion occurs.
17. The rate of the reaction expressed as the rate of change of concentration, dc/dt, depends upon the concentration of the reactants.
18. The discrepancies between the experimental data and the theoretical models are often due to nonspherical particles, a range in sizes, poor contact between reactants, formation of multiple products, and the dependency of the diffusion coefficient upon composition.
19. Arnold et al. [2.141] concluded that dynamic thermogravimetric studies provide insufficient data for calcula-

tion of reaction kinetics, that the data are influenced by the experimental procedures, and that the results are uncertain.

20. The enthalpy of the reaction is often sufficient to raise or lower the sample temperature by as much as 1000°C.
21. The flow of material by diffusion is proportional to the concentration gradient and is directed from the region of high concentration to one of low concentration.
22. In isometric crystals, the diffusion coefficient is isotropic, as it is in polycrystalline materials as long as no preferred orientation exists.
23. In most real cases, the diffusion coefficient can vary with time, temperature, composition, or position along the sample, or any combination of these.
24. Silica-forming reactions are the most desirable for protection against oxygen diffusion.
25. A diffusion flux will set up a thermal gradient in an isothermal system.

2.11 ADDITIONAL RELATED READING

Vetter, K.J. *Electrochemical Kinetics*; Academic Press, New York, 1967.

Bockris, O.M.; Reddy, A.K.N. *Modern Electrochemistry*; Plenum Press, New York, 1970; Vol. 2.

Shaw, D.J. Charged Interfaces. *Introduction to Colloid and Surface Chemistry*, 3rd Ed.; Butterworths, London, 1980; 148–182. Chp. 7.

Marshall, C.E. *The Physical Chemistry and Mineralogy of Soils: Soils in Place*; Wiley & Sons: New York, 1977; Vol. II.

Reviews in Mineralogy: Mineral–Water Interface Geochemistry; Hochella, M.F. Jr., White A.F., Eds.; Mineral. Soc. Am., Washington, DC, 1990; Vol. 23.

Burns, R.G. *Mineralogical Applications of Crystal Field Theory*; Cambridge University Press: Cambridge, 1970.

Shackelford J.F., Ed.; *Bioceramics, Applications of Ceramic and Glass Materials in Medicine*; Trans Tech Publications: Switzerland, 1999.

Reviews in Mineralogy: Health Effects of Mineral Dusts. Guthrie, G.D. Jr.; Mossman B. T., Eds.; Mineral. Soc. Am., Washington, DC, 1993; Vol. 28.

P.G. Shewmon. *Diffusion in Solids*. J. Williams Book Co., Jenks, OK, 1983.

2.12 EXERCISES, QUESTIONS, AND PROBLEMS

1. Discuss the reaction products that may form and how they may relate to any interfacial reaction layer formed.
2. If a "unified theory of corrosion of ceramics" were to be developed, what structural characteristic would be included and why?
3. Look up the vapor pressure of several materials to confirm the concept that covalent materials vaporize more quickly than ionic materials due to their higher vapor pressure.
4. Why does the corrosion rate decrease when a thermal gradient is present?
5. The Arrhenius equation has been used to represent the temperature dependence of corrosion. Discuss when this equation is most appropriate and why.
6. Discuss the difference between direct and indirect dissolution. What other terms are used to describe these types of dissolution?
7. What is the most predominant parameter in the equation for corrosion rate under free convection? Why is this parameter more predominant than the others?
8. Discuss the various problems relating to the experimental verification of the galvanic corrosion of ceramics.
9. Describe how the cross-linking of silica tetrahedra affect corrosion in silicates by aqueous solutions.
10. How does pH affect the corrosion of crystalline ceramics and how does this relate to isoelectric point (IEP)?

11. Discuss the difference between electrochemical and chemical dissolution. What material parameters are important in each type?
12. Describe how one tells whether solid–solid corrosion occurs by bulk, grain boundary, or surface diffusion.

REFERENCES

2.1. *Phase Diagrams for Ceramists*. Vol. I–XII, Am. Ceram. Soc., Westerville, Ohio.

2.2. Cooper, A.R. The use of phase diagrams in dissolution studies. In *Refractory Materials*; Alper, A.M., Ed.; Academic Press: New York, 1970; Vol. 6-III, 237–250.

2.3. Kramer, D.P.; Osborne, N.R. Effects of atmosphere and dew point on the wetting characteristics of a glass-ceramic on two nickel-based superalloys. In *Ceramic Engineering and Science Proceedings*; Smothers, W.J., Ed.; Am. Ceram. Soc. Westerville, Ohio, 1983; *4* (9–10), 740–750.

2.4. Noyes, A.A.; Whitney, W.R. Rate of solution of solid materials in their own solutions. (Ger) Z. Physik. Chem. 1897, *23*, 689–692.

2.5. Nernst, W. Theory of reaction velocities in heterogeneous systems. (Ger) Z. Physik. Chem. *47*, 52–55.

2.6. Berthoud, A. Formation of crystal faces. J. Chem. Phys. *10*, 624–635, 1912.

2.7. Prandtl, L. NACE Tech. Memo. No. 452, 1928.

2.8. Levich, B.G. *Physicochemical Hydrodynamics*; Prentice Hall: Englewood Cliffs, NJ, 1962.

2.9. Levich, B.G. Theory of concentration polarization. Discuss. Faraday Soc. *1*, 37–43, 1947.

2.10. Cooper, A.R., Jr.; Kingery, W.D. Dissolution in ceramic systems: I, Molecular diffusion, natural convection, and forced convection studies of sapphire dissolution in calcium aluminum silicate. J. Am. Ceram. Soc. 1964, *47* (1), 37–43.

2.11. Samaddar, B.N.; Kingery, W.D.; Cooper, A.R., Jr. Dissolution in ceramic systems: II, Dissolution of alumina, mullite, anorthite, and silica in a calcium–aluminum–silicate slag. J. Am. Ceram. Soc. 1964, *47* (5), 249–254.

2.12. Oishi, Y.; Copper, A.R., Jr.; Kingery, W.D. Dissolution in ceramic systems: III, Boundary layer concentration gradients. J. Am. Ceram. Soc. 1965, *48* (2), 88–95.

2.13. Hrma, P. Contribution to the study of the function between the rate of isothermal corrosion and glass composition. (Fr) Verres Refract. 1970, *24* (4–5), 166–168.

2.14. Lakatos, T.; Simmingskold, B. Influence of constituents on the corrosion of pot clays by molten glass. Glass Technol. 1967, *8* (2), 43–47.

2.15. Lakatos, T.; Simmingskold, B. Corrosion effect of glasses containing Na_2O–CaO–MgO–Al_2O_3–SiO_2 on tank blocks Corhart ZAC and sillimanite. Glastek. Tidskr. 1967, *22* (5), 107–113.

2.16. Lakatos, T.; Simmingskold, B. Influence of viscosity and chemical composition of glass on its corrosion of sintered alumina and silica glass. Glastek. Tidskr. 1971, *26* (4), 58–68.

2.17. Chung, Y.-.D.; Schlesinger, M.E. Interaction of CaO–FeO–SiO_2 slags with partially stabilized zirconia. J. Am. Ceram. Soc. 1994, *77* (3), 612.

2.18. Pons, A.; Parent, A. The activity of the oxygen ion in glasses and its effect on the corrosion of refractories. (Fr) Verres Refract. 1969, *23* (3), 324–333.

2.19. Blau, H.H.; Smith, C.D. Refractory problems in glass manufacture. Bull. Am. Ceram. Soc. 1950, *29* (1), 6–9.

2.20. Woolley, F.E. Prediction of refractory corrosion rate from glass viscosity and composition. In *UNITECR '89 Proceedings*; Trostel, L.J., Jr., Ed.; Am. Ceram. Soc. Westerville, OH, 1989, 768–779.

2.21. Fox, D.S.; Jacobson, N.S.; Smialek, J.L. Hot corrosion of silicon carbide and nitride at 1000°C. In *Ceramic Transactions: Corrosion and Corrosive Degradation of Ceramics*; Tressler, R.E., McNallan, M., Eds.; Am. Ceram. Soc. Westerville, OH, 1990; Vol. 10, 227–249.

2.22. Jacobson, N.S.; Stearns, C.A.; Smialek, J.L. Burner rig corrosion of SiC at 1000°C. Adv. Ceram. Mater. 1986, *1* (2), 154–161.

2.23. Cook, L.P.; Bonnell, D.W.; Rathnamma, D. Model for molten salt corrosion of ceramics. In *Ceramic Transactions: Corrosion and Corrosive Degradation of Ceramics*; Tressler, R.E., McNallan, M., Eds.; Am Ceram. Soc. Westerville, OH, 1990; Vol. 10, 251–275.

2.24. Gordon, S.; McBride, B.J. Computer Program for Calculation of Complex Chemical Equilibrium Compositions, Rocket Performance, Incident & Reflected Shocks, and Chapman–Jongnet Detonations. NASA SP-273, US Printing Office: Washington, DC, 1971.

2.25. Borom, M.P.; Arendt, R.H.; Cook, N.C. Dissolution of oxides of Y, Al, Mg, and La by molten fluorides. Ceram. Bull. 1981, *60* (11), 1168–1174.

2.26. Le Clerc, P.; Peyches, I. Polarization of refractory oxides immersed in molten glass. (Fr) Verres Refract. 1953, *7* (6), 339–345.

2.27. Godrin, Y. Review of the Literature on Electrochemical Phenomena. International Commission on Glass: Paris, 1975.

2.28. Vetter, K.J. *Electrochemical Kinetics*; Academic Press: New York, 1967.

2.29. Wall, F.D.; Taylor, S.R.; Cahen, G.L. The simulation and detection of electrochemical damage in BMI/graphite fiber composites using electrochemical impedance spectroscopy. In *High Temperature and Environmental Effects on Polymeric Composites, STP 1174*; Harris, C.E., Gates, T.S., Eds.; ASTM: Philadelphia, PA, 1993; 95–113.

2.30. Lindsay, J.G.; Bakker, W.T.; Dewing, E.W. Chemical resistance of refractories to Al and Al–Mg alloys. J. Am. Ceram. Soc. 1964, *47* (2), 90–94.

2.31. Busby, T. Hotter refractories increase the risk of downward drilling. Glass Ind. 1992, *73* (1), 20, 24.

2.32. Lasaga, A.C. Atomic treatment of mineral–water surface reactions. In *Reviews in Mineralogy, Mineral–Water Interface Geochemistry*; Hochella, M.F., Jr., White, A.F., Eds.; Mineral. Soc. Am. Washington, DC, 1990; Vol. 23, 17–85. Chp. 2.

2.33. Marshall, C.E. *The Physical Chemistry and Mineralogy of Soils: Soils in Place*; Wiley & Sons: New York, 1977; Vol. II.

2.34. Huang, P.M. Feldspars, olivines, pyroxenes, and amphiboles. In *Minerals in Soil Environments*; Dinauer, R.C., Ed.; Soil Sci. Soc. Am. Madison, WI, 1977, 553–602. Chp. 15.

2.35. Casey, W.H.; Bunker, B. Leaching of mineral and glass surfaces during dissolution. In *Reviews in Mineralogy, Vol. 23: Mineral–Water Interface Geochemistry*; Hochella, M.F., Jr., White, A.F., Eds.; Mineral Soc. Am. Washington, DC, 1990; Vol. 23, 397–426. Chp. 10.

2.36. Borchardt, C.A. Montmorillonite and other smectite minerals. In *Minerals in Soil Environments*; Dinauer, R.C., Ed.; Soil Sci. Soc. Am. Madison, WI, 1977, 293–330. Chp. 9.

2.37. Schnitzer, M.; Kodama, H. Reactions of minerals with soil humic substances. In *Minerals in Soil Environments;* Dinauer, R.C., Ed.; Soil Sci. Soc. Am. Madison, WI, 1977, 741–770. Chp. 21.

2.38. Jennings, H.M. Aqueous solubility relationships for two types of calcium silicate hydrate. J. Am. Ceram. Soc. 1986, *69* (8), 614–618.

2.39. Marshall, C.E. *The Physical Chemistry and Mineralogy of Soils: Soil Materials*; Krieger Publishing Company: Huntington, NY, 1975; Vol. I.

2.40. Elmer, T.H. Role of acid concentration in leaching of cordierite and alkali borosilicate glass. J. Am. Ceram. Soc. 1985, *68* (10), C273–C274.

2.41. Burns, R.G. *Mineralogical Applications of Crystal Field Theory*; Cambridge University Press: London, 1970; 162–167.

2.42. Hawkins, D.B.; Roy, R. Distribution of trace elements between clays and zeolites formed by hydrothermal alteration of synthetic basalts. Geochim. Cosmochim. Acta 1963, *27* (165), 785–795.

2.43. Shaw, D.J. Charged interfaces. *Introduction to Colloid and Surface Chemistry*, 3rd Ed.; Butterworths: London, 1980; 148–182. Chp. 7.

2.44. Brady, P.V.; House, W.A. Surface-controlled dissolution and growth of minerals. In *Physics and Chemistry of Mineral Surfaces*; Brady, P.V., Ed.; CRC Press: New York; 1996, 225–305. Chp. 4.

2.45. Parks, G.A. The isoelectric points of solid oxides, solid hydroxides, and aqueous hydroxo complex systems. Chem. Rev. 1965, *65* (2), 177–198.

2.46. Diggle, J.W. Dissolution of oxide phases. In *Oxides and Oxide Films*; Diggle, J.W., Ed.; Marcel Dekker: New York, 1973; Vol. 2, 281–386. Chp. 4.

2.47. Bright, E.; Readey, D.W. Dissolution kinetics of TiO_2 in HF–HCl solutions. J. Am. Ceram. Soc. 1987, *70* (12), 900–906.

2.48. Hulbert, S.F.; Bokros, J.C.; Hench, L.L.; Wilson, J.; Heimke, G. In *High Tech Ceramics*; Vincenzini, P., Ed.; Elsevier Science Pub. B.V.: Amsterdam, 1987; 180–213.

2.49. Hench, L.L.; Wilson, J. Introduction. *An Introduction to Bioceramics,* Advanced Series in Ceramics. World Scientific Publishing Co. Ltd.: Singapore, 1993; Vol. 1, 1–24.

2.50. *Reviews in Mineralogy, Health Effects of Mineral Dusts.*; Guthrie, G.D., Jr., Mossman, T., Eds.; Mineral. Soc. Am. Washington, DC, 1993; Vol. 28.

2.51. Nolan, R.P.; Langer, A.M. Limitations of the Stanton hypothesis. In *Health Effects of Mineral Dusts*; Guthrie, G.D., Jr., Mossman, T., Eds.; Reviews in Mineralogy. Min. Soc. Am. Washington, DC, 1993; Vol. 28. Chp. 9.

2.52. Correns, C.W. Growth and dissolution of crystals under linear pressure. Discuss. Faraday Soc. No. 5; 1949, 267–271.

2.53. Winkler, E.M. Salt action on stone in urban buildings. In *Application of Science in Examination of Works of Art*; Joung, W.J., Ed.; Museum of Fine Arts: Boston, 1973.

2.54. Skoulikidis, T.N. Atmospheric corrosion of concrete reinforcements, limestones, and marbles. In *Atmospheric Corrosion*; Ailor, W.H., Ed.; John Wiley & Sons: New York, 1982; 807–825.

2.55. Amoroso, G.G.; Fassina, V. *Stone Decay and Conservation*; Elsevier: Amsterdam, 1983; 12.

2.56. Hoffmann, M.R. Fog and cloud water deposition. In *Materials Degradation Caused by Acid Rain*; ACS Symposium Series 318; Baboian, R., Ed.; Am. Chem. Soc. Washington, DC, 1986; 64–91.

2.57. Mulawa, P.A.; Cadle, S.H.; Lipari, F.; Ang, C.C.; Vandervennet, R.T. Urban dew: Composition and influence on dry deposition rates. In *Materials Degradation Caused by Acid Rain*, ACS Symposium Series 318; Baboian, R., Ed.; Am. Chem. Soc. Washington, DC, 1986; 61–91.

2.58. Semonin, R.G. Wet deposition chemistry. In *Materials Degradation Caused by Acid Rain,* ACS Symposium Series 318; Baboian, R., Ed.; Am. Chem. Soc. Washington, DC, 1986; 23–41.

2.59. Neal, K.M.; Newnam, S.H.; Pokorney, L.M.; Rybarczyk, J.P. Elemental analysis of simulated acid rain stripping of Indiana limestone, marble, and bronze. In *Materials Degradation Caused by Acid Rain*, ACS Symposium Series 318; Baboian, R., Ed.; Am. Chem. Soc. Washington, DC, 1986; 285–300.

2.60. Reddy, M.M.; Sherwood, S.I.; Doe, B.R. Limestone and marble dissolution by acid rain: An onsite weathering experiment. In *Materials Degradation Caused by Acid Rain*, ACS Symposium Series 318; Baboian, R., Ed.; Am. Chem. Soc. Washington, DC, 1986; 226–238.

2.61. Kobussen, A.G. Corrosion in condensing gas-fired central heating boilers. In *Dewpoint Corrosion*; Holmes, D.R., Ed.; Ellis Horwood Ltd.: Chichester, UK, 1985; 179–190.

2.62. Penkett, S.A. Chemical changes in the air. *SCI Sulfur Symposium*; May 1979; 109–122.

2.63. Cox, W.M.; Farrell, D.M; Dawson, J.L. Corrosion monitoring for process control. In *Dewpoint Corrosion*; Holmes, D.R., Ed.; Ellis Horwood Ltd.: Chichester, UK, 1985; 191–217.

2.64. Cussler, E.L.; Featherstone, J.D.B. Demineralization of porous solids. Science Aug. 1981, *213*, 1018–1019.

2.65. Yoshimura, M.; Hiuga, T.; Somiya, S. Dissolution & reaction of yttria-stabilized zirconia single crystals in hydrothermal solutions. J. Am. Ceram. Soc. 1986, *69* (7), 583–584.

2.66. Sato, T.; Ohtaki, S.; Shimada, M. Transformation of yttria partially stabilized zirconia by low temperature annealing in air. J. Mater. Sci. 1985, *20* (4), 1466–1470.

2.67. Janowski, K.R.; Rossi, R.C. Mechanical degradation of MgO by water vapor. J. Am. Ceram. Soc. 1968, *51* (8), 453–455.

2.68. White, W.B. Glass structure and glass durability. In *Materials Stability and Environmental Degradation, Materials Research Society Symposium Proceedings*; Barkatt, A., Verink, E.D., Jr., Smith, L.R., Eds.; Mater. Res. Soc. Pittsburgh, PA, 1988; Vol. 125, 109–114.

2.69. Wald, J.W.; Messier, D.R.; DeGuire, E.J. Leaching behavior of Si–Y–Al–O–N glasses. Int. J. High Technol. Ceram. 1986, *2* (1), 65–72.

2.70. Douglas, R.W.; El-Shamy, T.M.M. Reaction of glass with aqueous solutions. J. Am. Ceram. Soc. 1967, *50* (1), 1–8.

2.71. Jantzen, C.M. Thermodynamic approach to glass corrosion. In *Corrosion of Glass, Ceramics, & Ceramic Superconductors*; Clark, D.E., Zoitos, B.K., Eds.; Noyes Publications: Park Ridge, NJ, 1992; 153–217. Chp. 6.

2.72. Newton, R.G.; Paul, A. A new approach to predicting the durability of glasses from their chemical compositions. Glass Technol. 1980, *21* (6), 307–309.

2.73. Pourbaix, M. *Atlas of Electrochemical Equilibria in Aqueous Solution*; NACE: Houston, TX, 1974. Eng Trans by J.A. Franklin.

2.74. Garrels, R.M.; Christ, C.L. *Solutions, Minerals, and Equilibria*; Harper and Row: New York, 1965.

2.75. Hench, L.L.; Clark, D.E. Physical chemistry of glass surfaces. J. Non-Cryst. Solids 1978, *28*, 83–105.

2.76. McVay, G.L.; Peterson, L.R. Effect of gamma radiation on glass leaching. J. Am. Ceram. Soc. 1981, *64* (3), 154–158.

2.77. Hogenson, D.K.; Healy, J.H. Mathematical treatment of glass corrosion data. J. Am. Ceram. Soc. 1962, *45* (4), 178–181.

2.78. Budd, S.M. The mechanism of chemical reaction between silicate glass and attacking agents; Part 1. Electrophilic and nucleophilic mechanism of attack. Phys. Chem. Glasses 1961, *2* (4), 111–114.

2.79. Budd, S.M.; Frackiewicz, J. The mechanism of chemical reaction between silicate glass and attacking agents; Part 2. Chemical equilibria at glass–solution interfaces. Phys. Chem. Glasses 1961, *2* (4), 115–118.

2.80. Simmons, C.J.; Simmons, J.H. Chemical durability of fluoride Glasses: I, Reaction of fluorozirconate glasses with water. J. Am. Ceram. Soc. 1986, *69* (9), 661–669.

2.81. Thomas, W.F. An investigation of the factors likely to affect the strength and properties of glass fibers. Phys. Chem. Glasses 1960, *1* (1), 4–18.

2.82. Wojnarovits, I. Behavior of glass fibers in strong acidic and alkaline media. J. Am. Ceram. Soc. 1983, *66* (12), 896–898.

2.83. Hench, L.L. Bioactive glasses help heal, repair and build human tissue. Glass Res. 2002–2003, *12* (1–2), 18.

2.84. White, J.E.; Day, D.E. Rare earth aluminosilicate glasses for *in vivo* radiation delivery. In *Rare Elements in Glasses*; Key Engineering Materials; Shelby, J.E., Ed.; Trans Tech Pub., Switzerland, 1994; Vols. 94–95, 181–208.

2.85. Oda, K.; Yoshio, T. Properties of Y_2O_3–Al_2O_3–SiO_2 glasses as a model system of grain boundary phase of Si_3N_4 ceramics, Part 2: Leaching characteristics. J. Ceram. Soc. Jpn. 1991, *99* (11), 1150–1152.

2.86. Erbe, E.M.; Day, D.E. In *Proceedings: Science & Tech of New Glasses*; Sakka, S., Soga, N., Eds.; Ceram. Soc. Japan: Tokyo, 1991.

2.87. Conzone, C.D.; Brown, R.F.; Day, D.E.; Ehrhardt, G.J. *In vitro* and *in vivo* dissolution behavior of a dysprosium lithium borate glass designed for the radiation synovectomy treatment of rheumatoid arthritis. J. Biomed. Mater. Res. 2002, *60* (2), 260–268.

2.88. Day, D.E. Reactions of bioactive borate glasses with physiological liquids. Glass Res. 2002–2003, *12* (1–2), 21–22.

2.89. Bauer, J.F. Corrosion and surface effects of glass fiber in biological fluids. Glass Res. 2000, *9* (2), 4–5.

2.90. Kubaschewski, O.; Hopkins, B.E. *Oxidation of Metals and Alloys*; Butterworths: London, 1962.

2.91. Readey, D.W. Gaseous corrosion of ceramics. In *Ceramic Transactions, Corrosion and Corrosive Degradation of Ceramics*; Tressler, R.E., McNallan, M., Eds.; Am. Ceram. Soc., Westerville, OH, 1990; Vol. 10, 53–80.

2.92. Yokokawa, H.; Kawada, T.; Dokiya, M. Construction of chemical potential diagrams for metal–metal–nonmetal systems: Applications to the decomposition of double oxides. J. Am. Ceram. Soc. 1989, *72* (11), 2104–2110.

2.93. Grimley, R.T.; Burns, R.P.; Inghram, M.G. Thermodynamics of vaporization of Cr_2O_3: Dissociation energies of CrO, CrO_2 and CrO_3. J. Chem. Phys. 1961, *34* (2), 664–667.

2.94. Graham, H.C.; Davis, H.H. Oxidation/vaporization kinetics of Cr_2O_3. J. Am. Ceram. Soc. 1971, *54* (2), 89–93.

2.95. Pilling, N.B.; Bedworth, R.E. The oxidation of metals at high temperature. J. Inst. Met. 1923, *29*, 529–591.

2.96. Jorgensen, P.J.; Wadsworth, M.E.; Cutler, I.B. Effects of oxygen partial pressure on the oxidation of silicon carbide. J. Am. Ceram. Soc. 1960, *43* (4), 209–212.

2.97. Engell, H.J.; Hauffe, K. Influence of adsorption phenomena on oxidation of metals at high temperatures. Metall 1952, *6*, 285–291.

2.98. Langmuir, I. Evaporation of small spheres. Phys. Rev. 1918, *12* (5), 368–370.

2.99. Tichane, R.M.; Carrier, G.B. The microstructure of a soda–lime glass surface. J. Am. Ceram. Soc. 1961, *44* (12), 606–610.

2.100. Simpson, H.E. Study of surface structure of glass as related to its durability. J. Am. Ceram. Soc. 1958, *41* (2), 43–49.

2.101. Tichane, R.M. Initial stages of the weathering process on a soda–lime glass surface. Glass Technol. 1966, *7* (1), 26–29.

2.102. Burggraaf, A.J.; van Velzen, H.C. Glasses resistant to sodium vapor at temperatures to 700°C. J. Am. Ceram. Soc. 1969, *52* (5), 238–242.

2.103. Johnston, W.D.; Chelko, A.J. Reduction of ions in glass by hydrogen. J. Am. Ceram. Soc. 1970, *53* (6), 295–301.

2.104. Gibson, A.S.; LaFemina, J.P. Structure of mineral surfaces. In *Physics and Chemistry of Mineral Surfaces*; Brady, P.V., Ed.; CRC Press: New York, 1996; 1–62.

2.105. Hochella, M.F. Jr.; White, A.F. *Mineral–Water Interface Geochemistry;* Rev. Mineral; 23 pp.

2.106. Washburn, E.W. Note on a method of determining the distribution of pore sizes in a porous material. Proc. Natl. Acad. Sci. 1921, *7*, 115–116.

2.107. Smithwick, R.W.; Fuller, E.L. A generalized analysis of hysteresis in mercury porosimetry. Powder Technol. 1984, *38*, 165–173.

2.108. Conner, W.C. Jr.; Blanco, C.; Coyne, K.; Neil, J.; Mendioroz, S.; Pajares, J. Measurement of morphology of high surface area solids: Inferring pore shape characteristics. In *Characterization of Porous Solids*; Unger, K.K., et al., Ed.; Elsevier Science Publishers: Amsterdam, 1988; 273–281.

2.109. Moscou, L.; Lub, S. Practical use of mercury porosimetry in the study of porous solids. Powder Technol. 1981, *29*, 45–52.

2.110. Lapidus, G.R.; Lane, A.M.; Ng, K.M.; Conner, W.C. Interpretation of mercury porosimetry data using a pore-throat network model. Chem. Eng. Commun. 1985, *38*, 33–56.

2.111. Conner, W.C.; Lane, A.M. Measurement of the morphology of high surface area solids: Effect of network structure on the simulation of porosimetry. J. Catal. 1984, *89*, 217–225.

2.112. Van Brakel, J.; Modry, S.; Svata, M. Mercury porosimetry: State of the art. Powder Technol. 1981, *29*, 1–12.

2.113. Rootare, H.M.; Nyce, A.C. The use of porosimetry in the measurement of pore size distribution in porous materials. Int. J. Powder Metall. 1971, *7* (1), 3–11.

2.114. Smith, C.S. Grains, phases, and interpretation of microstructure. Trans. AIME 1948, *175* (1), 15–51.

2.115. White, J. Magnesia-based refractories. In *High Temperature Oxides, Part 1: Magnesia, Lime and Chrome Refractories*. Alper, A.M., Ed.; *Refractory Materials: A Series of Monographs*; Margrave, J.L., Ed.; Academic Press: New York, 1970; Vol. 5-1, 77–141.

2.116. Inomata, Y. Oxidation resistant Si-impregnated surface layer on reaction sintered articles. Yogyo Kyokaishi 1975, *83* (1), 1–3.

2.117. Messier, D.R. Use of Ti to enhance wetting of reaction-bonded Si_3N_4 by Si. In *Ceramic Engineering and Science Proceedings*; Smothers, W.J., Ed.; Am. Ceram. Soc. Westerville, OH, 1980, *1* (7–8B), 624–633.

2.118. Puyane, R.; Trojer, F. Refractory wear and wettability by glass at high temperatures. Glass 1980, *57* (12), 5–8.

2.119. Carre, A.; Roger, F.; Varinot, C. Study of acid/base properties of oxide, oxide glass, and glass-ceramic surfaces. J. Colloid Interface Sci. 1992, *154* (1), 174–183.

2.120. Gaskell, D.R. *Introduction to Metallurgical Thermodynamics*, 2nd Ed.; McGraw-Hill: New York, 1981.

2.121. Swalin, R.A. *Thermodynamics of Solids*. Wiley & Sons Inc.: New York, 1962.

2.122. Bent, H.A. *The Second Law*. Oxford University Press: New York, 1965.

2.123. Chase, M.W., Jr.; Davies, C.A.; Downey, J.R., Jr.; Frurip, D.J.R.; McDonald, R.A.; Syverud, A.N. J. Phys. Chem. Reference Data, Vol. 14, Suppl. No. 1, *JANAF Thermochemical Tables*, 3rd Ed., Parts I & II, Am. Chem. Soc. & Am. Inst. Phys., 1985.

2.124. Kubaschewski, O.; Evans, E.L.; Alcock, C.B. *Metallurgical Thermodynamics*. Pergamon Press: Oxford, 1967.

2.125. Krupka, K.M.; Hemingway, B.S.; Robie, R.A.; Kerrick, D.M. High temperature heat capacities and derived thermodynamic properties of anthophyllite, diopside, dolomite, enstatite, bronzite, talc, tremolite, and wollastonite. Am. Mineral. 1985, *70*, 261–271.

2.126. Eriksson, G. Thermodynamic studies of high temperature equilibria. XII. SOLGASMIX, a computer program for calculation of equilibrium compositions in multiphase systems. Chem. Scr. 1975, *8*, 100–103.

2.127. Latimer, W.M. *The Oxidation States of the Elements and Their Potentials in Aqueous Solutions*; Prentice-Hall: Englewood Cliffs, NJ, 1952.

2.128. Brenner, A. The Gibbs–Helmoltz equation and the EMF of galvanic cells, II. Precision of its application to concentration cells. J. Electrochem. Soc. 1975, *122* (12), 1609–1615.

2.129. Livey, D.T.; Murray, P. The stability of refractory materials. In *Physicochemical Measurements at High Temperatures*; Bockris, J. O'M., et al., Ed.; Butterworths Scientific Publications: London, 1959; 87–116.

2.130. Luthra, K.L. Chemical interactions in ceramic and carbon–carbon composites. In *Materials Research Society Symposium Proceedings: Materials Stability and Environmental Degradation*; Barkatt, A. Verink, E.D., Jr., Smith, L.R., Eds.; Mat. Res. Soc.: Pittsburgh, PA, 1988; Vol. 125, 53–60.

2.131. Ellingham, H.J.T. Reducibility of oxides and sulfides in metallurgical processes. J. Soc. Chem. Ind. 1944, *63*, 125.

2.132. Richardson, F.D.; Jeffes, J.H.E. The thermodynamics of substances of interest in iron and steel making from 0°C to 2400°C; I Oxides. J. Iron Steel Inst. 1948, *160*, 261.

2.133. Darken, L.S.; Gurry, R.W. *Physical Chemistry of Metals*; McGraw-Hill: New York, 1953; 348–349.

2.134. Lou, V.L.K.; Mitchell, T.E.; Heuer, A.H. Review—Graphical displays of the thermodynamics of high-temperature gas–solid reactions and their application to oxidation of metals and evaporation of oxides. J. Am. Ceram. Soc. 1985, *68* (2), 49–58.

2.135. Quets, J.M.; Dresher, W.H. Thermochemistry of the hot corrosion of superalloys. J. Mater. 1969, *4* (3), 583–599.

2.136. Barret, P., Ed.; *Reaction Kinetics in Heterogeneous Chemical Systems*; Elsevier: Amsterdam, 1975.

2.137. Sharp, J.H.; Brindley, G.W.; Narahari Achar, B.N. Numerical data for some commonly used solid state reaction equations. J. Am. Ceram. Soc. 1966, *49* (7), 379–382.

2.138. Frade, J.R.; Cable, M. Reexamination of the basic theoretical model for the kinetics of solid-state reactions. J. Am. Ceram. Soc. 1992, *75* (7), 1949–1957.

2.139. Freeman, E.S.; Carroll, B. The application of thermoanalytical techniques to reaction kinetics. The thermogravimetric evaluation of the kinetics of the decomposition of calcium oxalate monohydrate. J. Phys. Chem. 1958, *62* (4), 394–397.

2.140. Sestak, J. Errors of kinetic data obtained from thermogravimetric curves at increasing temperature. Talanta 1966, *13* (4), 567–579.

2.141. Arnold, M.; Veress, G.E.; Paulik, J.; Paulik, F. The applicability of the Arrhenius model in thermal analysis. Anal. Chim. Acta. 1981, *124* (2), 341–350.

2.142. Sestak, J. *Thermophysical Properties of Solids. Thermal Analysis*; Wendlandt, W.W., Ed.; Part D; Svehla, G., Ed.; *Comprehensive Analytical Chemistry*; Svehla, G., Ed.; Vol. XII; Elsevier: Amsterdam, 1984.

2.143. Holman, J.P. *Heat Transfer*. McGraw-Hill: New York, 1963.

2.144. Courtright, E.L. Engineering limitations of ceramic composites for high performance and high temperature applications. In *Proc. 1993 Conf. on Processing, Fabrication and Applications of Advanced Composites, Long Beach, CA, Aug 9–11*; Upadhya, K., Ed.; ASM: Ohio, 1993; 21–32.

2.145. Tripp, W.C.; Davis, H.H.; Graham, H.C. Effects of SiC additions on the oxidation of ZrB_2. Ceram. Bull. 1973, *52* (8), 612–616.

2.146. Oishi, Y.; Kingery, W.D. Self-diffusion in single crystal and polycrystalline aluminum oxide. J. Chem. Phys. 1960, *33*, 480.

2.147. Sucov, E.W. Diffusion of oxygen in vitreous silica. J. Am. Ceram. Soc. 1963, *46* (1), 14–20.

2.148. Kingery, W.D.; Pappis, J.; Doty, M.E.; Hill, D.C. Oxygen ion mobility in cubic $Zr_{0.85}Ca_{0.15}O_{1.85}$. J. Am. Ceram. Soc. 1959, *42* (8), 393–398.

2.149. Paladino, A.E.; Kingery, W.D. Aluminum ion diffusion in aluminum oxide. J. Chem. Phys. 1962, *37* (5), 957–962.

2.150. Rhodes, W.H.; Carter, R.E. Ionic self-diffusion in calcia stabilized zirconia, 64th Annual Mtg Abstracts. Am. Ceram. Soc. Bull. 1962, *41* (4), 283.

2.151. Lindner, R.; Parfitt, G.D. Diffusion of radioactive magnesium in magnesium oxide crystals. J. Chem. Phys. 1957, *26*, 182.

2.152. Lindner, R.; Akerstrom, A. Self-diffusion and reaction in oxide and spinel systems. Z. Phys. Chem. 1956, *6*, 162.

2.153. Lindner, R.; Hassenteufel, W.; Kotera, Y. Diffusion of radioactive lead in lead metasilicate glass. Z. Phys. Chem. 1960, *23*, 408.

2.154. Shewmon, P.G. *Diffusion in Solids*; J. Williams Book Co.: Jenks, OK, 1983.

2.155. Crank, J. *The Mathematics of Diffusion*; Oxford University Press: Fair Lawn, NJ, 1956.

3

Methods of Corrosion Analysis

One must learn by doing the thing; for though you think you know it, you have no certainty until you try.

SOPHOCLES

3.1 INTRODUCTION

According to Weisser and Bange [3.1] it was Lavoisier who in 1770 first recorded the aqueous corrosion of a silicate glass predominantly by use of an analytical balance. The analysis of corrosion has been changing over the years with the greatest changes probably taking place within the last 25 years. These changes have been due mostly to the availability of sophisticated computerized analytical tools. It has taken many years for investigators to become familiar with the results obtained and how to interpret them. In some cases, special sample preparation techniques had to be perfected. Although one could conceivably employ all the various characterization methods described below, in most cases, only a few are needed to obtain sufficient information to solve a particular problem. The determination of the overall mechanism of corrosion requires a thorough detailed investigation using several characterization methods. Many times, though, the investigator has a limited amount of time and/or funds to obtain data and thus must rely on a few well-chosen tools. It should be obvious that considerable thought should be given to the selection of samples, test conditions, characterization methods, and interpretation of the results, especially if the data are to be used for prediction of lifetimes in actual service conditions. The reader is referred to the book by Wachtman [3.2] for a review of the principles involved in the various characterization techniques.

3.2 LABORATORY TEST VS. FIELD TRIALS

There are two general ways to approach a corrosion problem: either to conduct some laboratory tests to obtain information as to how a particular material will behave under certain conditions, or to perform a postmortem examination of field trial samples. It is best to perform the laboratory test first to aid in making the proper selection of materials for a particular environment and then perform the field trial. Laboratory tests, however, do not always yield the most accurate information since they rely on the investigator for proper setup; however, they are easier to control. The investigator must have a thorough understanding of the environment where the ceramic is to be used and must select the portions of the environment that may cause corrosion. For example, it is not sufficient to know that a furnace for firing ceramicware is heated by fuel oil to a temperature of 1200°C. One must also know what grade fuel oil is used and the various impurities contained in the oil and at what levels. In addition, parameters such as partial pressure of oxygen, moisture content, etc. may be important. Once all these various parameters are known, the investigator can set up an appropriate laboratory test.

One must also understand all the various things that cannot be scaled down to a laboratory test, such as viscosity of liquids, time, temperature, etc. Care must be exercised when attempting to perform an accelerated laboratory test, which is usually accomplished by raising the temperature or increasing the concentration of the corrosive medium or both. Since the mechanism of corrosion in the accelerated test may not be the same (generally, it is not the same) as that under actual service conditions, erroneous conclusions and inaccurate predictions may be obtained. The mechanisms must be the same for accurate application of laboratory test results to actual service conditions. Sample size is one parameter that is easily scaled; however, this can also cause problems. For example, when testing the corrosion of a ceramic by a liquid, the ratio of liquid volume present to the surface area of the exposed ceramic is very important. The investigator must remember that corrosion is controlled predominantly by thermodynamics and kinetics. Assuming that the proper laboratory tests have been conducted, the probability that any problems will arise is minimal.

The only way to analyze corrosion accurately is to conduct a field trial. This entails placing selected materials in actual service conditions, generally for an abbreviated time, and then collecting samples for analysis along with all the operational data of the particular environment. The size and amount of material or samples placed into actual service conditions for a field trial can be as little as one small laboratory test bar, or, for example, as large as a complete wall in a large industrial furnace. The larger the installation for the field trial, the more confidence one must have in the selection of materials. The larger installations are generally preceded by several laboratory tests and possibly a small-scale field trial. Abbreviated times may be as long as several years or as short as several days.

Data such as temperature and time are the obvious ones to collect, but there exists a large amount of other data that should be examined. Many times, however, some of the more important data do not exist for one reason or another. For example, maybe the oxygen partial pressure was not determined during the duration of the service life of the ceramic. In some cases, it may be impossible to collect certain pieces of data during the operation of the particular piece of equipment. At these times, a knowledge of phase equilibria, thermodynamics, and kinetics can help fill in the gaps or at least give an indication as to what was present.

3.3 SAMPLE SELECTION AND PREPARATION

It should be obvious that powders will present a greater surface area to corrosion and thus will corrode more rapidly than a solid sample. One may think this to be a good way to obtain a rapid test, but saturation of the corroding solution may cause corrosion to cease, or even cause a reverse reaction (i.e., crystal growth), giving misleading results. This points to the extreme importance of the surface area to volume ratio (SA/V) of the ceramic to the corroding solution. Another factor related to this is that during corrosion, the surface may change, altering the SA/V ratio effect. Surface areas during dissolution have been reported to increase presumably due to opening of etch pits, microfissures, etc. [3.3].

Selecting samples for analysis provides another challenge to the investigator. Foremost in the selection process is selecting an area for analysis that is representative of the overall corrosion process. If this cannot be done, then many samples must be analyzed. Much of the modern analytical equipment necessitates the analysis of very small samples, thus one must be very sensitive to the selection of representative samples or at least evaluate multiple samples.

Much care must be given to preparing samples that contain an adherent reaction product surface layer. It is best to select a sample that is many times larger than required by the final technique and then mounting this in some metallurgical mountant (e.g., epoxy). After the larger sample has been encased, then smaller samples can be safely cut from the larger piece.

Solid samples, when prepared for laboratory tests, should be cleaned in a noncorrosive solution to remove any loose particles adhering to the surface and any extraneous contamination. Brady and House [3.3] have reported that initially, an accelerated, nonlinear dissolution may occur from high-energy sites caused by grinding and incomplete removal of ultrafine particles. Best results are obtained if the cleaning is done in an ultrasonic cleaner. These cleaning solutions can be obtained from any of the metallographic supply companies. If the sample is mounted into one of the epoxy-type metallographic mountants, one must be aware that some cleaning solutions will react with the mountant. It is best to use supplies from one manufacturer to avoid these problems.

If as-manufactured samples are used for corrosion tests, one should remove a thin surface layer by grinding and cleaning before performing the corrosion tests. In this way, remnants from such things as powder beds or encapsulation media used in the production of the material can be eliminated and therefore not interfere with the corrosion process.

Quite often, the as-manufactured surface of a ceramic will have a different microstructure or even chemistry than the bulk. This often manifests itself as a thin surface layer (as much as several millimeters thick) that contains smaller grain sizes (more grain boundaries) and possibly a lower porosity. If the corrosion test corrodes only this thin surface layer, again, misleading results will be obtained. One way to solve this problem is to remove the surface layer by grinding. Grinding, however, must be done with some thought to the final surface roughness since, again, this will affect the SA/V ratio. Diamond-impregnated metal grinding disks should be used rather than silicon carbide paper disks or silicon carbide loose grit. Loose grit and the grinding media from paper disks have a tendency to become lodged within the pores and cracks of the sample being prepared. The final grinding media grit size should be no greater than 15 μm. It is best to clean samples after each grinding step in an ultrasonic cleaner with the appropriate cleaning solution.

Surface roughness of solid samples is an item that is often overlooked. Not only does a rough surface increase the area exposed to corrosion, but it may also lead to problems with some analytical techniques. For example, when the surface roughness is on the order of the reaction layer thickness caused by corrosion, errors will be present in the depth profiles obtained by secondary ion mass spectroscopy (SIMS). In those cases when surface analyses are planned, one should prepare solid samples to at least a 5-μm finish.

Grinding and polishing of samples that contain a reaction product surface layer should be done so that the reaction layer is not damaged, or the interface obscured. If part of the sample is metal, then polishing should be done in the direction ceramic toward metal to eliminate smearing the metal over the ceramic. If very thin reaction layers are present, one can prepare taper sections to increase the area that is examined. Sample preparation of composites presents some additional problems since materials of very different characteristics will be presented at the surface being polished. Chanat [3.4] of Buehler Ltd. has offered some good advice for mounting, sectioning, and polishing. The most effective method involves the use of diamond abrasives with low nap cloths for FRCMC*. High nap cloths can induce excessive relief at boundaries of different materials.

Many tips on how to prepare samples can be obtained by reading the various technical journals published by the manufacturers of consumable grinding and polishing supplies. One particular article that offers some new ideas was that of Damgaard and Geels [3.5]. They emphasized the importance of polishing disk diameter and velocity, indicating that both are directly proportional to material removal rates. Although this

*FRCMC = fiber-reinforced ceramic matrix composite.

may be true, one must be very aware of the amount of lubricant used, the pressures applied, and the area of the sample being polished. If the lubricant supply rate is constant, which is generally the case, the material removal rate will peak at about 300–350 rpm. Thus if one is thinking of purchasing automatic grinding and polishing equipment, he should look for something that has automatic lubrication flow rate control. Although many advances have been made in the grinding and polishing of ceramics, this area still is very much an art. Subtle changes in the procedure can make a major difference in the final finish of the sample*.

3.4 SELECTION OF TEST CONDITIONS

Although the selection of appropriate samples can be a major problem, the selection of the appropriate test conditions is an even more difficult task. The goal of the industrial corrosion engineer in selecting test conditions is to simulate actual service conditions. Selection of test conditions is much easier for the scientist, who is attempting to determine mechanisms. The major problem in attempting to simulate service conditions is the lack of detailed documentation. This is caused by not knowing the importance of such data in the corrosion of ceramics, the cost of collecting the data, or both. Thus if one wants to perform meaningful laboratory corrosion studies, it is imperative that the industrial environment of interest be accurately characterized.

When conducting laboratory oxidation studies, a convenient way to obtain a range of oxygen partial pressures is desirable. Very low partial pressures are never attained in practice by the use of a vacuum system. Instead, a mixture of gases in which oxygen is a component is used to establish the low partial pressure. The most important mixtures that are

*One should also remember that a perfectly polished surface, although excellent for reports, is not necessary to obtain sufficient data to solve a corrosion problem.

used are CO_2 + CO and H_2O + H_2. Since the oxygen pressures are obtained through the equilibrium reactions:

$$CO_2 \leftrightarrow CO + 1/2\ O_2 \tag{3.1}$$

and

$$H_2O \leftrightarrow H_2 + 1/2\ O_2 \tag{3.2}$$

the partial pressure of oxygen is given by:

$$pO_2 = k_1 \left(\frac{pCO_2}{pCO}\right)^2 \text{ and } pO_2 = k_2 \left(\frac{pH_2O}{pH_2}\right)^2 \tag{3.3}$$

where k_1 and k_2 are the equilibrium reaction constants. For constant ratios, the partial pressure of oxygen is independent of the total pressure. Thus these gas mixtures provide a means to obtain a range of oxygen pressures. Several techniques to mix these gases are discussed by Macchesney and Rosenberg [3.6].

In the study of corrosion in coal gasification atmospheres, gas mixtures such as $CH_4 + H_2$ and $H_2S + H_2$ become important along with the ones listed above. As the gas mixture becomes more complex, the number of equations that must be solved to obtain the equilibrium gas composition at elevated temperatures and pressures also increases, making it convenient to use a program such as SOLGASMIX [3.7] for the calculations. One should not make the erroneous assumption that gas mixtures are the same at all temperatures since the equilibrium mixture is dependent upon the equilibrium constant, which is temperature-dependent.

3.5 CHARACTERIZATION METHODS

3.5.1 Microstructure and Phase Analysis

Visual Observation

The most obvious method of analysis is that of visual observation. The human eye is excellent at determining differences between a used and an unused ceramic. Such things as variations in color, porosity, and texture should be noted. If no obvious changes have taken place, one should not assume that no alteration has occurred. Additional examination on a much finer scale is then required. Many times, visual observation can be misleading. For example, a sample may exhibit a banded variation in color, indicating a possible chemical variation. On closer examination, however, the color differences may be due only to porosity variation. An aid to visual observation is the dye penetration test. In this method, a sample is immersed into a solution such as methylene blue and then examined under a stereomicroscope.

Optical Microscopy

A compliment to visual observation is that of optical microscopy. Many people have devoted their entire lives to the study of ceramic microstructures through the examination of various sample sections and the use of some very sophisticated equipment. A preliminary examination should be conducted with a stereomicroscope and photographs taken. It is sometimes difficult to remember what a particular sample looked like after it has been cut into smaller pieces and/or ground to a fine powder for further analysis. A photographic record solves that problem.

The ceramics community has fallen into the habit of making only polished sections for observation by reflected light, when a tremendous amount of information can be obtained by observing thin sections with transmitted light. This trend has been brought about by the presence of many other pieces of equipment. Polished sections must be supplemented by x-ray diffractometry and also energy dispersive spectroscopy and/or scanning electron microscopy to obtain a full identification. A full identification can be made, however, with the use of thin sections. The only drawback is that an expert microscopist is required who understands the interaction of polarized and unpolarized light with the various features of the sample. It is true that the preparation of a thin section is more tedious than that of a polished section, but with today's automatic equipment, there is not much difference. In addition, a thin section does not require the fine polishing (generally down to submicron grit sizes) that a polished section does. The problem of pullouts does not interfere with the interpretation of the microstructure in transmitted light like it does in reflected light. The major drawback of a thin section, which should be on the order of 30 μm thick, is that it must be not greater than one crystal thick. With today's advanced ceramics being produced from submicron-sized powders, many products do not lend themselves to thin section examination. In those cases, polished sections must suffice.

One major advantage of the light microscope over electron microscopes is the ability to observe dynamic processes. Time-lapse video microscopy can be used to follow real-time corrosion processes. Obviously, room-temperature processes and those in aqueous media are the easiest to observe. Much of the latest work in the area of video microscopy has taken place in cell biology. Anyone interested in additional reading in this area should read the book by Cherry [3.8].

X-ray Diffractometry

Phase analysis is normally accomplished through the use of x-ray diffractometry (XRD), although optical microscopy can also be used. X-ray diffractometry is generally best done on powdered samples; however, solid flat surfaces can also be evaluated. Generally, a sample of about 1.5 g is necessary, but sample holder designs vary considerably and various sample sizes can be accommodated. Solid flat samples should be on the order of about 0.5 in. square. Powder camera techniques are available that can be used to identify very small quantities of powders. In multiphase materials, the minor components must be present in amounts greater than about 1–2 wt.% for identification. Once the mineralogy of the corroded ceramic is known, a comparison with the original uncorroded material can aid in the determination of the mechanism of corrosion.

Although quantitative XRD can be performed, the accuracy is dependent upon the sample preparation (crystal orientation plays a major role), the quality of the standards used, and the care taken in reducing various systematic and random errors. Several articles have been published in the literature that the interested reader may want to consult before taking on the task of quantitative XRD [3.9–3.12]. The one by Brime [3.12] is especially good since it compares several techniques.

Although the author is unaware of the use of high-temperature XRD in the evaluation of corrosion, there is no technical reason why it could not be useful. The major problem with high-temperature XRD is the identification of multiple phases at temperatures where the peaks become sufficiently broadened to obscure one another*.

Scanning Electron Microscopy/Energy Dispersive Spectroscopy

If an evaluation of the corroded surface is required and one does not want to destroy the sample totally, then an examination by scanning electron microscopy/energy dispersive spectroscopy (SEM/EDS) can yield valuable information. With most ceramics, however, the sample requires a conductive coating of carbon or gold before examination. If the same sample is to be used for both optical reflected light microscopy and SEM, the optical work should be done first. Quite often, the polished section prepared for optical examination is too large for the SEM, and the coating required for SEM may interfere with optical examination.

Chemical analysis by EDS can be quite useful in identifying phases observed in reflected light optical microscopy. Although the resolution of topographic features can be as good as several hundred angstroms in the SEM, the resolution of the EDS data is generally on the order of 1 μm. The EDS data also come from a small volume of sample and not just the surface. This may lead to the EDS signal originating from several

*High-temperature XRD also has many problems related to sample holders, sample temperature determination, and thermal expansion effects. Anyone considering this technique should have ample time to obtain results.

overlapping features and not just what one observes from the topographic features. Although SEM can be performed on as-received or rough surfaces, EDS is best performed on polished or flat surfaces. The analysis by SEM/EDS in combination with XRD and optical microscopy is a powerful tool in the evaluation of corrosion. See Fig. 5.3, which shows optical, SEM/EDS, and XRD data for the corrosion of a mullite refractory, and the corresponding text for an example of the use of EDS in phase identification.

Transmission Electron Microscopy

Transmission electron microscopy (TEM) can be used to evaluate the corrosion effects upon grain boundary phases. Transmission electron microscopy used in this way can be very useful; however, it is a very time consuming method, and, quite often, the samples are not representative due to their small size (several millimeters or less) and the thinning process. Transmission electron microscopy does not lend itself to the observation of porous samples and thus is confined to observation of dense regions of corroded samples.

3.5.2 Chemical Analysis

Bulk Analysis

The bulk chemical analysis of a corroded material is also a widely used tool in the evaluation of corrosion. In most cases, it is the minor constituents that will be most important. It may even be necessary to examine the trace element chemistry. When corrosion has taken place through reaction with a liquid, it is important to analyze the chemistry of the liquid. In this way, it is possible to establish whether it is the bulk or the bonding phases that are being corroded. Schmidt and Rickers [3.13] determined the concentration of chemical species in corroding fluids and melts by synchrotron radiation x-ray fluorescence (SR-XRF). The studies of Schmidt and Rickers are quite interesting since they were performed in situ at pressures up to 1.1 GPa and temperatures up to 800 °C in a hydrothermal diamond-anvil cell.

A chemical analysis that is normally not done is that of the gaseous phases produced during corrosion. This is not an easy task for large-scale experiments but can be accomplished on the microscale, such as that done with the aid of a thermobalance (TG) connected to a gas chromatograph (GS), mass spectrometer (MS), or infrared absorption spectrometer (FTIR).

Surface Analysis

Since corrosion takes place through reaction with the surface of a material, it is easier to determine mechanisms when the chemistry of the surfaces involved is analyzed. In this way, one may no longer be confronted with evaluation of minor constituents and trace elements since the corrosive reactants and products are more concentrated at the surface. The only drawback to surface analysis is that of the cost of the equipment and the necessity of a skilled technician. Secondary ion mass spectroscopy (SIMS) is a technique that currently receives wide use since it provides element detection limits in the subparts per million range and very good spatial resolution. Profiling of the various elements, another form of surface analysis, in question can be a very enlightening experiment. In this way, the depth of penetration can be determined and the elements that are the more serious actors can be evaluated. Lodding [3.14] has provided an excellent review of the use of SIMS to the characterization of corroded glasses and superconductors.

Determination of surface structures of ceramics for corrosion studies is most likely best accomplished by techniques such as Auger photoelectron diffraction (APD), x-ray photoelectron diffraction (XPD), or atomic force microscopy (AFM). Other techniques are available (e.g., LEED*), but they are better suited to other materials or suffer from various limitations. Gibson and LaFemina [3.15] offer an excellent discussion of how the various surface analysis techniques are used to characterize mineral surface dissolution.

*LEED is the acronym for low-energy electron diffractometry.

3.5.3 Physical Property Measurement

Gravimetry and Density

The evaluation of weight change during a reaction in many cases is sufficient to determine that corrosion has taken place. Weight change in itself, however, is not always detrimental. In the case of passive corrosion, a protective layer forms on the exposed surface. This would indicate that corrosion had taken place, but it is not necessarily detrimental since the material is now protected from further corrosion.

If at all possible, one should perform weight change experiments in a continuous manner on an automated thermal analyzer rather than performing an interrupted test where the sample is removed from the furnace after each heat treatment and weighed. In the interrupted test, one runs the risk of inaccurate weight measurements due to handling of the sample.

Density measurements are another form of gravimetry, but in this case, the volume change is also measured. Many times, volumetric changes will take place when a material has been held at an elevated temperature for an extended time. This implies that additional densification or expansion has taken place. Additional densification, although not necessarily a form of corrosion, can cause serious problems in structural stability. Expansion of a material generally implies that corrosion has taken place and that the reactions present involve expansion. Again, these may not be degrading chemically to the material but may cause structural instability.

One must exercise care in comparing density data obtained by different methods. Generally, the apparent density obtained from helium pycnometry is slightly higher than that obtained from water absorption*. For example, the data

*Helium is more penetrating than water and thus yields a smaller volume determination. This is dependent upon the pore size distribution.

for a sample of fusion cast α/β alumina gave 3.47 g/mL by water absorption compared to 3.54 g/mL by helium pycnometry. Helium pycnometry lends itself to the determination of densities of corroded samples.

Porosity-Surface Area

The evaluation of the porosity of a corroded sample generally presents the investigator with a rather difficult task. Most often, the best method is a visual one. Determination of the variations in pore size distribution in different zones of the sample may be a significant aid to the analysis. With modern computerized image analysis systems, one has the capability of evaluating porosity and pore size distributions rather easily [3.16]. One must be aware of the fact that sample preparation techniques can greatly affect the results obtained by image analysis.

The determination of the porosity of an uncorroded specimen, however, is extremely important in determining the surface area exposed to corrosion. Two samples identical in every way except porosity will exhibit very different corrosion characteristics. The one with the higher porosity or exposed surface area will exhibit the greater corrosion. This is therefore not a true test of corrosion but is valuable in the evaluation of a particular as-manufactured material. Not only is the value of the total volume of porosity important, but the size distribution is also important.

The porosity test by water absorption is not sufficient since the total porosity available for water penetration is not equivalent to the total porosity available for gaseous penetration. Although water absorption is a convenient method to determine porosity, it yields no information about pore size, pore size distribution, or pore shape. Mercury intrusion, however, does yield information about pore size distribution in the diameter range between 500 and 0.003 μm. One must remember that the size distribution obtained from mercury intrusion is not a true size distribution but one calculated from an equivalent volume. By assuming the pores to be cylindrical, one can calculate an approximate surface area from the total volume intruded by the mercury. A sample that has been used for mercury intrusion should not be subsequently used for corrosion testing since some mercury remains within the sample after testing. For applications involving gaseous attack, a method that measures gas permeability better evaluates the passage of gas through a material. Permeability tests, however, are not as easy to perform as porosity tests. A major problem with the permeability test is sealing the edges of the sample against gas leakage.

Determination of the surface area directly by gas adsorption (BET*) or indirectly by mercury intrusion may not correlate well with the surface area available to a corrosive liquid since the wetting characteristics of the corrosive liquid are quite different from that of an adsorbed gas or mercury. Thus one should exercise caution when using data obtained by these techniques.

Mechanical Property Tests

Probably the most widely used mechanical property test is that of modulus of rupture (MOR). One generally thinks of corrosion as lowering the strength of a material; however, this is not always the case. Some corrosive reactions may, in fact, raise the strength of a material. This is especially true if the MOR test is done at room temperature. For example, a high-temperature reaction may form a liquid that more tightly bonds the material when cooled to room temperature. A method that is often used is first soaking the samples in a molten salt and then performing a MOR test. This evaluates both the high-temperature strength and the effects of corrosion upon strength. Long-term creep tests or deformation under load tests can yield information about the effects of alteration upon the ability to resist mechanical deformation. For a more detailed discussion of the effects of corrosion upon mechanical properties, see Chap. 8.

---

*BET is an acronym for the developers of the technique, Brunauer, Emmett, and Teller.

3.6 DATA REDUCTION

The corrosion data that have been reported in the literature have been in many forms. This makes comparison between various studies difficult unless one takes the time to convert all the results to a common basis. Those working in the area of leaching of nuclear waste glasses have probably made the most progress in standardizing the reporting of data; however, a major effort is still needed to include the entire field of corrosion of ceramics. The work and efforts of organizations like ASTM can aid in providing standard test procedures and standard data reporting methods. These are briefly described in Chap. 4.

3.7 ADDITIONAL RELATED READING

Riga A.T., Patterson G.H., Eds.; *Oxidative Behavior of Materials by Thermal Analytical Techniques*; ASTM STP 1326, ASTM: West Conshohocken, Pa., 1997; 247 pp.

Gibson, A.S.; LaFemina, J.P. Structure of Mineral Surfaces. In *Physics and Chemistry of Mineral Surfaces*; Brady, P.V., Ed.; CRC Press, NY, 1996, 1–62.

Zipperian, D.C. Microstructural Analysis Using Image Analysis. In *Ceramic Transaction, Advances in Ceramic–Matrix Composites*; Bansal, N.P., Ed.; Am. Ceram. Soc., Westerville, OH, 1993; Vol. 38, 631–651.

Mason, C.W.; *Handbook of Chemical Microscopy*, 4th Ed.; John Wiley Sons, Inc.: New York, 1983; Vol. 1.

Cherry, R.J., Ed.; *New Techniques of Optical Microscopy and Microspectroscopy, Topics in Molecular and Structural Biology*; Neidle, S., Fuller, W., Series Eds.; CRC Press: Boca Raton, FL, 1991.

Chinn, R.E., Ed.; *Ceramography: Preparation and Analysis of Ceramic Microstructures*. ASM International & The Amer. Ceram. Soc. 2002, 214 pp.

3.8 EXERCISES, QUESTIONS, AND PROBLEMS

1. List all the possible techniques that one may use to analyze a corroded sample and the type of information obtained.
2. Describe the differences between laboratory tests and field trials.
3. List the various parameters of a laboratory test that can be scaled from the actual environment and list those that cannot. How will this affect the overall interpretation of the results of a lab test?
4. Discuss the errors that may arise when performing an accelerated laboratory test. In addition, what characteristics of a small lab sample lead to errors compared to the full-size installation?
5. Calculate the increased interface surface exposed by polishing a sample at a 45° taper, if the original perpendicular cross section had a 1-μm thick interface.
6. What parameters are important in the grinding and polishing of a sample and how do they affect the final result?
7. Discuss the information that one may obtain by examining a corroded sample with the unaided human eye.
8. Discuss the importance of the surface area of the corroded sample to the volume of the corroding liquid.
9. How does an interrupted weight change test vs. temperature interfere with the results? How can this problem be overcome?

REFERENCES

3.1. Weisser, M.; Bange, K. Sophisticated methods available to analyze glass corrosion. Glass Res. 2000, *9* (2), 16–17, 21.

3.2. Wachtman, J.B. *Characterization of Materials*; Butterworth-Heinemann: Boston, 1993.

3.3. Brady, P.V.; House, W.A. Surface-controlled dissolution and growth of minerals. In *Physics and Chemistry of Mineral Surfaces*; Brady, P.V., Ed.; CRC Press: New York, 1996; 225–305. Chp. 4

3.4. Chanat, S. Preparation techniques for analysis of fiber reinforced ceramic matrix composites. In *Ceramic Transaction, Advances in Ceramic–Matrix Composites*; Bansal, N.P., Ed.; Am. Ceram. Soc.: Westerville, OH, 1993; Vol. 38, 603–615.

3.5. Damgaard, M.J.; Geels, K. High capacity materialographic specimen preparation. Struers J. Materialogr. 2001; 7–11. Structure 38.

3.6. Macchesney, J.B.; Rosenberg, P.E. The methods of phase equilibria determination and their associated problems. In *Phase Diagrams: Materials Science and Technology*; Alper, A.M., Ed.; *Refractory Materials*; Margrave, J.L., Ed.; Academic Press: New York, 1970; Vol. 6-1, 113–165. Chp. 3.

3.7. Eriksson, G. Thermodynamic studies of high temperature equilibria. XII. SOLGASMIX, A computer program for calculation of equilibrium compositions in multiphase systems. Chem. Scr. 1975, *8*, 100–103.

3.8. Cherry, R.J., Ed.; *New Techniques of Optical Microscopy and Microspectroscopy*; Topics in Molecular and Structural Biology; Neidle, S., Fuller, W., Series Eds.; CRC Press: Boca Raton, FL, 1991.

3.9. Alexander, L.; Klug, H.P. Basic aspects of X-ray absorption. Anal. Chem. 1948, *20*, 886–889.

3.10. Chung, F.H. Quantitative interpretation of X-ray diffraction patterns of mixtures: I. Matrix-flushing method for quantitative multicomponent analysis. J. Appl. Cryst. 1974, *7*, 519–525.

3.11. Dickson, M.J. The significance of texture parameters in phase analysis by X-ray diffraction. J. Appl. Cryst. 1969, *2*, 176–180.

3.12. Brime, C. The accuracy of X-ray diffraction methods for determining mineral mixtures. Mineral. Mag. 1985, *49* (9), 531–538.

3.13. Schmidt, C.; Rickers, K. In-situ determination of mineral solubilities in fluids using a hydrothermal diamond-anvil cell and SR-XRF: Solubility of AgCl in water. Am. Mineral. 2003, *88* (2–3), 288–292.

3.14. Lodding, A. Characterization of corroded ceramics by SIMS. In *Corrosion of Glass, Ceramics and Ceramic Superconductors*; Clark, D.E., Zoitos, B.K., Eds.; Noyes Publications: Park Ridge, NJ, 1992; 103–121. Chp. 4.

3.15. Gibson, A.S.; LaFemina, J.P. Structure of mineral surfaces. In *Physics and Chemistry of Mineral Surfaces*; Brady, P.V., Ed.; CRC Press: New York, 1996; 1–62.

3.16. Exner, H.E., Hougardy, H.P., Eds.; *Quantitative Image Analysis of Microstructures*; DGM Informationsgesellschaft mbH: Germany, 1988, 235 pp.

4

Corrosion Test Procedures

> When you can measure what you are speaking about and express it in numbers you know something about it; but when you cannot measure it, when you cannot express it in numbers, your knowledge is of a meager and unsatisfactory kind.
>
> LORD KELVIN

4.1 INTRODUCTION

The American Society for Testing and Materials (ASTM) was formed in 1898 through the efforts of Andrew Carnegie and the chief chemist of the Pennsylvania Railroad, Charles Dudley, who were both convinced that a solution was necessary to the unexplainable differences of testing results that arose between their laboratories. These early efforts were focused upon improving the understanding between seller and buyer of the quality of their products. Although ASTM and other organizations have made considerable progress in eliminating the unexplainable differences in testing results between laboratories, new materials and new applications continue to present new and exciting challenges to the corrosion engineer. These challenges, however, are ones that must be overcome if there is to be honest competition in the world market of materials.

Many of us have fallen into the habit of performing a test only once and believing the results. This is probably one of the most important things not to do when evaluating a particular material for use under a certain set of conditions. The results of a test will generally vary to a certain degree and can vary considerably. It is up to the testing engineer to know or determine the test method variation. All ASTM standards now contain a statement of precision and bias to aid the test engineer in determining how his test fits into the overall imprecision of the procedure developed by the standards committee. In the development of an ASTM standard, a ruggedness test (ASTM Standard E-1169) is performed to determine the major sources of variation. This test should be performed for any laboratory test that one might conduct to minimize the major sources of error. The idea of the ruggedness test is to determine the major sources of variation of a procedure and then minimize those variations to within acceptable limits.

Many standard tests have been developed through ASTM to evaluate the corrosion resistance of various ceramic materials. These various tests have been listed in Tables 4.1 and 4.2 and can be found in the Annual Book of ASTM Standards, volumes 2.05, 4.01, 4.02, 4.05, 12.01, 14.04, 15.01, and 15.02. A brief summary of each of these is given below. Standards that are in the process of being developed have not been listed in Tables 4.1 and 4.2. These draft standards can be found on the ASTM web site.* ASTM designates some procedures as standard test methods and others as standard practices. The distinction between these two is best given by their definitions.

* The ASTM web site can be found at www.astm.org.

ASTM defines test method as a definitive procedure for the identification, measurement, and evaluation of one or more qualities, characteristics, or properties of a material, product, system, or service that produces a test result, and practice as a definitive procedure for performing one or more specific operations or functions that does not produce a test result [4.1]. Standard practices provide the user with accepted procedures for the performance of a particular task. Test methods provide the user with an accepted procedure for determination of fundamental properties (i.e., density, viscosity, etc.). These standards must be updated or reapproved by the end of the 8th year after the last approval. If not reapproved, the standard is then withdrawn.

The Materials Characterization Center* (MCC) is another organization that has developed standard test procedures [4.2]. Several of these tests have been used extensively by those investigating the leaching of nuclear waste glasses. Test MCC-1 involves a procedure for testing the durability of monolithic glass samples in deionized or simulated groundwater at 40°C, 70°C, and 90°C for 28 days. One disadvantage of this test is that no standard glass is used, thus eliminating corrections for bias. It does, however, require the reporting of mass loss normalized to the fraction of the element leached in the glass sample allowing one to make comparisons between glasses. Test MCC-3, in contrast, evaluates an agitated crushed glass sample to maximize leaching rates. Test temperatures are extended to 110°C, 150°C, and 190°C. Again, a standard glass is not used. Both of these tests have now been developed into ASTM standard test methods, C1220 and C-1285, respectively.

With the global economy of today, the engineer must be familiar with standards from countries other than the United States. In addition to the individual countries that maintain standards, there are also the International Organization for Standardization (ISO) and the International Electrotechnical

* The MCC was created in 1980 by the U.S. Department of Energy and is operated for the DOE by the Pacific Northwest Laboratories of the Battelle Memorial Institute in Richland, WA.

TABLE 4.1 ASTM Test Methods Related to Corrosion of Ceramics

ASTM designation	Vol. no.	Title
C-151	4.01	Autoclave Expansion of Portland Cement
C-157	4.02	Length Change of Hardened Hydraulic-Cement Mortars and Concrete (reapproved 1999)
C-225	15.02	Resistance of Glass Containers to Chemical Attack (reapproved 1999)
C-267	4.05	Chemical Resistance of Mortars, Grouts, and Monolithic Surfacings (reapproved 1997)
C-282	2.05	Acid Resistance of Porcelain Enamels (reapproved 1999)
C-283	2.05	Resistance of Porcelain Enameled Utensils to Boiling Acid (reapproved 1997)
C-288	15.01	Disintegration of Refractories in an Atmosphere of Carbon Monoxide (reapproved 1997)
C-370	15.02	Moisture Expansion of Fired Whiteware Products (reapproved 1999)
C-413	4.05	Absorption of Chemical-Resistant Mortars, Grouts, and Monolithic Surfacings (reapproved 1996)
C-452	4.01	Potential Expansion of Portland Cement Mortars Exposed to Sulfate (reapproved 1995)
C-456	15.01	Hydration Resistance of Basic Brick and Shapes (reapproved 1998)
C-492	15.01	Hydration of Granular Dead-Burned Refractory Dolomite (reapproved 1998)
C-544	15.01	Hydration of Magnesite or Periclase Grain (reapproved 1998)

TABLE 4.1 Continued

ASTM designation	Vol. no.	Title
C-556	15.02	Resistance of Overglaze Decorations to Attack by Detergents (withdrawn 1994)
C-577	15.01	Permeability of Refractories (reapproved 1999)
C-614	2.05	Alkali Resistance of Porcelain Enamels (reapproved 1999)
C-620	15.01	Hydration Resistance of Pitch-Bearing Basic Refractory Brick (withdrawn 1996)
C-621	15.01	Isothermal Corrosion Resistance of Refractories to Molten Glass (reapproved 1995)
C-622	15.01	Corrosion Resistance of Refractories to Molten Glass Using the Basin Furnace (withdrawn in 2000)
C-650	15.02	Resistance of Ceramic Tile to Chemical Substances
C-675	15.02	Alkali Resistance of Ceramic Decorations on Returnable Beverage Glass Containers (reapproved 1996)
C-676	15.02	Detergent Resistance of Ceramic Decorations on Glass Tableware (reapproved 1999)
C-724	15.02	Acid Resistance of Ceramic Decorations on Architectural Type Glass (reapproved 2000)
C-735	15.02	Acid Resistance of Ceramic Decorations on Returnable Beer and Beverage Glass Containers (reapproved 1999)
C-738	15.02	Lead and Cadmium Extracted from Glazed Ceramic Surfaces (reapproved 1999)

TABLE 4.1 Continued

ASTM designation	Vol. no.	Title
C-777	15.02	Sulfide Resistance of Ceramic Decorations on Glass (reapproved 1999)
C-863	15.01	Evaluating Oxidation Resistance of Silicon Carbide Refractories at Elevated Temperatures
C-872	2.05	Lead and Cadmium Release from Porcelain Enamel Surfaces (reapproved 1999)
C-895	15.02	Lead and Cadmium Extracted from Glazed Ceramic Tile (reapproved 1998)
C-927	15.02	Lead and Cadmium Extracted from Lip and Rim Area of Glass Tumblers Externally Decorated with Ceramic Glass Enamels (reapproved 1999)
C-1012	4.01	Length Change of Hydraulic-Cement Mortars Exposed to a Sulfate Solution
C-1034	15.02	Lead and Cadmium Extracted from Glazed Ceramic Cookware (withdrawn in 2001)
C-1106	4.05	Chemical Resistance and Physical Properties of Carbon Brick
C-1203	15.02	Quantitative Determination of Alkali Resistance of a Ceramic-Glass Enamel (reapproved 1996)
C-1370	4.05	Determining the Chemical Resistance of Aggregates for Use in Chemical-Resistant Sulfur Polymer Cement Concrete and Other Chemical-Resistant Polymer Concretes

TABLE 4.2 ASTM Practices Related to Corrosion of Ceramics

ASTM designation	Vol. no.	Title
C-454	15.01	Disintegration of Carbon Refractories by Alkali (reapproved 1998)
C-768	15.01	Drip Slag Testing Refractory Brick at High Temperature
C-874	15.01	Rotary Slag Testing of Refractory Materials
C-987	15.01	Vapor Attack on Refractories for Furnace Superstructures
G-7	14.04	Atmospheric Environmental Exposure Testing of Nonmetallic Materials
G-90	14.04	Accelerated Outdoor Weathering of Nonmetallic Materials Using Concentrated Natural Sunlight

Commission (IEC). A list of the various acronyms used for these various organizations is given in Table 4.3. The development of international standards is indispensable for the future progress of international trade.

4.2 ASTM STANDARDS

4.2.1 Autoclave Expansion of Portland Cement, C-151

Samples of portland cement are exposed to water vapor at 2 MPa and 23°C for 3 hr in an autoclave. The test evaluates the potential for delayed expansion caused by the hydration of CaO or MgO or both. The percent linear expansion change is reported.

4.2.2 Length Change of Hardened Hydraulic-Cement Mortars and Concrete, C-157

Samples of hardened cement or concrete are tested in lime-saturated water at 23°C for 15 or 30 min depending upon

TABLE 4.3 Standards Organizations and Their Acronyms

ACNOR	Association Canadienne de Normalisation
AENOR	Asociacion Espanola de Normalizacion y Certificacion
AFNOR	Association Française de Normalisation
ANSI	American National Standards Institute
ASA	Australian Standards Association
ASTM	American Society for Testing and Materials
BDS	Bulgarian Institute for Standardization
BNSI	Barbados National Standards Institution
BOBS	Botswana Bureau of Standards
BPS	Bureau of Philippine Standards
BSI	British Standards Institution
BSN	Badan Stardardisasi Nasional (Indonesia)
BSTI	Bangladesh Standards and Testing Institute
CAS	China Association of Standardization
CEN	Comite Europeen de Normalisation
CCN	Conseil Canadien des Normes
CSA	Canadian Standards Association
CSNI	Czech Standards Institute
DGN	Direccion General de Normes (Mexico)
DIN	Deutsches Institut für Normung (German)
DS	Danish Standards Bureau
DSM	Department of Standards of Malaysia
ENS	European Standards
EOS	Egyptian Organization for Standardization
GOST	Gosudarstvenii Stardart (Russian Federation)
GSB	Ghana Standards Board
HSI	Hungarian Standards Institution
IAN	Institute Algerien de Normalisation
IBN	Institute Belge de Normalisation (Belgium)
INEN	Instituto Ecuatoriano de Normalizacion (Ecuador)
IIRS	Irish Institute for Industrial Research and Standards
ISI	Indian Standards Institute
ISO	International Organization for Standardization
JBS	Jamaica Bureau of Standards
JSA	Japanese Standards Association

TABLE 4.3 Continued

JISC	Japanese Industrial Standards Committee
KATS	Korean Agency for Technology and Standards
KEBS	Kenya Bureau of Standards
MCC	Materials Characterization Center (USA)
MSB	Mauritius Standards Bureau
NC	Oficina Nacional de Normalizacion de Cuba
NIST	National Institute for Standards and Technology (USA)
NSF	Norges Standardiseringsforbund (Norway)
ON	Osterreichisches Normungsinstitut (Austria)
SABS	South African Bureau of Standards
SAC	Standards Association of China
SANZ	Standards Association of New Zealand
SCC	Standards Council of Canada
SIS	Shanghai Institute of Standardization (China)
SIS	Swedish Institute for Standardization
SNV	Schweizerische Normenvereinigung (Swiss)
SZS	Savezni Zavod za Standardizaciju (Yugoslavia)
TISI	Thai Industrial Standards Institute
TSE	Turk Standardslari Enstitusu
TTBS	Trinidad & Tobago Bureau of Standards
UNI	Ente Nationale Italiano di Unificazione
VSC	Vietnam Standards Centre

sample size. The samples are then dried at 23°C and a relative humidity of 50%. The length change is recorded after 4, 7, 14, and 28 days and 8, 16, 32, and 64 weeks.

4.2.3 Resistance of Glass Containers to Chemical Attack, C-225

Attack by dilute sulfuric acid (representative of products with pH less than 5.0) or distilled water (representative of products with pH greater than 5.0) on glass bottles and the attack by pure water upon powdered glass (for containers too small to test solubility by normal methods) all at 121°C is covered in this standard test method.

4.2.4 Chemical Resistance of Mortars, Grouts, and Monolithic Surfacings, C-267

This method tests the resistance of resin, silica, silicate, sulfur, and hydraulic materials, grouts, and monolithic surfacings to a simulated service environment. Any changes in weight, appearance of the samples or test medium, and the compressive strength are recorded.

4.2.5 Acid Resistance of Porcelain Enamels, C-282

This test method was developed to test the resistance of porcelain enamel coatings on stoves, refrigerators, table tops, sinks, laundry appliances, etc. to 10% citric acid at 26°C. Several drops of acid solution are placed onto a flat area about 50 mm in diameter. After 15 min, the samples are cleaned and evaluated for changes in appearance and cleanability.

4.2.6 Resistance of Porcelain Enameled Utensils to Boiling Acid, C-283

Test samples 82 mm in diameter make up the bottom of glass tube that is filled with 150 mL of a solution prepared from 6 g of citric acid in 94 g of distilled water. The test cell is placed onto a hot plate and the solution is allowed to boil for 2 1/2 hr. The results are reported as the change in weight.

4.2.7 Disintegration of Refractories in an Atmosphere of Carbon Monoxide, C-288

Providing a higher than expected amount of carbon monoxide normally found in service conditions, this method can be used to obtain the relative resistance of several refractory products to disintegration caused by exposure to CO. Samples are heated in nitrogen to the test temperature of 500°C then held in an atmosphere of 95% CO for times sufficient to produce complete disintegration of half the test samples.

4.2.8 Moisture Expansion of Fired Whiteware Products, C-370

Unglazed, rod-shaped samples are tested for their resistance to dimensional changes caused by water vapor at elevated temperatures and pressures. Five samples are placed into an autoclave for 5 hr in an atmosphere of 1 MPa of steam. The amount of linear expansion caused by moisture attack is then recorded.

4.2.9 Absorption of Chemical-Resistant Mortars, Grouts, and Monolithic Surfacings, C-413

Silica and silicate samples, in addition to other materials, are tested for absorption in boiling xylene after 2 hr. The percent absorption is recorded.

4.2.10 Potential Expansion of Portland-Cement Mortars Exposed to Sulfate, C-452

Samples of portland cement are mixed with gypsum and then immersed in water at 23°C for 24 hr and 14 days or more. The change in linear expansion is recorded.

4.2.11 Disintegration of Carbon Refractories by Alkali, C-454

Carbon cubes with a hole drilled into them to form a crucible are used as the samples to test their resistance to attack from molten potassium carbonate at approximately 1000°C for 5 hr. The results of this standard practice are reported as visual observations of the degree of cracking. Variations of this procedure have been used by many to investigate the resistance of refractories to attack by molten metals and molten glasses.

4.2.12 Hydration Resistance of Basic Brick and Shapes, C-456

One-inch cubes cut from the interior of basic brick are tested in an autoclave containing sufficient water to maintain a pressure of 552 kPa at 162°C for 5 hr. This test is repeated for successive 5-hr periods to a maximum of 30 hr or until the samples disintegrate. The results are reported as visual observations of hydration and cracking.

4.2.13 Hydration of Granular Dead-Burned Refractory Dolomite, C-492

A 100-g dried powder sample of dolomite that is coarser than 425 μm is tested by placing it into a steam-humidity cabinet that is maintained at 71°C and 85% humidity for 24 hr. The sample is then dried at 110°C for 30 min, and the amount of material passing a 425-μm sieve is determined.

4.2.14 Hydration of Magnesite or Periclase Grain, C-544

A carefully sized material that is between 425 μm and 3.35 mm is tested by placing a dried 100-g sample into an autoclave maintained at 162°C and 552 kPa for 5 hr. The sample is then weighed after removal from the autoclave and dried at 110°C. The hydration percentage is calculated from the weight difference between the final dried weight and the weight of any material coarser than 300 μm.

4.2.15 Resistance of Overglaze Decorations to Attack by Detergents, C-556; Withdrawn 1994

Overglaze decorations on pieces of dinnerware are tested by submerging the samples into a solution of sodium carbonate and water at a temperature of 95°C. Samples are removed after 2, 4, and 6 hr and rubbed with a muslin cloth. The results are reported as visual observations of the degree of material removed by rubbing.

4.2.16 Permeability of Refractories, C-577

Although not a corrosion test, C-577 is important in determining the ease of flow of various gases through a material. This test method is designed to determine the unidirectional rate of flow of air or nitrogen through a 2-in. cube of material at room temperature.

4.2.17 Alkali Resistance of Porcelain Enamels, C-614

The coatings on washing machines, dishwashers, driers, etc. are tested for their resistance to solution containing 260 g of tetrasodium pyrophosphate dissolved in 4.94 L of distilled water. The loss in weight is determined after exposure for 6 hr at 96°C.

4.2.18 Hydration Resistance of Pitch-Bearing Refractory Brick, C-620

Full-sized pitch-containing bricks are placed into a steam-humidity cabinet and tested for 3 hr at 50°C and 98% humidity. The test is repeated for successive 3-hr periods until visually affected. The results are reported as visual observations of hydration and disintegration.

4.2.19 Isothermal Corrosion Resistance of Refractories to Molten Glass, C-621

This method compares the corrosion resistance of various refractories to molten glass under static, isothermal conditions. Samples approximately 1/2 in. square by 2 in. long are immersed into molten glass, then heated to a temperature that simulates actual service conditions. The duration of the test should be sufficient to produce a glass-line cut of 20–60% of the original sample thickness. After the test, samples are cut in half lengthwise and the width or diameter is measured at the glass line and halfway between the glass line and the bottom of the sample before testing.

4.2.20 Corrosion Resistance of Refractories to Molten Glass Using the Basin Furnace, C-622; Withdrawn in 2000

This standard practice determines the corrosion of refractories by molten glass in a furnace constructed of the test blocks with a thermal gradient maintained through the refractory. Because of the cooling effects of the thermal gradient, the duration of this test is 96 hr. Since the glass is not replaced during the test, solution products may modify the results of the test. The depth of the glass-line cut is determined across the sample, and the volume corroded is determined by filling the corroded surface with zircon sand and determining the volume of sand required.

4.2.21 Resistance of Ceramic Tile to Chemical Substances, C-650

This method is designed to test plain colored, glazed, or unglazed impervious ceramic tile of at least 4 1/4 × 4 1/4 in. to the resistance against attack by any chemical substance that may be of interest. The test conditions may be any combination of time and temperature deemed appropriate for the expected service conditions. Hydrochloric acid or potassium hydroxide at 24°C for 24 hr is the recommended exposure. The results are reported as visually affected or not affected, and also the calculated color difference may be reported.

4.2.22 Alkali Resistance of Ceramic Decorations on Returnable Beverage Glass Containers, C-675

Two ring sections cut from each container and representative of the label to be evaluated are placed into the test solution at 88°C of sodium hydroxide, trisodium phosphate, and tap water for successive 2-hr intervals. The results are reported as the time required for 90% destruction of the label. A variation of this method conducted at 60°C for 24 hr in a mixture of sodium hydroxide, trisodium phosphate, and distilled water determines the reduction in thickness of the label.

4.2.23 Detergent Resistance of Ceramic Decorations on Glass Tableware, C-676

In this standard method, glass tableware with ceramic decorations is immersed into a solution of sodium pyrophosphate and distilled water at 60°C for successive 2-hr periods. The samples are then rubbed with a cloth under flowing water, dried, and evaluated as to the degree of loss of gloss up to complete removal of the decoration.

4.2.24 Acid Resistance of Ceramic Decorations on Architectural Type Glass, C-724

A citric acid solution is placed onto the ceramic decoration of the architectural glass for 15 min at 20°C, and the degree of attack after washing is determined visually.

4.2.25 Acid Resistance of Ceramic Decorations on Returnable Beer and Beverage Glass Containers, C-735

Representative containers are immersed into hydrochloric acid solution such that half the decoration is covered for 20 min at 25°C. The results are reported as the visually observed degree of attack.

4.2.26 Lead and Cadmium Extracted from Glazed Ceramic Surfaces, C-738

This standard method determines quantitatively by atomic absorption the amount of lead and cadmium extracted from glazed ceramic surfaces when immersed into 4% acetic acid solution at 20–24°C for 24 hr.

4.2.27 Drip Slag Testing Refractory Brick at High Temperature, C-768

Test samples of this standard practice are mounted into the wall of a furnace such that their top surface slops down at a 30° angle. Rods of slag are placed through a hole in the furnace wall such that when the slag melts, it will drip and fall 2 in. to the surface of the refractory test piece. Slag is fed continuously to maintain consistent melting and dripping onto the sample. Test temperatures are about 1600°C and the duration of the test is from 2 to 7 hr. The volume of the corroded surface is determined by measuring the amount of sand required to fill the cavity. In addition, the depth of penetration of slag into the refractory is determined by cutting the sample in half.

4.2.28 Sulfide Resistance of Ceramic Decorations on Glass, C-777

Decorated ware is immersed into a solution of acetic acid, sodium sulfide, and distilled water at room temperature for 15 min such that only half the decoration is covered by the test solution. The results are reported as visually observed deterioration of the decoration.

4.2.29 Evaluating Oxidation Resistance of Silicon Carbide Refractories at Elevated Temperatures, C-863

The volume change of one-fourth of a 9-in. straight is evaluated in an atmosphere of steam and at any three temperatures of 800°C, 900°C, 1000°C, 1100°C, and 1200°C. The duration of the test is 500 hr. In addition to the average volume change of three samples, any weight, density, or linear changes are also noted in this standard method.

4.2.30 Lead and Cadmium Release from Porcelain Enamel Surfaces, C-872

Samples cut from production parts or prepared on metal blanks under production conditions are exposed to 4% acetic acid at 20–24°C for 24 hr. Samples 26 cm^2 are placed into a test cell similar to the one used in C-283 and covered with 40 mL of solution for each 6.45 cm^2 of exposed surface area. The Pb and Cd released into solution are determined by atomic absorption spectrophotometry.

4.2.31 Rotary Slag Testing of Refractory Materials, C-874

This standard practice evaluates the resistance of refractories to flowing slag by lining a rotary furnace, tilted at 3° axially toward the burner, with the test samples. The amount of slag used and the temperature and duration of the test will depend upon the type of refractory tested. The results are reported as the percent area eroded.

4.2.32 Lead and Cadmium Extracted from Glazed Ceramic Tile, C-895

This standard method determines quantitatively by atomic absorption the amount of lead and cadmium extracted from glazed ceramic tile when immersed into 4% acetic acid solution at 20–24°C for 24 hr.

4.2.33 Lead and Cadmium Extracted from Lip and Rim Area of Glass Tumblers Externally Decorated with Ceramic-Glass Enamels, C-927

This standard method determines quantitatively by atomic absorption the amount of lead and cadmium extracted from the lip and rim area of glass tumblers when immersed into 4% acetic acid solution at 20–24°C for 24 hr.

4.2.34 Alkali Vapor Attack on Refractories for Glass-Furnace Superstructures, C-987

This standard practice evaluates the resistance to alkali attack of refractories by placing a 55-mm square by 20-mm-thick sample over a crucible containing molten reactant such as sodium carbonate at 1370°C. A duration at test temperature of 24 hr is recommended, although other times can be used to simulate service conditions. The results are reported as visual observations of the degree of attack.

4.2.35 Length Change of Hydraulic-Cement Mortars Exposed to a Sulfate Solution, C-1012

Samples are tested in a solution of Na_2SO_4 or $MgSO_4$ in water (50 g/L) at 23°C for times initially ranging from 1 to 15 weeks. Extended times may be used if required. The percent linear expansion is recorded.

4.2.36 Lead and Cadmium Extracted from Glazed Ceramic Cookware, C-1034; Withdrawn in 2001

This standard test method determines quantitatively by atomic absorption the amount of lead and cadmium extracted from glazed ceramic cookware when immersed into boiling 4% acetic acid solution for 2 hr.

4.2.37 Chemical Resistance and Physical Properties of Carbon Brick, C-1106

At least three 2-in. cubes per test medium and per test temperature are immersed into approximately 150 mL of the desired test liquid. The closed containers are placed into a constant temperature oven or bath and then examined after 1, 7, 14, 28,

56, and 84 days. The samples are evaluated for weight change and compressive strength change.

4.2.38 Quantitative Determination of Alkali Resistance of a Ceramic-Glass Enamel, C-1203

The chemical dissolution of a ceramic-glass enamel-decorated glass sample is determined by immersing it into a 10% alkali solution near its boiling point (95°C) for 2 hr. The dissolution is determined by calculating the difference in weight losses between the decorated sample and an undecorated sample, normalized for the differences in areas covered and uncovered by the decoration.

4.2.39 Determining the Chemical Resistance of Aggregates for Use in Chemical-Resistant Sulfur Polymer Cement Concrete and Other Chemical-Resistant Polymer Concretes, C-1370

This standard test method determines the chemical resistance of at least three 200-gm samples of aggregate immersed into 400 mL of the desired solution, covered, and held at 60°C for 24 hr. The resistance to attack is determined by the change in weight during the test.

4.2.40 Atmospheric Environmental Exposure Testing of Nonmetallic Materials, G-7

This standard practice evaluates the effects of climatic conditions upon any nonmetallic material. Samples are exposed at various angles to the horizon and generally are faced toward the equator. It is recommended that temperature, humidity, solar radiation, hours of wetness, and presence of contaminants be recorded.

4.2.41 Performing Accelerated Outdoor Weathering of Nonmetallic Materials Using Concentrated Natural Sunlight, G-90

This standard practice describes the use of a Fresnel-reflector to concentrate sunlight onto samples in the absence of moisture. A variation in the procedure allows the spraying of purified water at regular intervals on the samples.

4.3 NONSTANDARD TESTS

Many individual laboratories use test procedures that are similar to ASTM standard procedures; however, they have been modified to suit their own particular needs or capabilities. Although a particular ASTM test was developed for a certain material under specific conditions, it does not imply that other materials cannot be tested in the same manner. For example, C-621 for corrosion of refractories by molten glass could be used to test nonrefractories by various other liquids. A variation of this test has been used by some glass technologists where the refractory samples are rotated to simulate a forced convection situation. The real problem with this test is that one generally does not know the glass velocity distribution along the sample with sufficient accuracy to extrapolate laboratory results to commercial furnaces. A more appropriate test to evaluate forced convection upon dissolution is the rotating disk test, shown in Fig. 4.1. In this setup, the diffusion boundary layer across the lower disk face has a constant value for any experimental temperature and rotational velocity. The dissolution of the solid disk is therefore constant, a situation that does not occur in the finger test (see also Chap. 2, page 15 on Attack by Molten Glasses). Any test that is used should be subjected to ruggedness testing first to determine the important variables.

It is almost impossible to test the corrosion of ceramics and maintain all samples equivalent since variations in density and porosity are generally present. Thus it is important to test more

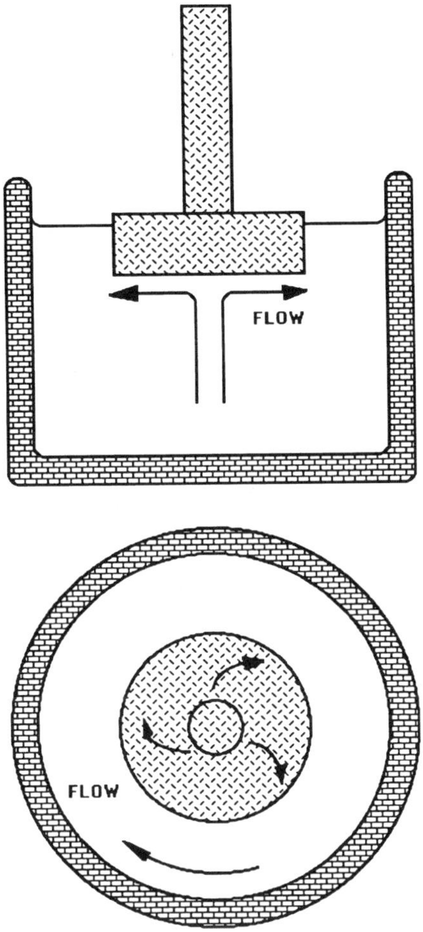

FIGURE 4.1 Rotating disk setup.

than one sample under a particular set of conditions and average the results or normalize the test results to constant porosity.

4.4 ADDITIONAL RELATED READING

Youden, W.J. Experimental design and ASTM committees, Mater. Res. Stand. 1961, *1* (11), 862–867.

Wernimont, G. Ruggedness evaluation of test procedures. ASTM Standard. News March 1977, 13–16.

Smith G.L.; Marschman, S.C. Nuclear waste analytical round robins 1–6 summary report, Mater. Res. Soc. Symp. Proc. 1994, *333*, 461–472.

Sterling, J. The importance of international standards. ASTM Standard. News June 2001, 24, 27 pp.

4.5 EXERCISES, QUESTIONS, AND PROBLEMS

1. Why is it important to determine the factors that cause variation during testing? What ASTM standard addresses these factors?
2. What is the difference between a standard method and a standard practice?
3. Are there situations when a nonstandard test may be used? If yes, what precautions should be taken?
4. Discuss the importance of developing International Standards related to corrosion of ceramic materials.

REFERENCES

4.1. *Form and Style for ASTM Standards*, 7th Ed. ASTM: Philadelphia, March 1986.

4.2. Mendel, J.E. (compiler). *Nuclear Waste Materials Handbook-Waste Form Test Methods*, Materials Characterization Center, Pacific Northwest Laboratories, Richland, WA, U.S. DOE Report DOE/TIC-11400, 1981.

5

Corrosion of Specific Crystalline Materials

The most beautiful thing we can experience is the mysterious. It is the source of all true art and science.

ALBERT EINSTEIN

5.1 ATTACK BY LIQUIDS

5.1.1 Attack by Glasses

In the indirect corrosion of oxides by glasses, the crystalline phase that forms at the interface is dependent upon the glass composition and the temperature. Various interface phases that form in some silicate melts are listed in Table 5.1. Whether the system is under forced convection or not will also play an important role in the formation of a crystalline interface phase. An excellent study of the effects of forced convection is that by

TABLE 5.1 Interfacial Reaction Products Caused by Molten Liquid Attack

Oxide	Liquid[a]	Interface[a]	Ref.
Al_2O_3	CAS	CA_2 and CA_6	5.2
Al_2O_3	Coal Slag	Mixed spinel	5.3
Al_2O_3	CMAS	$MgAl_2O_4$	5.4,5.5
Al_2O_3	S	A_3S_2	5.6
Al_2O_3–Cr_2O_3	CMAS	Mixed spinel	5.1,5.5
Al_2O_3–Cr_2O_3	Coal Slag	Mixed spinel	5.3
AZS	Coal Slag	CA_6 and C_2AS	5.3
AZS	NCS	NAS_2 and Z	5.6
AZS	KPS	KAS_2 and Z	5.7
CaO	CAS	C_2S and C_3S	5.8
CaO	CFS	C_2S and C_3S	5.8
Cr_2O_3 spinel	Coal Slag	Mixed spinel	5.3
Fused SiO_2	CAS	Cristobalite	5.9
MgO	CAS	C_2MS_2 and M	5.9
MgO	CFS	MF solution	5.10
$MgAl_2O_4$	CAS	C_2AS or CAS_2	5.9
$Al_6Si_2O_{13}$	NCS	NAS_2 and A	5.6
$Y_3Al_5O_{12}$	CAS	C_2AS	5.9
$ZrSiO_4$	KPS	KZS_3 and Z	5.11

[a] A = Al_2O_3, C = CaO, F = FeO, K = K_2O, M = MgO, N = Na_2O, P = PbO, S = SiO_2, Z = ZrO_2.

Sandhage and Yurek [5.1], who, in their studies of the indirect dissolution of chrome–alumina crystalline solution materials in CaO–MgO–Al_2O_3–SiO_2 melts at 1550°C, reported that the reaction layer thickness of the spinel that formed decreased with increasing rotational rpm, but did not change with time at constant rpm. The reaction layer was an order-of-magnitude thinner (30 vs. 300 μm) at 1200 rpm when compared to the case with no forced convection.

The investigator must be careful in his interpretation of the crystalline phases present after an experiment has been completed, so that he does not confuse phases that precipitate during cooling with those that were present during the experiment. The reader, if interested in a particular system, should examine the original articles of those listed in Table 5.1 to determine the exact experimental conditions. The following sections describe some of the more important systems that have been investigated but no attempt has been made for an exhaustive survey.

Alumina-Containing Materials

The corrosion of multicomponent materials proceeds through the path of least resistance. Thus, those components with the lowest resistance are corroded first. This is really a form of selective corrosion and may proceed through either the direct or indirect corrosion process. The corrosion of a fusion cast alumina–zirconia–silica (AZS) refractory will be used as an example of a case when selective direct corrosion is operative. This particular material is manufactured by fusing the oxides, casting into a mold, and then allowing crystallization to occur under controlled conditions. The final microstructure is composed of primary zirconia, alumina, alumina with included zirconia, and a glassy phase that surrounds all the other phases (Fig. 5.1). The glassy phase (about 15% by volume) is necessary for this material to provide a cushion for the polymorphic transformation of zirconia* during cooling and subsequent use. This material is widely used as a basin-wall material in soda–lime–silica glass furnaces. The corrosion proceeds by the diffusion of sodium ions from the bulk glass into the glassy phase of the refractory. As sodium ions are added to this glass, its viscosity is lowered and it becomes corrosive toward the refractory. The corrosion next proceeds by solution of the

*Tetragonal zirconia transforms to monoclinic zirconia athermally upon cooling at about 1000°C accompanied by an expansion of about 5 vol.%.

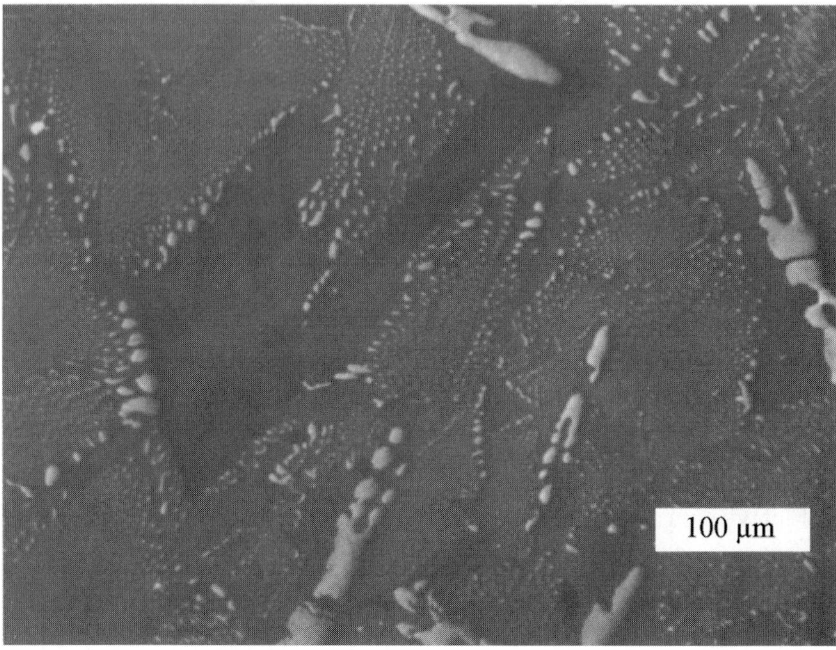

FIGURE 5.1 Microstructure of an alumina–zirconia–silica refractory. Reflected oblique illumination (magnification 200×). Brightest areas are ZrO_2, next darker areas are Al_2O_3, next darker areas are silicate glass (diagonal area across middle) and the few darkest spots are pores.

alumina and finally by partial solution of the zirconia. Under stagnant conditions, an interface of zirconia embedded in a high-viscosity, alumina-rich glass is formed (Fig. 5.2). If the diffusion of sodium ions into the glassy phase is sufficient, the glassy phase may contain sufficient sodium so that upon cooling, nepheline ($Na_2Al_2Si_2O_8$) crystals precipitate, or if the temperature is proper, the nepheline may form in service. The presence of nepheline has been reported by several investigators [5.6,5.12,5.13]. In actual service conditions, however, the convective flow of the bulk glass erodes this interface,

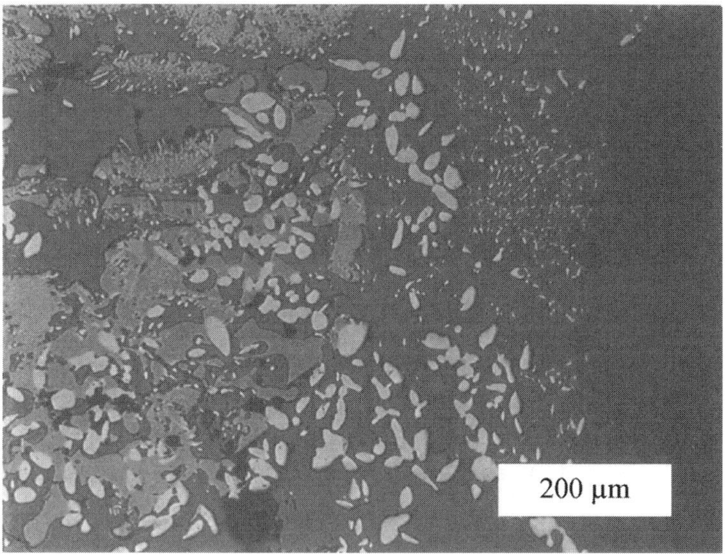

FIGURE 5.2 ZrO_2 interface on AZS refractory corroded by a soda–lime–silica glass at 1450°C for 7 days. Reflected light illumination. (Courtesy of Corning, Inc.)

allowing continuous corrosion to take place until the refractory is consumed. This type of corrosion can take place in any multicomponent material where the corroding liquid diffuses into a material that contains several phases of varying corrosion resistance.

Hilger et al. [5.7] reported the corrosion of an AZS refractory by a potassium–lead–silicate glass at 1200°C to be very similar to that discussed above. In this case, the potassium ions diffused into the glassy phase of the refractory, dissolved the alumina of the refractory, and formed a glassy phase with a composition very similar to leucite ($K_2Al_2Si_4O_{12}$). Actual crystals of leucite were found upon examination of used blocks. It was interesting that very little lead diffused into the refractory.

In these refractory materials containing ZrO_2, one should note that the ZrO_2 is very insoluble in soda–lime–silicate and potassium–lead–silicate glasses. Thus the corrosion of AZS refractories in these glasses is very similar to that which occurs in alumino-silicate (e.g., mullite) refractories. The major difference being the skeletal interface layer of undissolved ZrO_2 that forms on the AZS materials. The presence of lead in the corroding glass acts predominantly to lower the viscosity, with increasing lead contents producing more severe corrosion [5.11].

Lakatos and Simmingskold [5.14] studied the effects of various glass constituents upon the corrosion of two pot clays, one with 21% alumina and one with 37% alumina. Their silicate glasses contained K_2O, Na_2O, CaO, and PbO in varying amounts. They found that PbO had no significant effect upon corrosion, that Na_2O was 2–3 times more corrosive than K_2O, and that CaO followed a cubic function. As their tests were conducted at 1400°C, it should be obvious that the glass viscosities varied considerably. They concluded that 95–96% of the total variance in corrosion was a result of viscosity differences, and that the specific chemical effects existed only to a small extent.

Lakatos and Simmingskold [5.15] later found in isoviscosity tests that the corrosion of alumina depended upon the lime and magnesia content of the glass, whereas the corrosion of silica depended upon the alkali content.

During the testing of refractories for resistance toward coal-ash slags, Bonar et al. [5.3] determined that AZS type refractories exhibited complete dissolution at the slag line, alumina exhibited significant corrosion, and a chrome-spinel refractory exhibited negligible attack at 1500°C and 10^{-3} Pa oxygen pressure for 532 hr. These results were consistent with the determined acid/base ratios of the slags and what one would predict knowing the acid or base character of the refractories.

Fig. 5.3 shows the results of a mullite refractory that was removed from the regenerator division wall of a soda–lime–silicate container glass furnace. The sample was in service for 1 year at a temperature of approximately 1480°C. The attacking glass was from batch particulate carryover and condensation of

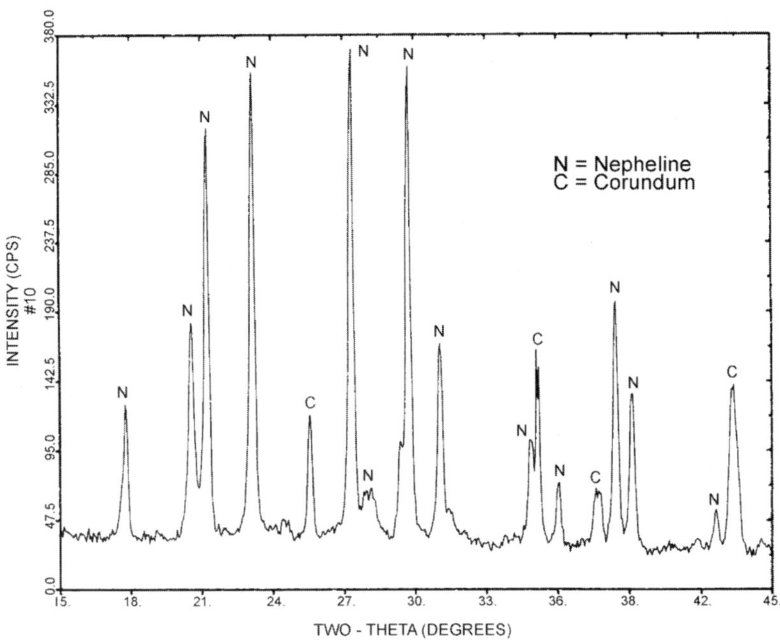

(a)

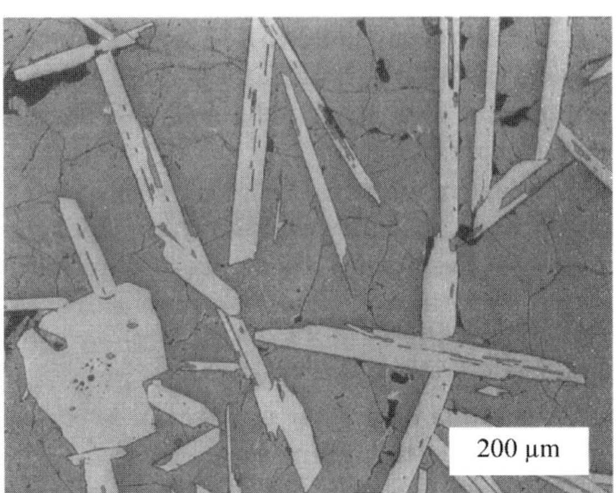

(b)

FIGURE 5.3 Corrosion of mullite refractory: (a) XRD pattern, (b) reflected light optical micrograph (magnification 100×, lighter areas are corundum and darker areas are nepheline), and (c) EDS spot maps for several elements.

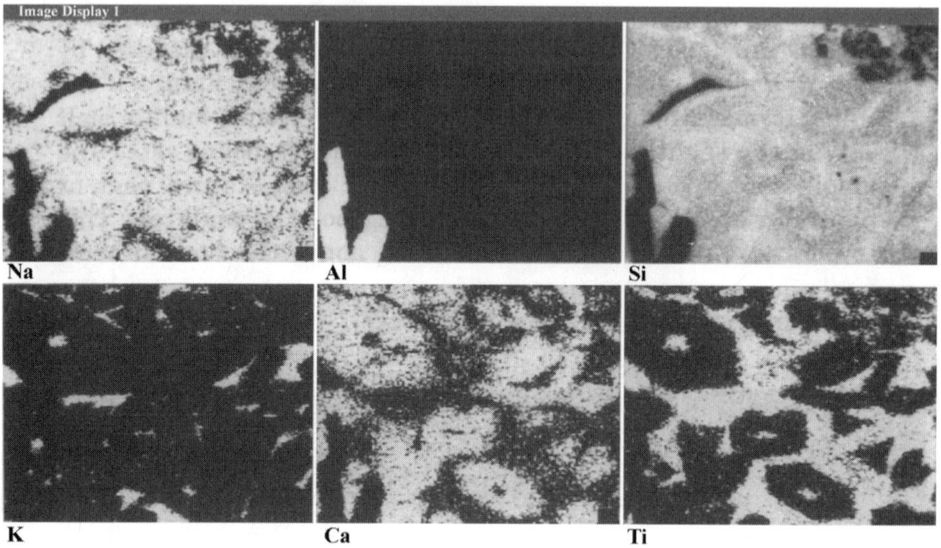

(c)

FIGURE 5.3 Continued.

volatiles. A small amount of convective flow down the vertical face of the wall was present due only to gravity. The alteration of the refractory due to corrosion occurred to a depth of about 25 mm. As can be seen from Fig. 5.3, the mullite has completely converted to predominantly corundum and nepheline. Fig. 5.3a is the X-ray diffraction (XRD) pattern supporting the presence of only corundum and nepheline. The optical micrograph shown in Fig. 5.3b indicated the presence of an additional phase. Upon examination of elemental maps via scanning electron microscopy (SEM)/energy dispersive spectroscopy (EDS) shown in Fig. 5.3c, it was determined that the nepheline contained a reasonable amount of dissolved calcium and that the crystalline nepheline was embedded in a matrix of vitreous potassium–titanium–silicate. The potassium diffused into the refractory from glass batch impurities and the titanium was present in the original refractory in minor amounts.

Orlova et al. [5.16] reported that divalent cation additions to alkaline silicate melts decreased the dissolution temperature of mullite while increasing the intensity of dissolution in the order CdO > ZnO > BaO. Additional crystalline phases form depending on the divalent cation added: corundum (CdO), spinel (ZnO), or celsian (BaO).

There are times when the microstructure of the resultant corrosion product can offer information useful in determining the cause of the deterioration. Fig. 5.4 shows a sample taken from surface runnage on a corundum refractory that was attacked by silica in a glass furnace. The dendritic and fibrous nature of the mullite formed were indicative of crystallization from a mullite melt containing a slight excess of silica and a variable cooling rate. The sample shown had apparently been at a temperature near or slightly in excess of 1850°C (the melting point of mullite). This temperature was approximately 300°C above the normal operating temperature for this furnace. The mullite identification for the dendrites was confirmed by SEM/EDS.

Zircon

The attack of zircon by soda–lime–silicate glasses is similar to that of AZS materials in that an interface of zirconia crystals embedded in a highly viscous glass is formed. The difference is the lack of alumina, which keeps nepheline from forming and the viscous glass is now a silicious glass as opposed to an alumina-rich glass. Thomas and Brock [5.17] reported that as the sodium content of the attacking glass decreased, the thickness of the zirconia layer decreased. The attack of zircon by E-glass (a borosilicate) that contains only about 0.5% Na_2O exhibited no observable alteration. Zircon has been successfully used in contact with high-temperature lithium–alumino-silicate glasses. A protective layer of zirconia crystals suspended in a very viscous glass is formed by the leaching of silica from the zircon. Because these glasses are melted at temperatures above 1700°C, it is quite possible that the zircon dissociates into zirconia and silica, with the silica then going into solution

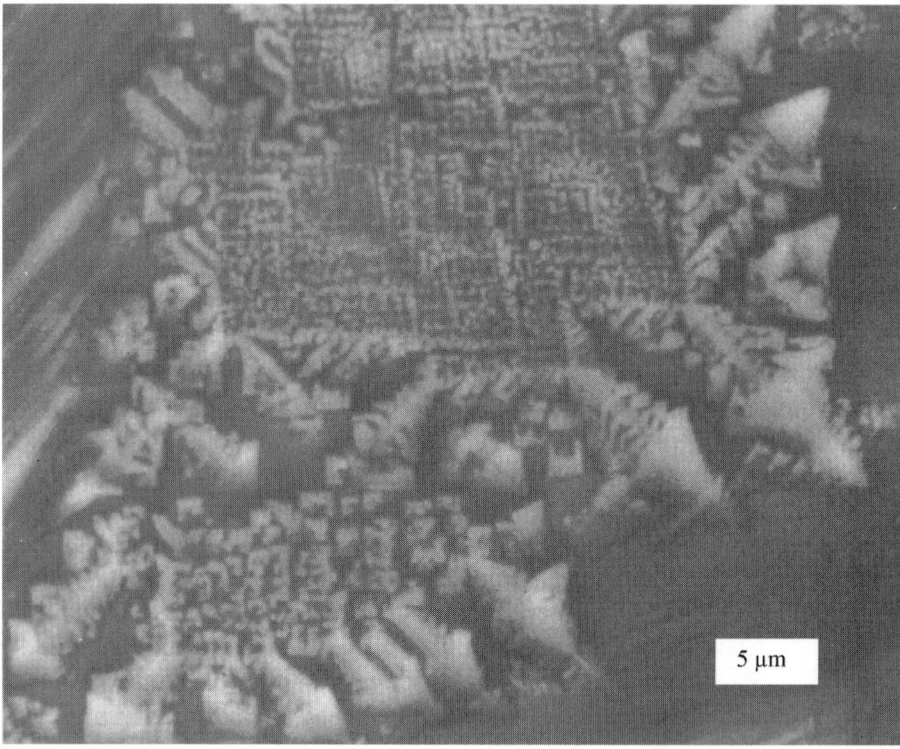

FIGURE 5.4 Dendritic and fibrous mullite formation caused by silica attack of a corundum refractory. Transmitted light illumination.

rather than the silica being leached from the zircon. As long as thermal cycling does not occur, this protective layer remains intact.

Zirconia

Because of the polymorphic transitions associated with zirconia, it has not been widely used as material for furnace linings except when combined with other materials, such as the AZS refractories described above, or when stabilized in the cubic phase. Stabilization of the cubic phase by incorporation of CaO,

MgO, or Y_2O_3 has been successful. These oxides, however, are susceptible to leaching when in contact with liquids. Chung and Schlesinger [5.18] studied the effects of molten calcia–ferrosilicate slags upon the stability of zirconia stabilized by magnesia and calcia. Three slags were chosen: a basic, an acid, and a high iron slag. They found that low basicity slag attacked magnesia-stabilized zirconia by leaching the magnesia. This subsequently allowed the more soluble monoclinic phase to form. Higher iron content slags had lower viscosities and thus were more corrosive. High basicity slag was found to attack magnesia stabilized zirconia by direct dissolution of the cubic phase. Calcia stabilized zirconia samples were completely destroyed by these slags even after short time periods.

Carbides and Nitrides

Silicon carbide and nitride are relatively inert to most silicate liquids as long as they do not contain significant amounts of iron oxide. The reaction:

$$SiC + 3FeO \rightarrow SiO_2 + 3Fe + CO \tag{5.1}$$

can occur and becomes destructive at temperatures above about 1100°C [5.19].

The dissolution of Si_3N_4 by glass is important not only in evaluating attack by various environments, but also for gaining an understanding of the operative mechanism in liquid phase sintering and solution/precipitation creep phenomena occurring in materials that contain a glassy bonding phase. Tsai and Raj [5.20] studied the dissolution of β-silicon nitride in a Mg–Si–O–N glass, which they reported separated into SiO_2-rich, and MgO- and N-rich regions. They concluded that the dissolution of β-silicon nitride into the glass at 1573–1723 K occurred in three steps, with precipitation of Si_2N_2O:

1. β-Si_3N_4 dissolved into the melt as Si and N.
2. This Si and N diffused through the melt toward Si_2N_2O.
3. The Si and N then attached to the growing Si_2N_2O.

Ferber et al. [5.21] reported that the corrosion of α-SiC at 1175–1250°C when coated with a static layer of a basic coal slag involved at least three reaction mechanisms. These were:

1. Oxidation of SiC with the formation of silica between the slag and SiC
2. Dissolution of the silica by the slag
3. Localized formation of Fe–Ni silicides at the SiC surface due to reaction of the SiC with the slag

Which of these predominated was dependent upon the thickness of the slag layer, which in turn, determined the local partial pressure of oxygen available by diffusion through the layer. When the slag thickness was <100 μm, passive oxidation occurred with the formation of SiO_2. As the slag thickness increased, the available oxygen at the surface was insufficient for SiO_2 formation causing SiO to form instead. This active oxidation, forming the gaseous phases of SiO and CO, disrupted the silica layer allowing the slag to come in contact with the SiC thus forming iron and nickel silicides.

McKee and Chatterji [5.22] have reported a similar effect upon the corrosion mechanism of SiC, where a molten salt layer provided a barrier to oxygen diffusion promoting the formation of SiO gas.

Based upon the work of Deal and Grove [5.23], Ferber et al. [5.21] gave the following equation for calculating the oxygen partial pressure at the SiC/slag interface:

$$pO_2 = RTC^*/(1 + 2X_1/A) \tag{5.2}$$

where:

R = gas constant
T = temperature
C^* = equilibrium concentration of oxygen in slag
X_1 = slag layer thickness
A = constant determined from kinetics

For oxidation in air only, estimating A as 0.31 μm at 1250°C and taking C^* as 0.086 mol/m^3 (oxygen concentration in silica), the critical slag layer thickness for the change from passive to active oxidation was calculated as 155 μm. Based upon the various assumptions involved in the calculation, this is very close to the experimentally determined value of 100 μm.

Reaction bonded SiC (RB SiC) is produced by the reaction of either liquid or gaseous silicon or SiO with carbon in a silicon carbide/carbon compact. This results in a porous body with continuous silicon carbide phase; however, these pores can be filled with nonreacted Si (2–10%) yielding a dense product that results in excellent mechanical properties. The excellent wettability between Si and SiC allows this to be carried out for RB SiC but not for RBSN (silicon nitride). This interpenetrating grain boundary phase of silicon metal limits the high-temperature mechanical properties to the melting point of Si (1410°C). When exposed to aggressive environments, the silicon may be attacked relatively easily, leading to degradation of properties.

5.1.2 Attack by Aqueous Solutions

The resistance to attack by aqueous solutions can be very important for many applications and especially where the shape-forming step involves slip casting of powders suspended in slurries. In Chapter 2, it was pointed out that a tremendous amount of literature is available concerning the dissolution in aqueous media of soil minerals. Some of these are mentioned below; however, the area of soils dissolution is too extensive to warrant an exhaustive review as is the area of cement/concrete chemistry and the dissolution of the various cement phases. Both of these areas are important ones to consider for those interested in hazardous waste disposal. Anyone interested in the hydrous and anhydrous cement phase chemistry should see Lea [5.24]. Calcia–silica–water chemistry has been discussed by Taylor [5.25] and Jennings [5.26]. Hydration of dicalcium silicate has been discussed by McConnell [5.27]. The area of dissolution studies related to nuclear waste disposal for at least the last 20 years has appeared in a series of symposia proceedings published by the Materials Research Society under the series title *Scientific Basis for Nuclear Waste Management* [5.28] and by the American Ceramic Society under the series title *Nuclear Waste Management* [5.29–5.32].

Alumina

Alumina has been shown by Sato et al. [5.33] to dissolve in aqueous solution at 150–200°C containing NaOH by the following reactions:

$$Al_2O_3 + OH^- + 2H_2O \rightarrow Al(OH)_4^- + AlOOH \quad (5.3)$$

$$AlOOH + OH^- + H_2O \rightarrow Al(OH)_4^- \quad (5.4)$$

with the second reaction being the faster of the two. Although no surface interfacial layers were reported, AlOOH solid formed as part of the overall reaction as shown above. The rate of dissolution was linearly proportional to the NaOH concentration. Because the samples of Sato et al. were impure, containing a silicate grain boundary phase (7% and 0.5%), the grain boundaries exhibited enhanced corrosion.

The ligand-enhanced dissolution of alumina was discussed by Brady and House [5.34]. They reported that the dissolution was dependent upon the structure of the ligand; the oxalate five-membered ring was the most corrosive, with the malonate six-membered ring being intermediate, and the succinate seven-membered ring being the least corrosive. These three ligands are all adsorbed through two functional groups (i.e., bidentate adsorption) to a single site. Benzoate, which is adsorbed through only one functional group (i.e., unidentate adsorption), was even less corrosive than those adsorbed through two functional groups.

Silica and Silicates

The dissolution at room temperature and pH = 7 of the various forms of silica has been reported to be a function of the silica tetrahedra packing density by Wilding et al. [5.35]. Thus the dissolution increases in the sequence: quartz, cristobalite, opal, amorphous silica. A wide variation in the solubility data has been reported in the literature, which has been attributed to the various investigators using different test conditions—pH, temperature, particle size, silica surface condition, and various components dissolved in the water. Quartz is not attacked by HCl, HNO_3, or H_2SO_4 at room temperature; however, it is slowly attacked by alkaline solutions. At elevated temperatures, quartz is readily attacked by NaOH, KOH, Na_2CO_3, Na_2SiO_3, and $Na_2B_4O_7$. The presence of organics dissolved in the water has been shown to increase greatly the solubility of silica with the formation of Si–organic molecular complexes (see Ref. [5.36] for comparison to silicates). Acetic and oxalic acids have been reported to increase the rate of mineral breakdown, accelerating their solubility [5.37]. Various chemisorbed metallic ions (especially Al^{3+}) have been reported to inhibit dissolution with the formation of relatively insoluble silicates.

Fig. 5.5 shows a comparison of the dissolution of quartz at 25 and 60°C in 0.1 M NaCl solution with that of fused silica* at 65°C. These data exhibited several important points—(1) that dissolution rates increased with temperature, (2) that the dissolution rates were relatively constant below pH = 7 and increased after that, and (3) that the crystalline form of silica dissolved at a slower rate than the amorphous form. The difference in rates between crystalline and amorphous silica was confirmation of the effects of tetrahedra packing density. Although the temperature of the tests between crystalline and amorphous silica were different, Brady and House [5.34] concluded that this difference was probably insignificant.

The dissolution of quartz in 49% HF solutions has been reported to vary depending upon the crystallographic face being attacked. Liang and Readey [5.38] reported that the rate

*The fused silica was tested in a buffered solution of distilled, deionized water. Different inorganic and organic buffers were used for each pH range.

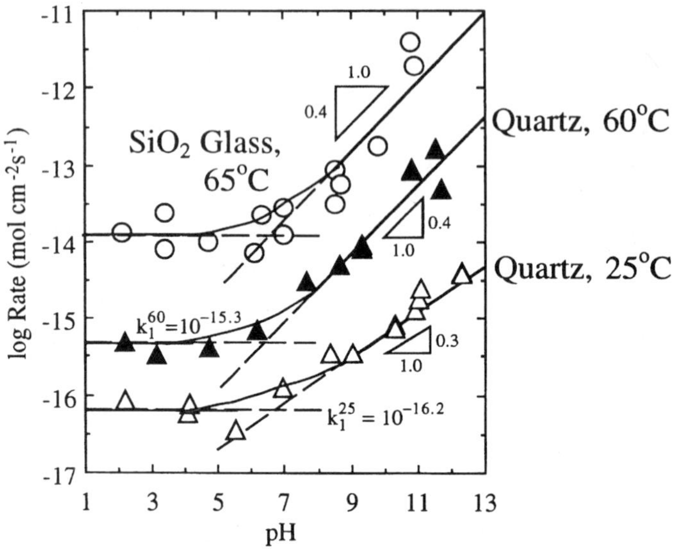

FIGURE 5.5 Dissolution rates of quartz at 25 and 60 °C, and SiO_2 glass at 65 °C. (From Ref. 5.34. Copyright CRC Press, Boca Raton, FL.)

of dissolution of *X*-cut quartz was about twice as high for the positive end vs. the negative end, whereas for *Y*- and *Z*-cut samples both ends exhibited similar rates. Both the *X*- and *Y*-cut ends exhibited dissolution rates much lower than the rate for *Z*-cut samples. The rates of dissolution were reported to be HF concentration dependent and surface reaction controlled. The substitution of HF molecules in solution with a surface complex ion was suggested as the surface reaction. The differences among the various *X*-, *Y*-, or *Z*-cut samples was attributed to a difference in the number of reactive or kink sites, rather than a difference in structure for the various crystallographic faces.

An area of extreme importance that has received little attention until recently is the relationship between the surface activity of minerals and their toxicity to humans. Most of the biological studies of toxicity examine only the effects of particle size and shape and the mass concentration. An excellent review of the effects of inhaled minerals was recently given by Guthrie [5.39], who pointed out the need for collaborative studies between the health and mineral scientists.

Because the inhaled minerals will undergo some form of alteration during the time they remain within the human body, it is of interest to study the biodurability of these minerals as a factor in mineral dust-related diseases. Hume and Rimstidt [5.40] have studied the dissolution of chrysotile in an effort to develop a general test for mineral dust biodurability. At $pH<9$, the reaction:

$$Mg_3Si_2O_5(OH)_4 + 6H^+ \rightarrow 3Mg^{2+} + 2H_4SiO_4 + H_2O \quad (5.5)$$

describes the dissolution. Based upon some reported concentration levels of Mg^{2+}, H^+, and silica in lung tissue fluids, Hume and Rimstidt determined that chrysotile would be in equilibrium with these fluids at a pH of 8. However, body fluids never reach pH = 8, thus creating an environment for continuous dissolution. Its persistence is attributable to a very slow dissolution rate. Dissolution occurs in two steps: first, Mg is leached, and then the silica matrix dissolves. Thus the lifetime of chrysotile is determined by the silica dissolution. Hume and Rimstidt gave the following equation for calculating the lifetime of a chrysotile fiber:

$$t = (3/4)/(d/V_m k) \quad (5.6)$$

where:

d = fiber diameter (m)
k = rate constant for silica dissolution (mol/m^2s)
V_m = volume of 1 mol of silica in chrysotile (equal to $5.4 \times 10^{-5}\ m^3/mol$)

A 1-μm-diameter chrysotile fiber will take approximately 9 months to dissolve (based upon the silica concentration of the fluid at a pH of 4–7 and 37°C). This lifetime is an order-of-magnitude less than the time required for the onset of diseases' symptoms. Thus, any biological model must explain this difference. By comparison, a fused silica fiber of 1-μm diameter under the same conditions takes 438 years to dissolve completely. One should also remember that chrysotile being a sheet silicate with a misfit between the octahedrally coordinated magnesium-containing layer and the tetrahedrally coordinated silica layer curls up into a tube forming fibrils with the octahedral layer on the exterior (see Section 2.5.2 on Surface Area and Fig. 2.11). Thus the dissolution of a mineral of this structure can be quite different from one that forms a uniform chemistry across a solid fiber.

Concrete, Cement, Limestone, Marble, and Clay

All the building materials contain some lime, several being mostly calcium carbonate (limestone and marble). Thus these calcareous materials are attacked by acids. Generally, any aqueous liquid with a $pH < 6$ will exhibit some attack. The corrosion of concrete takes place by the leaching of water-soluble salts that are formed by the reaction of the acid and the calcium-containing compounds of the concrete [5.41]. For example, the reaction of calcium carbonate with SO_2 containing acids* forms calcium sulfate or gypsum. The gypsum, being much more soluble than the carbonate, is then washed away. Run-off water has been shown to be more important than the pH of the solution for $pH > 3$. This causes etching of the surface in addition to pitting and scaling. Webster and Kukacka gave the following mechanisms for the dissolution of concrete:

1. Dissolution of hydrated cement compounds
2. Dissolution of anhydrous cement compounds
3. Dissolution of calcareous aggregates in the mix
4. Deposition of soluble sulfate and nitrate salts
5. Formation of new solid phases within pores
6. Production of stresses from numbers 4 and 5

*Sulfuric acid has been reported to make up 60–70% of the acidity in acid rain for the northeastern United States.

Generally, a combination of the above parameters is operative. In addition, there is the deterioration of any reinforcing steel.

The most common soluble salts that are involved in efflorescence are the alkali and calcium and magnesium sulfates, chlorides, nitrates, and carbonates. Sodium sulfate generally leaves a fluffy deposit behind, whereas gypsum deposits are more adhesive and scaly. The chlorides are much more mobile in solution than the sulfates and therefore tend to penetrate deeper into the corroding material. The disruptive action of these soluble salts is mostly through hydration–dehydration–rehydration. Thus the damage caused by these salts results from their having several hydration states. Therefore, sodium chloride is not applicable to this type of disruption because it expands only during dehydration [5.42]. Sodium exists in two hydrate forms, the mono- and the deca-hydrate. Amoroso and Fassina [5.42] reported that the mono-hydrate exists only above 32°C, with the deca-hydrate being the stable form below that temperature. A hepta-hydrate may also exist at intermediate temperatures.

Amoroso and Fassina discussed the mechanism of limestone deterioration by sulfate pollutant attack. This deterioration starts by the deposition of either dry or wet SO_2. They discussed the effects of atmospheric conditions (i.e., wind, etc.) and the limestone surface characteristics upon the degree and rate of deterioration. At low relative humidity, the reaction favors the formation of calcium sulphite according the following equations:

$$CaCO_3 + SO_2 \rightarrow CaSO_3 \cdot 1/2H_2O + CO_2 \tag{5.7}$$

or

$$CaCO_3 + SO_2 \rightarrow CaSO_3 \cdot 2H_2O + CO_2 \tag{5.8}$$

When water is present, the sulfite is oxidized to the sulfate according to:

$$CaSO_3 \cdot 2H_2O + 1/2O_2 \rightarrow CaSO_4 \cdot 2H_2O \tag{5.9}$$

This later reaction is one thought to be not of major consequence since a large amount of water would not allow a sufficient deposition of dry SO_2 to occur. Oxidation of SO_2 in an atmospheric aqueous phase (i.e., rain, fog, etc.) to sulfuric acid that will then react with the limestone to form gypsum is most likely the predominant mechanism of deterioration.

Rain water can dissolve carbon dioxide of the atmosphere forming the weak carbonic acid:

$$CO_{2(gas)} + H_2O_{(liquid)} \rightarrow H_2CO_{3(liquid)} \tag{5.10}$$

This then becomes dissociated into $[HCO_3^-]$ and $[CO_3^{2-}]$, which is controlled by the carbonic acid and hydrogen ion concentrations. If the CO_2 concentration increases, increasing the concentration of carbonic acid, the hydrogen ion concentration must also increase. This, in turn, will alter the pH of the solution. The weak carbonic acid solution reacts with the carbonates of limestones, marbles, and mortars to form the more soluble bicarbonates. Amoroso and Fassina [5.42] have reported that calcium bicarbonate is about 100 times more soluble than the carbonate. The presence of impurity ions will influence what phase precipitates from the solution. Kitano [5.43,5.44] reported that the presence of Mg^{2+}, Sr^{2+}, and SO_4^{2-} ions supports the formation of aragonite rather than calcite.

Nitrogen oxides when in contact with water form nitrous and nitric acids. Although nitric acid is a weaker acid than sulfuric acid, it can cause more extensive damage to concrete because it reacts with $Ca(OH)_2$ to form the more soluble calcium nitrate.

The deterioration of brick masonry is a combination of corrosive attack upon the brick and the mortar that is used to bond the brick together. Mortars are generally of a calcareous type, and as such, their corrosion is very similar to that described above for concrete and limestone. The actual corrosion of the mortar depends upon the type used: clay-based, lime sand, or Portland cement. The most susceptible to acid rain are those containing calcium of one form or another. The formation of gypsum from the reaction of acid rain upon calcium carbonate produces local stresses sufficient mechanically to disrupt the mortar. Any calcium hydroxide present from the hydration of Portland cement is especially susceptible to acid attack. Additional reactions such as the reaction of calcium sulfate with tri-calcium aluminate forming ettringite ($3CaO{\cdot}Al_2O_3{\cdot}3CaSO_4{\cdot}32H_2O$) also produces large local stresses because of the associated expansive crystallization.

It has been noted that 12th century Venetian bricks were more durable than modern ones, although they were fired at lower temperatures. This has been attributed to a very uniform microstructure, the lack of a highly vitrified surface, and a larger pore size (avoiding the <1-μm critical pore size) when compared to modern bricks [5.45]. Bricks may also contain Na_2SO_4 if fired in a kiln using sulfur-rich fuel and at a sufficiently low temperature. The sodium sulfate, dissolved by water, can recrystallize after evaporation forming the anhydrous salt, thenardite, or the decahydrate, mirabilite. The specific environmental conditions of temperature and relative humidity will change the degree of hydration and the amount of mechanical disruption.

Zirconia-Containing Materials

The hydrothermal effect of water upon the dissolution of yttria (14 mol%)-stabilized zirconia (YSZ) single crystals was investigated by Yoshimura et al. [5.46]. They found four regimes of behavior for YSZ treated at 600°C and 100 MPa for 24 hr, depending upon the pH of the solution. In alkali solutions (those containing LiOH, KOH, NaOH, or K_2CO_3), partial decomposition and dissolution/precipitation were found, with yttria being the more soluble component. In acidic solutions (those containing HCl or H_2SO_4), rapid dissolution of yttria occurred forming an interface of polycrystalline monoclinic ZrO_2. In reactions with H_3PO_4 solution, the interface layer formed was ZrP_2O_7. In neutral solutions, the dissolution was minimal.

Superconductors

Murphy et al. [5.47] reported that reaction with water liberated oxygen, formed Y_2BaCuO_5 and CuO, in addition to barium hydroxide, and was a function of temperature and surface area. This is similar to the leaching of barium from perovskites in aqueous solutions when the pH is less than 11.5 reported by Myhra et al. [5.48].

Titanates and Titania

A crystalline titanate mineral assemblage called SYNROC* has been under investigation for many years as a possible encapsulant for high-level radioactive wastes. The titanates are commonly a mixture of perovskites, $CaTiO_3$ and $BaTiO_3$, zirconolite, $CaZrTi_2O_7$, and hollandite, $BaAl_2Ti_6O_{16}$. In a study of the dissolution of these titanates in CO_2 enriched (4 ppm) deionized water (pH = 5–6) at 300 and 350°C and 500 bars, Myhra et al. [5.48] reported the following reactions:

$$CaTiO_3 + CO_2 \rightarrow TiO_2 + CaCO_3 \quad (5.11)$$

$$BaTiO_3 + CO \rightarrow TiO_2 + BaCO_3 \quad (5.12)$$

$$CaZrTi_2O_7 + CO_2 \rightarrow 2TiO_2 + ZrO_2 + CaCO_3 \quad (5.13)$$

$$BaAl_2Ti_6O_{16} + CO_2 + H_2O \rightarrow 6TiO_2 + BaCO_3 + 2AlO(OH) \quad (5.14)$$

$$CaTiO_3 + H_2O \rightarrow TiO_2 + Ca(OH)_2 \quad (5.15)$$

$$BaTiO_3 + H_2O \rightarrow TiO_2 + Ba(OH)_2 \quad (5.16)$$

*SYNROCK was developed in 1978 at the Australian National University by Prof. T. Ringwood. Several compositional variants have been developed, called SYNROCK-C, -D, or -F. Additional information can be obtained from the website www.uic.com.au) of the Uranium Information Centre, Ltd. in Melbourne, Australia.

The dissolution mechanism proposed for these titanates was one involving initial selective leaching of the alkaline earth ions along with hydration of the titanate surface. This first step was rather rapid, but then overall dissolution slowed as the solution became saturated. When the solubility product was exceeded, precipitation and equilibration with CO_2 occurred. As the precipitate concentration increased, the dissolution rate decreased. Thus the overall dissolution of these titanates was dependent upon the solubility of the alteration products in the solution. In contrast, Kastrissios et al. [5.49] proposed that the calcium was not selectively leached from the perovskite but instead, the perovskite dissolved congruently forming an amorphous titanium-rich surface layer from which TiO_2 precipitated. This titania layer was not continuous and therefore did not protect the underlying material from continued corrosion.

Buykx et al. [5.50] gave a diagram of relative phase stability for various titanium-containing compounds, among others, for dissolution in water at 150°C for 3 days. No alteration was found for zirconolite–zirkelite ($CaZrTi_2O_7$). Some alteration and precipitation of TiO_2 was found for hollandite ($BaAl_2Ti_6O_{16}$), loveringite–landauite ($FeTi_3O_7$), pseudobrookite (Fe_2TiO_5), and rutile (TiO_2). Extensive replacement by TiO_2 was found for perovskite ($CaTiO_3$) and freudenbergite ($Na_2Ti_6 \cdot Fe_2O_{14}(OH)_4$). Complete and rapid dissolution was found for any glassy phases. The stoichiometries given above are only approximate, the complete analyzed stoichiometries for the compounds investigated are given in the original paper.

Titania was investigated by Bright and Readey [5.51] as the least complex titanate to evaluate the quantitative dependence of kinetics upon ambient conditions. Powdered anatase (~0.54 μm agglomerate size and ~0.13 μm crystallite size) was added to acid solutions of HF–HCl and stirred for several hours at temperatures ranging from 37.5 to 95.0°C. Although very little is known about the titanium species in HF–HCl solutions, it was believed that the most predominant complex was

$(TiF_6)^{2-}$. The rate-controlling step in the kinetics of dissolution was concluded to be the removal of the highly charged cations from kink sites on the surface. The average calculated initial (for the first hour) dissolution rate was 59.0 wt.% TiO_2 dissolved per hour.

Slightly reduced titania has been investigated for its use in electricity generation and for water decomposition [5.52]. In these applications, *n*-type semiconducting titania is used as a photoanode in an aqueous solution of 0.5 M H_2SO_4. The photo-generated positive holes in the valence band of illuminated n-type titania reacts with the solution according to the following equation:

$$SO_4^{2-} + p^+ \rightarrow SO_4^- \tag{5.17}$$

The $(SO_4)^-$ that forms is an active species that reacts with titania forming etch pits. This phenomenon is called *photo-electro-chemical aging*.

Transition Metal Oxides

The use of transition metal oxides (RuO_2, NiO, MoO_2, Mo_4O_{11}, Mo_8O_{23}, Mo_9O_{26}, and WO_2) as fuel cell electrocatalysts requires that they be stable in aqueous solution of 1 N H_2SO_4. These oxides are relatively stable in acid solutions but undergo redox reactions in the region of pH = 7. Horkans and Shafer [5.53] reported that Mo_4O_{11} exhibited anodic dissolution but that WO_2 did not; however, it did form a layer of WO_3 on its surface. They reported that MoO_3 was more soluble than WO_3 in acid solutions, whereas MoO_2 was more stable.

Horkans and Shafer [5.54] reported that the oxidized surface layers that formed were generally less conductive than the bulk reduced phase, that they were generally of a wide range of compositions, and that the actual composition of the reaction surface layer was highly dependent upon the electrode potential. They also found that MO_2 (M = Mo, Ru, W, Re, Os, and Ir) was substantially more stable in acid solutions than was indicated by their Pourbaix diagrams.

In a study of the corrosion of nuclear fuels, Clayton [5.55] investigated the effects of 633 K flowing water upon various thoria- and urania-containing materials. He found the following order of corrosion resistance: $ThO_2 > ThO_2{-}UO_2 > ZrO_2{-}UO_2 > ZrO_2{-}CaO{-}UO_2 > UO_2$. Corrosion in highly oxidative conditions (caused by the fission fragment radiolysis of the water) was attributed to the oxidation of uranium from the four-valent to the six-valent state. UO_2 corroded easily even in low-oxygen (5 ppm) water. The test conditions of pH, sample preparation, and compact attributes had no effect on corrosion resistance.

Carbides and Nitrides

The transition metal carbides and nitrides are chemically stable at room temperature, but exhibit some attack by concentrated acid solutions. The one exception to this is VC, which slowly oxidizes at room temperature.

Bowen et al. [5.56] reported the formation of $Al(OH)_3$ (bayerite) on AlN powder after 16 hr in contact with deionized water at 25°C. In the first 8 hr, growth of an amorphous hydrated layer occurred with a chemistry very close to AlOOH, while the pH of the solution drifted from 7 to 10 after 5.5 hr. The kinetics indicated a linear rate controlled by the surface reaction. Anyone involved in the aqueous processing of materials such as AlN should be aware of the potential reactions that can take place with the incorporation of oxygen into their product through the formation of hydrated surface layers.

The behavior of sintered SiC in 0.045 M Na_2SO_4 + 0.005 M H_2SO_4 and 0.1 M LiOH aqueous solutions at 290°C was studied by Hirayama et al. [5.57]. They examined weight losses for up to 200 hr in both oxygenated and deoxygenated solutions. Weight losses increased with increasing pH and were greater for oxygenated solutions. No surface silica layers were found, with dissolution progressing through SiC hydrolysis. The proposed reaction follows:

$$SiC + 4H_2O \rightarrow Si(OH)_4 + CH_4 \tag{5.18}$$

and

$$Si(OH)_4 \rightarrow H_3SiO_4^- + H^+ \rightarrow H_2SiO_4^{2-} + 2H^+ \tag{5.19}$$

where the $Si(OH)_4$ sol that forms immediately dissolves. The dissolution of the $Si(OH)_4$ in acidic solutions ($pH = 4$) was slower than that in alkaline solutions, and provided a small degree of protection, leading to a rate law that was approximately parabolic. In alkaline solutions ($pH = 10$), the rate law was linear.

The corrosive effect of HCl aqueous solutions at 70°C upon Si_3N_4 has been shown by Sato et al. [5.58] to be dependent upon the sintering aid used, or more specifically, the grain boundary phase present in hot isostatically pressed materials. In solutions of <1 M HCl, the corrosion was surface reaction controlled, whereas in solutions of >5 M HCl, the corrosion was controlled by diffusion through the interfacial reaction layer that formed (assumed to be silica). Corrosion occurred through dissolution of the Al and Y ions (Y ion dissolution was about twice that of Al) contained in the grain boundary phase, with dissolution decreasing as the degree of crystallinity increased for this phase. Negligible dissolution of silicon ions was reported. In contrast to the above dissolution, the corrosive effect of 0.1–10 *M* aqueous HF solutions between 50 and 80°C for HIP or hot-pressed Si_3N_4 containing Y_2O_3, Al_2O_3, and AlN additives has been shown by Sato et al. [5.59] to involve the selective dissolution of Si and Al ions but not Y ions. The Y ions instead formed insoluble YF_3.

Seshadri and Srinivasan [5.60] investigated the corrosion of a titanium diboride particulate reinforced silicon carbide at room temperature in several aqueous solutions (aqua regia, NaOH, and HF/HNO_3) for up to 500 hr. Aqua regia was the most corrosive and a 50% NaOH solution was the least.

Preferential leaching of the TiB_2 from the surface was reported to be the cause of decreasing weight loss with time. After about 100 hr, weight loss was stopped for the aqua regia and HF/ HNO_3 solutions, whereas it took approximately 250 hr in the 50% NaOH solution.

5.1.3 Attack by Molten Salts

Oxides

The importance of molten salt reactions is well known in alumina reduction cells for the production of aluminum metal (Hall–Heroult process). In this process, the electrolyte consists of a solution of alumina (<10 wt.%) dissolved into molten cryolite (Na_3AlF_6) [5.61]. Pure molten cryolite contains AlF_6^{3-}, AlF_4^-, F^-, and Na^+ ions. When alumina is added, the complex ion $AlOF_x^{(1-x)}$ ($x=3$–5) forms in addition to the others. In a study of the cryolite–mullite and cryolite/sodium fluoride–mullite systems, Siljan and Seltveit [5.62] reported that materials with high Si/Al ratios experienced high weight losses when in contact with NaF–cryolite eutectic melts because of the formation of gaseous SiF_4. They reported that mullite readily dissolved into cryolite and cryolite–NaF melts, and that NaF reacted with alumina to form β-alumina. Allaire [5.63] studied the resistance of 15 different commercial refractories with varying Al_2O_3/SiO_2 ratios in a simulated alumina reduction cell and found that the resistance to attack by molten metallic sodium and NaF increased with alumina content. He reported that above 700°C, an Al_2O_3/SiO_2 ratio greater than 0.90 should be used.

The corrosion of fused silica by molten sodium sulfate in atmospheres containing either 1% SO_2/O_2 or pure oxygen at 700 and 1000°C has been described by Lawson et al. [5.64] to take place according to the ease of sodium diffusion in the various phases that form. Sodium diffuses into the fused silica, leading to the nucleation of cristobalite. Once a continuous layer of cristobalite formed, sodium diffusion was minimized. The sodium at the cristobalite/fused silica interface then diffused further into the fused silica. The basicity of the reaction determined whether or not a cristobalite layer formed, with less cristobalite forming as the reaction became more acidic. Cristobalite globules, however, were reported to precipitate from the salt solution. Low partial pressures of SO_3 were reported to promote the fluxing action of the molten sulfate by increasing the activity of Na_2O.*

In the evaluation of cathode materials for molten carbonate fuel cells, Baumgartner [5.65] reported solubility data for NiO, CuO, ZnO, $LiFeO_2$, and $LaNiO_3$ in a molten binary carbonate of Li/K (62/38 molar ratio) between 823 and 1223 K. Both NiO and CuO exhibited dissolution (CuO being more soluble than NiO) into the molten carbonate and diffusion toward the anode until the local partial pressure of oxygen was sufficiently low for metal precipitation. At temperatures exceeding 1123 K, $LaNiO_3$ decomposed to La_2NiO_4 and NiO, which dissolved and reduced to metallic Ni. A similar situation was found for $LaCoO_3$, which decomposed to La_2CoO_4 and CoO at temperatures exceeding 1073 K. Dissolution of $LiFeO_2$ into the molten carbonate resulted in reduction at the anode to $LiFe_5O_8$, while ZnO at the anode became nonstoichiometric. The solubilities of these oxides were in the order $LaNiO_3 < NiO < LiFeO_2 < CuO < ZnO$ below 1023 K. Above this temperature, the relative solubilities of CuO and ZnO were reversed.

An example of when corrosion becomes beneficial is the removal of ceramic cores in the investment casting process. The new process of directional solidification and the new alloys involved (NiTaC) require contact between the molten metal and the core material for times up to 20 hr at temperatures as high as 1800°C [5.66]. The requirement that the core material must withstand these conditions and then be chemically removed is a contradiction in stability. Core removal requires high dissolution rates at low temperatures. Potential core

* Compare with the discussion in the next section involving equations 5.20 through 5.24 and the relationship of soda activity with SO_3 levels.

materials are Al_2O_3, Y_2O_3, $Y_3Al_5O_{12}$, $LaAlO_3$, and $MgAl_2O_4$, which all possess satisfactory resistance to the casting conditions as reported by Huseby and Klug [5.66]. These materials, except for Y_2O_3, are insoluble in aqueous acids or bases. The solvents used must be aggressive toward the core material but not toward the alloy. Borom et al. [5.67] reported the weakly basic or amphoteric oxides of Al_2O_3, Y_2O_3, and La_2O_3 can be dissolved by molten M_3AlF_6, M_3AlF_6+MF, M_3AlF_6+M′F_2, or M_3AlF_6+MCl, where M=Li, Na, or K, and M′=Mg, Ca, Ba, or Sr. The more acidic core materials, such as ZrO_2 or ThO_2, required alkali or alkaline earth oxide additions to make the molten salt more basic.

Another field of study where the solution in molten salts is beneficial is that of crystal growth. The solubility of Be_2SiO_4 and $ZrSiO_4$ in various solvents was studied by Ballman and Laudise [5.68]. Solvents studied included alkali vanadates and molybdates. Because of solvent volatility problems (more important for molybdates than vanadates), most of their data contained substantial error at higher temperatures. The reported ion solubilities were greater for Be_2SiO_4 ranging from 3 to 5 1/2 mol% in the 900–1000°C range than for $ZrSiO_4$, which ranged around 1 mol% in the 900–1400°C region. Except for the solution by $Na_2O{\cdot}3V_2O_5$, which reached as much as 6 mol% at 1400°C, the vanadates were more corrosive than the corresponding molybdates of those studied. The greater solubility of Be_2SiO_4 over that of $ZrSiO_4$ can also be predicted from acid/base theory, because BeO is a stronger base than ZrO_2.

Carbides and Nitrides

The normally protective layer of SiO_2 that forms on SiC and Si_3N_4 can exhibit accelerated corrosion when various molten salts are present. McKee and Chatterji [5.22] described several different modes of behavior of SiC in contact with gas–salt mixed environments relating to the formation of various interfacial reaction layers. Salt mixtures containing Na_2SO_4 and

Na_2CO_3, Na_2O, $NaNO_3$, Na_2S, or graphite were tested. McKee and Chatterji found that a SiO_2 protective layer corroded in a basic salt solution, but not in an acid salt solution. With low oxygen pressures, active corrosion took place by formation of SiO gas.

The activity of Na_2O has been shown to be an important parameter in the action of molten sodium salts by Jacobson and coworkers [5.69–5.74]. The higher this activity, the greater the potential reaction with silica. The relationship of soda activity and SO_3 partial pressure can be obtained from the following equation:

$$Na_2SO_4 \leftrightarrow Na_2O + SO_3 \quad (5.20)$$

where the equilibrium constant k (which can be written in terms of concentrations, activities, or partial pressures) is given by:

$$k = \frac{[Na_2O][SO_3]}{Na_2SO_4} = pSO_3 \quad (5.21)$$

Therefore, the highest Na_2O activity is related to the lowest partial pressure of SO_3. Jacobson [5.71] reported that at partial pressures of SO_3 greater than 0.1 Pa, no reaction occurred between SiC and Na_2SO_4 at 1000°C for at least up to 20 hr. It is assumed, as always, that Na_2O and Na_2SO_4 are chemically pure stoichiometric compounds and that SO_3 acts as an ideal gas. Experimentally, the Na_2O activity can be set by the appropriate partial pressure of SO_3.

The decomposition of sodium sulfate by reaction (5.20) is not something that takes place readily. Sodium sulfate melts at 884°C, and is relatively nonreactive toward silica, even at temperatures as high as 1400°C. To increase the reactivity, the sulfate must be reduced to some lower oxide. This has been known for many years by the manufacturers of soda–lime–silica glass. Sodium sulfate has been used not only as a source of sodium but also as a fining agent to remove the bubbles from the glass melt during processing. If the sulfate is not reduced, it either floats on the surface or forms lenticlular immiscible inclusions in the finished product. References in the old glass literature refer to *blocking the furnace*, a term used to describe the process of adding wooden blocks (i.e., carbon) to pools of nonreactive sodium sulfate floating on the surface of the molten glass. The carbon from the burning wood reduced the sulfate to a form reactive or at least miscible with the molten glass. This reaction, shown below:

$$2Na_2SO_4 + C \rightarrow 2Na_2SO_3 + CO + 1/2O_2 \qquad (5.22)$$

is controlled more scientifically in modern glass manufacture through the use of coal as a batch ingredient and precise control of the combustion system to control the partial pressure of oxygen above the melt, which in turn controls the SO_3 equilibrium through:

$$SO_2 + 1/2O_2 \leftrightarrow SO_3 \qquad (5.23)$$

and subsequently the soda (or some sodium sulfur containing compound) activity. The reaction of Na_2SO_3 with silica according to:

$$Na_2SO_3 + xSiO_2 \rightarrow Na_2O \cdot xSiO_2 + SO_2 \qquad (5.24)$$

is the one of importance in the dissolution of silica in the manufacture of glass and is most likely the one of major importance in the corrosion of the silica layer formed on SiC or Si_3N_4. Continued reduction* of the Na_2SO_3 to Na_2S, although still reactive with silica is not necessary for excessive dissolution of carbides and nitrides.

Jacobson and Smialek [5.69] found that the partial decomposition of Na_2SO_4 enhanced the oxidation of SiC, forming a layer of tridymite, a sodium silicate glass, and some Na_2SO_4. Any free carbon in the SiC enhanced the corrosion, because it aided in the reduction of the sulfate. This enhanced corrosion

*The more reduced forms of the sodium–sulfur compounds are the basis of the amber color formed in the manufacture of brown bottles.

resulted from the ease of diffusion of oxygen through the predominantly sodium silicate amorphous layer compared to that of a crystalline silica layer. A somewhat different mechanism has been proposed for the corrosion of SiC by potassium sulfate, although details of the behavior have not yet been reported [5.75]. In the case of the potassium salt, no silica layer formed on the carbide, because it was immediately dissolved by the sulfate according to:

$$16SiC + 13K_2SO_4 \rightarrow 4(K_2O \cdot 4SiO_2) + 16CO + 9K_2S_{1.44} \tag{5.25}$$

presumably due to dissolution being faster than oxidation.

Cree and Amateau [5.76], in their studies of the behavior of SiC in molten lithium salts, found that fracture strengths were reduced by a factor of 2 when tested above 600°C. In the mixed salts of LiS/Li/LiF and LiCl/Li/LiF (i.e., nonoxide), the corrosion was via grain boundary penetration, whereas in $LiSO_4$/Li/LiF the penetration was uniform. The decrease in fracture strength above 600°C was attributed to the large volume increase when lithia reacted with the silica surface coating on the SiC to form Li_2SiO_3.

The corrosion of hot-pressed silicon nitride (HPSN), reaction bonded silicon nitride (RBSN), and silicon carbide by molten sodium sulfate, sodium chloride and the eutectic composition between these two salts at temperatures from 670 to 1000°C for up to 120 hr was reported by Tressler et al. [5.77]. Molten sodium sulfate was the most corrosive, the eutectic composition was intermediate, and sodium chloride was the least effective in dissolving the silica surface layer present on these materials. HPSN was the most resistant, whereas silicon carbide completely dissolved in sodium sulfate at 1000°C within 20 min. The lower reactivity of Si_3N_4 compared to SiC with molten Na_2SO_4 was reported by Fox and Jacobson [5.72] to be a result of the formation of an inner protective layer of silica that stopped the continued reaction of Si_3N_4. The formation of this inner protective layer was highly dependent upon whether oxidation or dissolution was the faster mechanism. Sato et al. [5.75] reported that this inner protective layer of silica formed on pressureless sintered Si_3N_4 containing 5 wt.% Y_2O_3 and Al_2O_3 in contact with molten potassium sulfate at 1200°C, when the tests were conducted in air but not when conducted in nitrogen. This same situation was not true for attack by molten potassium carbonate at 1013°C. In this case, attack occurred in both nitrogen and air, with air causing a greater degree of reaction. Compared to studies performed in molten sodium and lithium sulfate and carbonates, Sato et al. found that the corrosion rate, in a nitrogen atmosphere, was independent of the alkali present, with the sulfates yielding an activation energy of 430 kJ/mol compared to that of the carbonates of 106 kJ/mol.

Superconductors

In an investigation of the molten-salt synthesis of YBa_2-Cu_3O_{7-x} (123), Raeder and Knorr [5.78] reported the stability of 123 against decomposition toward several molten salts at 1173 K. They concluded that 123 was not stable in molten LiCl or the dichlorides of Cu, Ca, Mg, or Ba, or their mixtures. However, minimum decomposition was found in the NaCl–KCl system. The mechanism of decomposition was postulated as being one of selective dissolution of the barium in the 123 forming $BaCl_2$ and causing the 123 to decompose into several oxide phases that were consistent with the phase diagram reported by Lee and Lee [5.79]. These oxide phases were generally CuO and $Y_2Cu_2O_5$, or CuO and Y_2BaCuO_5 depending upon the amount of barium in the initial mixture.

5.1.4 Attack by Molten Metals

The application of ceramics to withstand the attack by molten metals is a very large part of the ceramic industry. Refractories are used to line furnaces for the manufacture of steel and the nonferrous metals of which aluminum and copper are probably the most important. The steel and nonferrous metals industries consume approximately 60% of all refractories manufactured today. Thus an understanding of the potential problems one may encounter from metal attack is quite important.

The attack by molten metals generally involves mechanisms of corrosion other than those by liquids in combination with liquid attack. The actual process that occurs in a blast furnace, for example, is truly a combination of corrosion mechanisms. In many cases, small amounts of metal become oxidized and the corrosion is through essentially a molten slag process. An example of this is shown in Fig. 5.6, which illustrates the corrosion of a 60% MgO magnesite–chrome refractory from an electric furnace that was in contact with a high iron oxide content slag. Diffusion of the iron oxide into the refractory and reaction with the magnesia and chrome-containing spinel formed an interface of large iron-rich mixed spinel crystals. Diffusion of iron into the magnesia caused precipita-

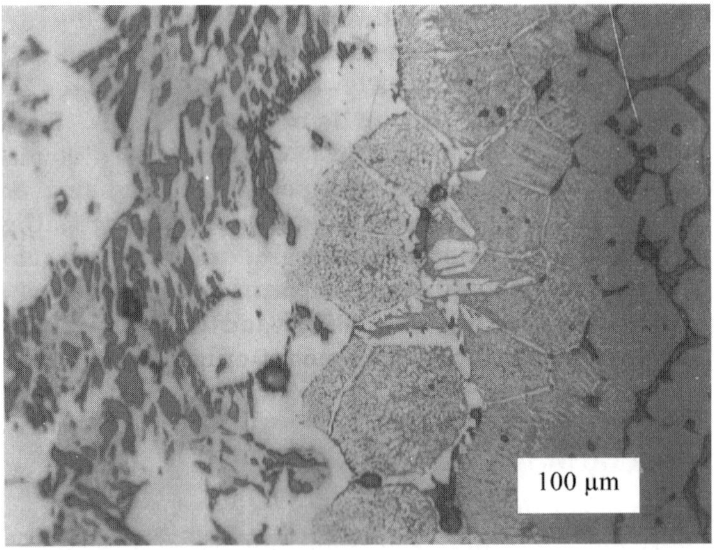

FIGURE 5.6 Corrosion interface between iron oxide-rich slag and a 60% MgO magnesite–chromite refractory (magnification 150×). Brightest regions are an iron-rich mixed spinel. (Courtesy of Harbison-Walker.)

tion of smaller iron-rich mixed spinel crystals within the magnesia and at the grain boundaries.

The reaction of silica-containing refractories with molten iron containing dissolved manganese has been known to be very deleterious. This reaction, however, is not only a reaction with a molten metal but also with an oxide of manganese (i.e., MnO). The initial reaction between SiO_2 and Mn forms MnO and Si metal. Although this reaction is thermodynamically unlikely, it has been reported by Kim et al. [5.80] to occur at 1600°C under an argon atmosphere. The subsequent reaction of MnO and silica can form one of two intermediate compounds, but more importantly, can form a eutectic liquid with a solidus temperature of 1250°C.

Many steel plant refractories are carbon-containing from manufacturing processes involving pitch or tar, organic resins, or graphite. Solid–solid corrosion through reduction by carbon takes place, or if the carbon becomes oxidized corrosion by molten metals is then very seldom a simple reaction involving only solid ceramic and molten metal.

The attack of molten aluminum upon materials containing silica follows the reaction:

$$4Al + 3SiO_2 \rightarrow 2Al_2O_3 + 3Si \tag{5.26}$$

which should be expected from examination of the free energy vs. temperature data of an Ellingham diagram. The alumina that forms in many cases provides an adherent protective layer against further corrosion [5.81]. This reaction is accompanied by a volume decrease of about 26% according to Siljan et al. [5.82]. Although many previous investigators have attributed the spalling of refractory linings to the volume expansion of this reaction, Siljan et al. have attributed any expansive spalling that may occur to the growth of corundum along thermal gradients and/or any contraction due to the volume decrease.

The action of molten aluminum upon any β-alumina contained in materials such as high alumina (70%) refractories produces metallic sodium [5.83]. The metallic sodium present can then lead to reduction of silica, and if oxidized, it can lead to the formation of $NaAlO_2$. The formation of $NaAlO_2$ is enhanced in the presence of a reducing atmosphere containing nitrogen by the intermediate formation of aluminum nitride according to the following reactions:

$$2Al_2O_3 + 2N_2 \rightarrow 4AlN + 3O_2 \tag{5.27}$$

$$2AlN + Na_2O + 3/2O_2 \rightarrow 2NaAlO_2 + N_2 \tag{5.28}$$

The difference in densities between alumina and sodium aluminate (3.96 vs. 2.69 g/cm^3) implies that a considerable volume expansion can take place during conversion of the original protective layer to a nonprotective aluminate, thus leading to continued corrosion.

According to Lindsay et al. [5.84], mullite, when attacked by molten aluminum, converted to silicon metal and alumina; when attacked by molten magnesium-containing aluminum alloys, mullite converted to spinel and magnesia. Brondyke [5.81] has shown that molten aluminum will reduce silica in all forms contained in alumino-silicate refractories.

According to Cornie et al. [5.85], the attack of SiC by molten aluminum at temperatures greater than 700°C can be represented by the following equation:

$$3SiC + 4Al \rightarrow Al_4C_3 + 3Si \tag{5.29}$$

Nickel-base eutectic alloys such as NiTaC provide very severe temperature (as high as 1800°C) requirements upon their containers. Huseby and Klug [5.66] studied the reactions of many oxides in contact with NiTaC-13 at 1700 and 1800°C, and found that only Al_2O_3, $Y_3Al_5O_{12}$, and $LaAlO_3$ formed no interfacial reaction layers.

In a study to evaluate sialon* crucible materials as candidates for containing molten silicon, Wills et al. [5.86] found that

*Sialon is an acronym used to represent solid solutions of silicon, aluminum, oxygen, and nitrogen.

the grain boundary chemistry is of utmost importance, with these phases being attacked first. Two sialon solid solutions (called O′ and β′) were evaluated at different compositions at 1450°C for 1–6 hrs under 0.84 MPa of argon.

5.2 ATTACK BY GASES

The corrosion of a ceramic by vapor attack is generally much more severe than that by liquids or solids. The major reason for the more severe attack is related to the increased surface that is available to gases as opposed to liquids or solids. The

TABLE 5.2 Interfacial Reaction Products Caused by Vapor Attack

Material	Vapor	Interface[a]	Ref.
Al_2O_3	Potassium	KA_{10}	5.89
Al_2O_3	Sodium	NA	5.178
Al_2O_3	Potassium	$KA_5 + Al$	5.90
Al_2O_3/TiC	Oxygen	AT	5.180
AlN	Oxygen	A	5.129
B_4C	Oxygen	B	5.145
$MgAl_2O_4$	Sodium	NA	5.178
$MgAl_2O_4$	Potassium	KA + M	5.178
Mg_2SiO_4	Sodium	$N_2M_2S_3$	5.178
Mg_2SiO_4	Potassium	KMS + M	5.178
$Al_6Si_2O_{13}$	Sodium	$NAS_2 + NA_{SS}$	5.178
$Al_6Si_2O_{13}$	Potassium	$KAS + KA_{SS}$	5.178
SiC	Oxygen	S	5.6
Si_3N_4	Oxygen	$S + Si_2O_2N$	5.6
Si_3N_4/R_2O_3	Oxygen	$RS_2 + S$	5.119
SiAlON	Oxygen	A_3S_2	5.125
TiB_2	Oxygen	B + T	5.160
TiC	Oxygen	T	5.156
TiN	Oxygen	T	5.179
$ZrSiO_4$	Potassium	$KZS_3 + Z$	5.178

[a] A = Al_2O_3, B = B_2O_3, K = K_2O, M = MgO, N = Na_2O, R = R_2O_3, (R = Y, Ce, La, Sm), S = SiO_2, T = TiO_2, Z = ZrO_2, subscript ss = solid solution.

various interfacial reaction products that may form as a result of attack by gases for several selected materials have been listed in Table 5.2. The interested reader should examine the original articles to determine the experimental conditions under which the various reaction products formed, and also to determine the exact nature of the ceramic tested. In the sections below, several selected materials are described in more detail.

Although oxidation is generally the most prevalent form of attack by gases, hydrogen reactions are becoming more important as a result of the development of miniature fuel cells for applications such as cell phones. Historically, hydrogen fuel cells were the realm of the space industry. According to Nelson [5.87], the corrosion of a material by hydrogen is dependent on the following parameters:

(a) ease of transport of H into the material
(b) type of reactions available between H and material
(c) design of the structure

Corrosion of ceramics by hydrogen is generally not a problem below about 1100°C, where the molecular form exists. Thermal dissociation begins to become significant above 1100°C, and transport rates to material surfaces increases rapidly.

5.2.1 Oxides

Alumina

In the ceramics community, alumina is considered one of the most inert materials toward a large number of environments. For this reason, alumina that is produced as a 95–100% Al_2O_3 material is used in many furnace applications. The one area where its reactivity is often overlooked is its application in laboratory furnaces. Most high-temperature laboratory furnaces use alumina as the standard lining. When the materials that are under investigation react to form gaseous species, and especially when the furnace atmosphere contains a low partial pressure of oxygen, one should be aware of the possible reactions that may occur with alumina.

The lining of a laboratory furnace can receive a much more severe usage than an industrial furnace. Generally, this is a result of repeated thermal cycling and to the investigation of a wide variety of materials that produce a wide range of corrosive environments. An example of the corrosion of sample crucible setter tiles is given below.

The test environment of a vertical tube furnace that used an alumina tube and horizontal alumina disks" on which to place alumina sample crucibles included an input atmosphere of hydrogen and methane that gave an oxygen partial pressure of 10^{-14} Pa at the test temperature of 1300°C. Samples being tested were various silicon nitride samples containing small quantities of MgO, Y_2O_3, Fe_2O_3, ZrO_2, Al_2O_3, and CaO. A discussion of the furnace set-up and operation can be found in Ref. [5.88].

The alumina disks in the upper portion of the furnace tube above the silicon nitride samples, along a temperature gradient that ranged from 1250 to 1185°C, exhibited a glazed surface layer of silicate glass containing crystals of cordierite and cristobalite (see Fig. 5.7). The formation of cordierite was caused by the active oxidation of the silicon nitride and the vaporization of magnesia contained within the nitride samples that subsequently reacted with the alumina to form cordierite. The following equations describe the reaction:

$$2Si_3N_4(s) + 3O_2 \rightarrow 6SiO(g) + 4N_2(g) \quad @\ 1300^\circ C \qquad (5.30)$$

$$SiO(g) + 1/2O_2 \rightarrow SiO_2(s) \quad @\ 1185\text{–}1200^\circ C \qquad (5.31)$$

$$MgO(s) \rightarrow MgO(g) \quad @\ 1300^\circ C \qquad (5.32)$$

$$2MgO(g) \rightarrow 2MgO(s) + 2SiO_2(s) + Al_2O_3(s)$$

$$\rightarrow 2MgO \cdot Al_2O_3 \cdot 2SiO_2(s) \quad @\ 1185-1250^\circ C \qquad (5.33)$$

Although Anderson [5.89] reported that oxygen was necessary for the formation of potassium β-alumina from sapphire

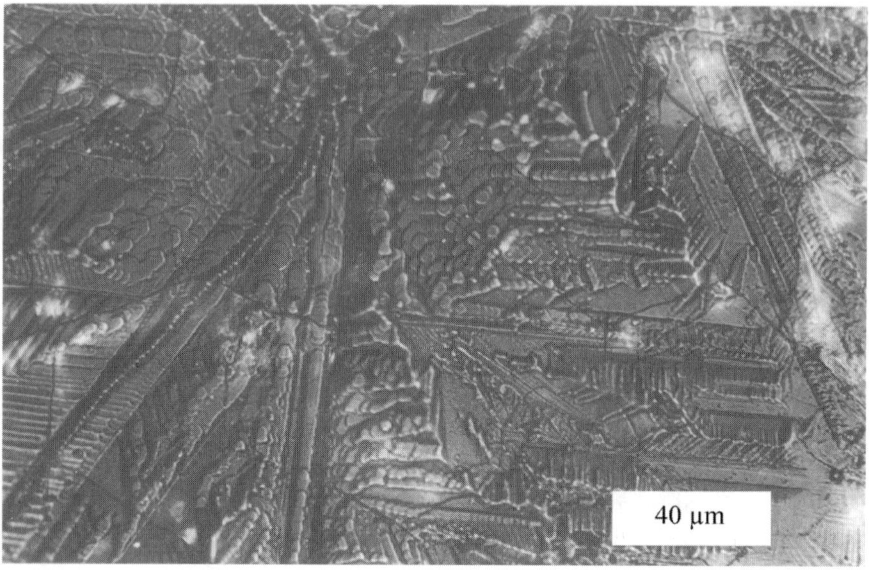

FIGURE 5.7 Cordierite and cristobalite formation on alumina. Reflected light, differential interference contrast micrograph (magnification 500×).

used in vapor arc lamps, van Hoek et al. [5.90] showed that potassium vapors (at 1 MPa) were able to reduce alumina at 1373 K in the absence of oxygen by the following reaction:

$$6K + 16Al_2O_3 \rightarrow 3(K_2O \cdot 5Al_2O_3) + 2Al \qquad (5.34)$$

forming a potassium β-alumina and metallic aluminum. Although approximately 3 wt.% aluminum should form, they detected no metallic aluminum by XRD in their samples. They suggested that the presence of metallic aluminum was confirmed by the blackening of samples due to the formation of finely divided metallic aluminum. They found that the β-alumina formed with the crystallographic *c*-axis parallel to the substrate surface. This created an easy diffusion path perpendicular to the surface for diffusion of potassium and continued corrosion. They also suggested that this oriented growth was not the result of epitaxial growth, because the starting alumina was a polycrystalline material. It is unfortunate that many references can be found in the literature that refer to oriented growth on polycrystalline substrates as epitaxy. Although localized crystallographic matching may occur on a grain-to-grain basis, forming an oriented polycrystalline layer on an oriented polycrystalline substrate, this was not the original meaning of epitaxy. It is enlightening that van Hoek et al. suggested another reason for oriented growth—the faster growth direction (easy diffusion path) being perpendicular to the reaction surface.

In a simulated industrial furnace atmosphere cycling between 8–10% combustibles (reducing) and 6–10% excess oxygen (oxidizing) at temperatures of 1260 and 1400°C, Mayberry et al. [5.91] showed that refractories containing chrome ore developed a permanent expansion and a loss in strength. This was the result of spinel solution into periclase, and then exsolution with the accompanying phase redistribution, recrystallization, and pore development during cycling. This example shows how a material may experience degradation as a result of atmosphere effects although it does not exhibit any signs of classic corrosion (weight gain or loss, reaction product surface layer formation, etc.).

In a study of UF_6 -fueled gas-core reactor systems, Wang et al. [5.92] investigated the effects of UF_6 gas (at pressures of 20.0–22.7 kPa) upon alumina at temperatures of 973, 1073, 1273, and 1473 K for times up to 4 hr. At the three lower temperatures, the following reaction was suggested to represent their findings:

$$6UF_6 + 2Al_2O_3 \rightarrow 4AlF_3 + 6UF_4 + 3O_2 \tag{5.35}$$

The AlF_3 formed on the alumina surface, whereas the UF_4 was found on the colder portions of the furnace chamber. At 1473 K, several oxides of uranium were found in the surface scale, and no AlF_3 was found because of its high vapor pressure (30.6 kPa) at that temperature. Weight gain was reported for the lower temperatures; however, at 1473 K, a large weight loss was exhibited as a result of vaporization of the AlF_3.

Wang et al. also performed a computer analysis of the expected reactions and found several differences between the experimental and calculated data. Although the differences that were found are not important, the cause of these differences was worth noting. The computer program used to predict chemical reactions at different combinations of temperature and pressure was dependent upon the database used, and thus could not predict products not contained in the database. The program used by Wang et al. could also analyze only a closed system at equilibrium. Any reactions (e.g., formation of interfacial layer) that may retard further reaction would prevent equilibration. Insufficient time for reactions to proceed to completion would also contribute to the differences, because the computer program based upon minimization of total free energy of formation cannot predict the kinetics of the reactions. Thus, one should remember that calculated reactions based upon thermodynamics is only a portion of any study, and only an indication of what should be expected during actual experimentation.

Alumino-Silicates

Arnulf Muan has provided a considerable amount of experimental data concerning the atmospheric effects upon the phase equilibria of refractory materials. One such article was reprinted in the *Journal of the American Ceramic Society* [5.93] as a commemorative reprint. This article stressed the importance of the oxygen partial pressure in determining the phases present in the reaction of iron oxides with alumino-silicate refractories. Under oxidizing conditions, large amounts of ferric iron can substitute for aluminum in the various aluminum-containing phases; however, under reducing conditions, this substitution is negligible. Large volume changes accompany some of the phase changes that occur with damaging results to the refractory. In addition, the temperature at which liquid phase develops decreases as the oxygen partial pressure decreases.

Reactions that have occurred between alumino-silicate refractories and the gaseous exhaust in glass furnace regenerators at about 1100–1200°C forming nepheline and noselite are shown below:

$$3Al_2O_3 \cdot 2SiO_2 + Na_2O \rightarrow Na_2O \cdot Al_2O_3 \cdot 2SiO_2 + 2Al_2O_3 \quad (5.36)$$

The nepheline formed then reacts with SO_3 and more Na_2O vapor of the exhaust forming noselite:

$$3(Na_2O \cdot Al_2O_3 \cdot 2SiO_2) + 2Na_2O + 2SO_3 \rightarrow 5Na_2O \cdot 3Al_2O_3 \cdot 6SiO_2 \cdot 2SO_3 \quad (5.37)$$

Although the precise mechanisms that take place have not been determined, most likely, the alumina and any silica available form more nepheline. Large volume expansions (10–15%) accompany these reactions resulting in spalling or shelling. Historically, these reactions have been a serious problem to the glass manufacturer, as they cause plugging of the regenerator and a less efficient combustion process. Various regenerator design and material changes have essentially eliminated this problem; however, it is a reaction that may still occur when the conditions are appropriate.

Magnesia-Containing Materials

McCauley and coworkers [5.94–5.97] have studied the effects of vanadium upon the phase equilibria in magnesia-containing materials. This work was initiated in an effort to understand the effects of vanadium impurities in fuel oils on basic refractories. It has been found that only small amounts of V_2O_5 are needed to alter the phase assemblages in high magnesia materials. The reactions that occur generally form low melting vanadates (i.e., tricalcium and trimagnesium vanadates with melting points of 1380 and 1145°C, respectively, and magnesium–calcium–vanadium garnet with its melting point of 1167°C) and, depending upon the exact compositions, can develop appreciable amounts of liquid at service temperatures.

Although the initial reaction is a gaseous phase reaction, it quickly converts to a liquid attack.

Ready [5.98] has shown that magnesia containing a small percentage of carbon black or graphite can form CO/CO_2 bubbles in oxidizing atmospheres at temperatures ranging from 1200 and 1500°C. These bubbles, which formed primarily at grain boundaries, were the source of intergranular fracture. For the unoxidized materials, the fracture was 100% transgranular, because no grain boundary separation (caused by the presence of bubbles) was present.

Zirconia

Lepistö et al. [5.99] studied the effects of humid conditions at 150°C for up to 1000 hrs upon the stability of metastable tetragonal phase in tetragonal zirconia polycrystals (TZP). Several different materials containing a small amount of yttria were tested, and all were found to contain increasing amounts of monoclinic zirconia as the exposure time increased. It was believed that surface finish along with the grain size and yttria content all had an effect upon the transformation of the tetragonal-to-monoclinic zirconia. In a later study, Lepistö and Mäntylä [5.100] concluded that the stability of yttria-containing TZP in humid atmospheres was through the dissolution of ZrO_2 at grain boundaries with the subsequent relief of localized stresses, followed by the transformation of the tetragonal-to-monoclininc phase. The proposed mechanism of dis- solution was through the dissociation of adsorbed water molecules on the zirconia surface. The oxygen ions formed were then proposed to anneal the oxygen vacancies present within the yttria-containing zirconia. As a result, water is formed, which then allowed continued dissolution. The water adsorption was greatest at defect sites, and thus was dependent upon the yttria concentration. Lepistö and Mäntylä found that both yttria and alumina increased the dissolution rate, whereas ceria did not, as it did not change the vacancy concentration. Thus the tetragonal phase stability toward humid atmospheres can be increased by using ceria as the dopant.

Along with alumina, Wang et al. [5.92] studied the effects of UF_6 gas upon partially stabilized zirconia containing 7 wt.% CaO. Temperatures of 873, 973, and 1073 K were investigated at times of up to 2 hr. The UF_6 gas pressure was maintained at 20–22.7 kPa. A weight increase was reported for only the lowest temperature. The surface scale was nonprotective for all temperatures. The compounds that formed on the sample surface after exposure at 873 K were ZrF_4, CaF_2, UO_3, and U_3O_8. Because of the hygroscopic nature of ZrF_4, this compound was identified as a hydrate. At 973 K, the same compounds were found, except for UO_3. At 1073 K, additional compounds (UF_4, UO_2F_2, and $UO_2F_2 \cdot 1.5H_2O$) were found as a result of the dissociation of UF_6 and reaction with atmospheric moisture. In addition, some zirconium oxyfluorides were found. As in the analysis of the alumina reactions, the computer predictions were different from the experimental results.

5.2.2 Nitrides and Carbides

None of the nitrides and carbides are thermodynamically stable in oxygen-containing environments. Under some conditions, some carbides and nitrides form a protective metal oxide layer that allows them to exhibit reasonably good oxidation resistance (e.g., $Si_3Al_3O_3N_5$ forms a protective layer of mullite). As can be seen from the examples given below, the corrosion of silicon nitride and carbide materials varies considerably based upon the characteristics of the individual material and the environment. Although many attempts have been made to determine an exact mechanism for this corrosion, there is still considerable disagreement, unless these materials are grouped according to their type and impurity level for each environment.

Silicon Nitride

Oxidation. The oxidation of Si_3N_4 is dependent upon the manufacturing process used: chemically vapor-deposited (CVD) materials exhibit the slowest and smallest amount of oxidation because of their purity, dense structure, and in some cases, larger grain size; hot-pressed (HP) and hot isostatically pressed (HIPed) materials exhibit an oxidation generally dependent upon the type and amount of additive used; and reaction sintered (RS) materials exhibit the most oxidation because of their large porosity. Materials containing more impurities, or at least more of those species that lower the viscosity of any silica reaction layer that may form, will provide a lower resistance to continued oxidation, because the diffusion of oxygen is easier through the lower viscosity coating.

The oxidation of Si_3N_4 has been described as occurring by either an active or a passive mechanism [5.101]. The active mechanism is one where the fugitive SiO forms in environments with low partial pressures of oxygen by the reaction:

$$2Si_3N_4(s) + 3O_2(g) \rightarrow 6SiO(g) + 4N_2(g) \tag{5.38}$$

Passive corrosion occurs in environments with high partial pressures of oxygen by the reaction:

$$Si_3N_4(s) + 3O_2(g) \rightarrow 3SiO_2(s) + 2N_2(g) \tag{5.39}$$

The SiO_2 that is produced forms a protective coating and further oxidation is limited. The pressure of N_2 formed at the interface can be large enough to form cracks or pores in the protective coating, which subsequently allow additional oxidation.

The active-to-passive transition, as determined by several investigators and compiled by Vaughn and Maahs [5.102], is shown in Fig. 5.8. The variations reported are a result of differences in the silicon nitride material tested and in the experimental conditions of the test.

Thermodynamically, silicon oxynitride (Si_2N_2O) should also form; however, it is further oxidized to SiO_2 according to:

$$2Si_3N_4 + 3/2O_2 \rightarrow 3Si_2N_2O + N_2 \tag{5.40}$$

$$Si_2N_2O + 3/2O_2 \rightarrow 2SiO_2 + N_2 \tag{5.41}$$

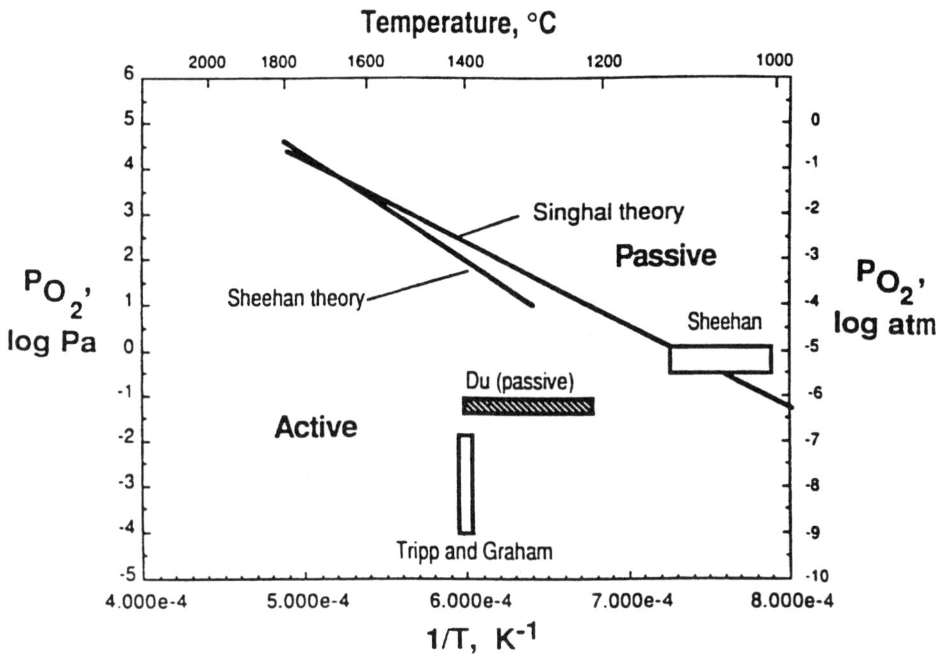

FIGURE 5.8 Literature data for active-to-passive oxidation transition for Si_3N_4. (From Ref. 5.102; reprinted with permission of The American Ceramic Society, www.ceramics.org. Copyright ©1990. All rights reserved.)

The reaction of SiO_2 with Si_3N_4 can also produce a loss in weight by the reaction:

$$Si_3N_4 + 3SiO_2 \rightarrow 6SiO + 2N_2 \tag{5.42}$$

As is quite often the case in corrosion of ceramics, the oxidation of silicon nitride initially follows a linear rate law, because the rate limiting step is the interface reaction. When the interface reaction layer becomes sufficiently thick, the kinetics shift to one limited by diffusion through the reaction surface layer, and thus obey a parabolic rate law. This change from linear to parabolic kinetics was reported to occur at an interface layer thickness of about 50 nm for the oxidation of CVD Si_3N_4 between 1000 and 1300°C by Choi et al. [5.103]. It was pointed out by Choi et al. that although two different materials (i.e., Si metal and Si_3N_4) exhibit the same major oxidation product (i.e., SiO_2) and follow the same oxidation kinetics (i.e., parabolic), the mechanism of oxidation is not necessarily the same. This was based upon the very different activation energies and rates of oxidation obtained for the two materials.

Although Choi et al. reported increased refractive index determinations for the SiO_2 reaction layer, consistent with significant nitrogen concentrations, they did not mention the formation of an oxynitride layer as did Du et al. [5.104]. Du et al. reported that the oxidation of pure CVD α-Si_3N_4 formed a double reaction layer; an inner zone of Si_2N_2O and an outer zone of SiO_2. A calculated plot of the thermodynamic stability fields for SiO_2, Si_2N_2O, and Si_3N_4 is shown in Fig. 5.9, which indicates a zone of Si_2N_2O separates the Si_3N_4 and SiO_2 throughout the temperature range examined (1000–2000 K). The assumption that Du et al. made in determining this was that no solutions existed among the three phases, which is not exactly correct. Indications are that the oxynitride is truly a solution of variable oxygen content with no distinct boundaries. Thus conventional thermodynamics cannot accurately predict the results. Du et al. also reported that the most probable rate-limiting step during oxidation was molecular oxygen diffusion through the inner zone of Si_2N_2O, which is structurally denser than SiO_2. In contrast to this molecular oxygen diffusion limited reaction, Luthra [5.105] has suggested that the reaction is controlled by a mixed process involving nitrogen diffusion and the reaction at the Si_2N_2O/SiO_2 interface. He also stated that the rate law for this mixed controlled process can be nearly parabolic, which is consistent with all the observations reported in the literature [5.106]. Luthra's assumption was that the Si_2N_2O interface thickness was too thin to provide a significant barrier to diffusion. Ogbuji [5.107], through some interesting experimentation, has given support to the diffusion barrier suggested by Du et al., although he did state that their model suffered from some inconsistencies as a result of incorrect assumptions. Thus it appears that an understanding of the

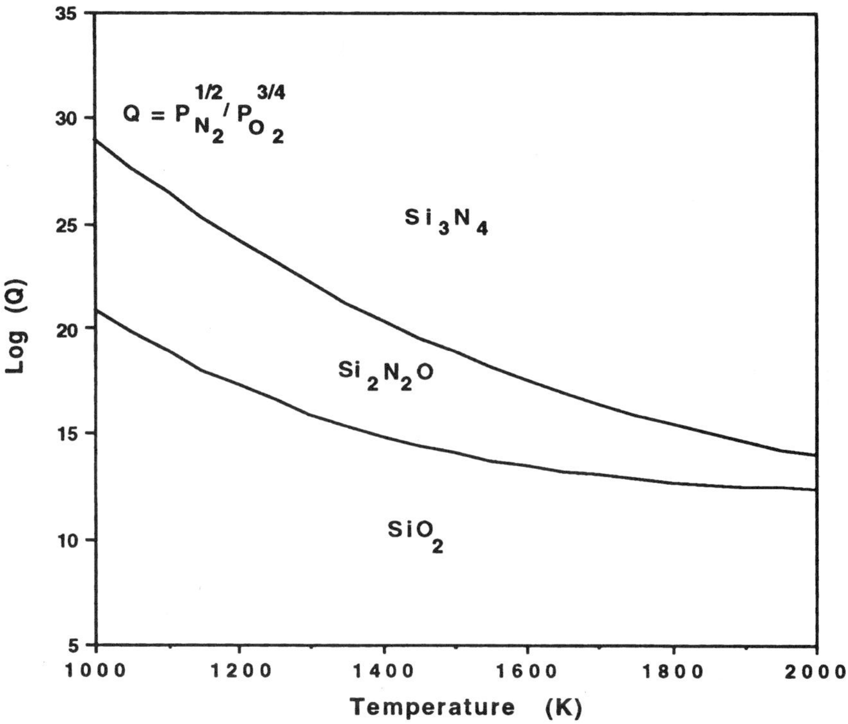

FIGURE 5.9 Thermodynamic stability fields for SiO_2, Si_2N_2O, and Si_3N_4. (Ref. 5.104; reproduced by permission of The Electrochemical Society, Inc.)

details of oxidation of Si_3N_4 is still incomplete. Ogbuji listed the areas of uncertainties as:

1. Formation sequence of Si_2N_2O and SiO_2
2. Oxygen diffusion routes
3. Effects of oxide layer crystallinity upon diffusion
4. Oxidation state of product gases
5. Effects of out-diffusion of nitrogen
6. Solubilities of O_2 and N_2 in the SiO_2 and Si_2N_2O phases

Both de Jong [5.108] and Joshi [5.109] have shown through binding energy calculations obtained by X-ray photoelectron spectroscopy (XPS) data that the oxidized double layer on commercially available as-received Si_3N_4 powders contained more oxygen for the inner layer than the stoichiometric Si_2N_2O. Their data match closely that which Bergstrom and Pugh [5.110] reported for Si_2ON. Joshi has suggested that the double layers consist of a thin outer layer of amorphous SiO_2 over an inner oxynitride layer that is amorphous near the silica outer layer and crystalline near the nitride, as shown in Fig. 5.10. He also suggested that the oxygen content varies through the oxynitride thickness as the silazane groups of the bulk are gradually replaced by siloxane groups near the surface.

As many silicon nitride materials contain various sintering aids and are therefore polyphase materials, their oxidation behavior is more complex than discussed above. The outward diffusion of the cations of the sintering aids along with various impurities that may be present, yield oxidation layers containing many mixed phases that depend upon what cations diffuse to the surface and how much of each is present. An indication of the equilibrium phases that should be present can be obtained from an examination of the appropriate phase diagram. The presence of any liquid-phase fields greatly increases the outward diffusion of the various cations (along with the inward

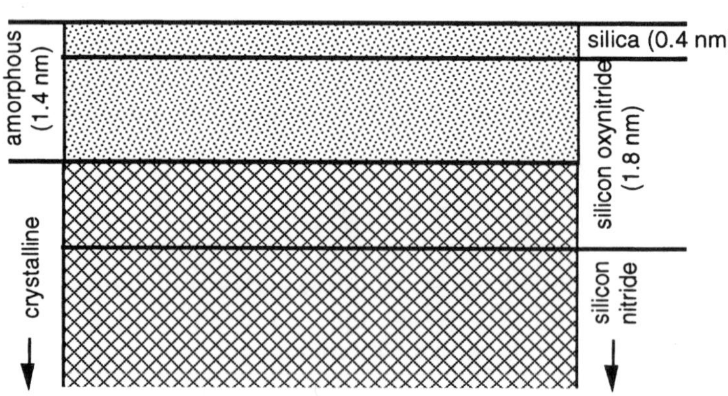

FIGURE 5.10 Typical oxide layer configuration for Si_3N_4 obtained from XPS and TEM data. Actual boundaries are diffuse. (From Ref. 5.109.)

diffusion of oxygen) and will also provide a means for the formation of a coherent layer, because a liquid can accommodate the volume changes that occur during the reaction of the various phases. In general, parabolic rate constants are obtained for the oxidation of these materials and these constants vary considerably from one material to the next.

Tripp and Graham [5.111] reported a marked change in the oxidation rate to higher values at about 1450°C for HP-Si_3N_4. This change in rate was related to a change in mechanism from passive-to-active oxidation at a partial pressure of oxygen of $10^{-7.3}$ atm at 1400°C. Both mechanisms were reported to occur between 10^{-7} and 10^{-10} atm. The passive-to-active oxidation transition was reported to occur at 1×10^{-4} MPa at 1400°C for a HP-Si_3N_4 containing 6 wt.% Y_2O_3 and 1.5 wt.% Al_2O_3 by Kim and Moorehead [5.112], which compares quite favorably with the calculated value of 2×10^{-4} MPa by Singhal [5.113].

The rate-determining step in HP-Si_3N_4 containing MgO was reported by Cubicciotti and Lau [5.114] to be the diffusion of MgO from the bulk material into the oxide surface layer. This surface layer was composed of SiO_2, $MgSiO_3$, or glass phase and some unoxidized Si_3N_4, and was porous due to released N_2. Similar results were reported by Kiehle et al. [5.115] for HP-Si_3N_4 containing impurities of magnesium, iron, aluminum, manganese, and calcium (all less than 0.6 wt.% each determined by emission spectroscopy). They found an amorphous silica film at temperatures as low as 750°C that converted to cristobalite at higher temperatures and/or times. Above 1000°C, sufficient migration of the impurities to the surface caused the formation of additional phases, such as enstatite ($MgSiO_3$), forsterite (Mg_2SiO_4), akermanite ($Ca_2MgSi_2O_7$), and diopside ($CaMgSi_2O_6$). Akermanite was the first silicate to crystallize, enstatite appeared only above 1350°C, and diopside appeared only after oxidation at 1450°C. Both akermanite and forsterite appeared only after longer heating times. Equilibrium phase assemblages occurred only at high temperatures and long times. It is quite interesting that Kiehle et al. reported neither a weight loss nor a weight gain (difference in weight before and after heating) for their oxidation tests. It was suggested that this was attributable to the simultaneous formation of both SiO_2 and the fugitive SiO. It is unfortunate that they did not report continuous weight change data, which may have shown an initial weight gain followed by a weight loss after the surface scale crystallized at about 1000°C.

Catastrophic oxidation at about 1000°C has been reported for HP-silicon nitride containing Y_2O_3, where the secondary phases are $Y_2Si_3O_3N_4$, $YSiO_2N$, or $Y_{10}Si_7O_{23}N_4$ [5.101]. If only Si_2N_2O and $Y_2Si_2O_7$ were present (true for materials containing less than about 6 wt.% Y_2O_3) as the secondary phases, the oxidation would be very low and well behaved.

In HIPed Si_3N_4 containing 5 wt.% Y_2O_3 oxidized over the temperature range 1200–1450°C for up to 100 hr, Plucknett and Lewis [5.116] found a variation in the scale microstructure from a phase mixture of small amounts of $Y_2Si_2O_7$ within larger amounts of amorphous silicate (containing some impurity cations) at short times and low temperatures, to a double layer scale of $Y_2Si_2O_7$ near the nitride and a semicontinuous SiO_2 outer layer at longer times and higher temperatures. As the oxidation times were increased, the amorphous silicate gradually converted to cristobalite.

Oxidation of RS-silicon nitride occurred through two steps: a fast internal oxidation of the open porosity until filled, and then a slower external oxidation of the surface [5.101]. In reaction sintered materials with surface areas of about 0.7 and 0.3 m^2/g, Gregory and Richman [5.117] reported that the surface pores were sealed by the oxidation product when the temperature of oxidation was above 1100°C. A plot of the weight gain for many RS materials as a function of the fractional open porosity after exposure to static air at 1400°C for 1000 hr is shown in Fig. 5.11 [5.118].

Nonporous sintered Si_3N_4 containing about 4–6 wt.% of Y_2O_3, Ce_2O_3, La_2O_3, or Sm_2O_3, in addition to about 4 wt.% SiO_2 as sintering aids, was investigated for oxidation resistance at 700, 1000, and 1370°C in air for up to 200 hr by Mieskowski and

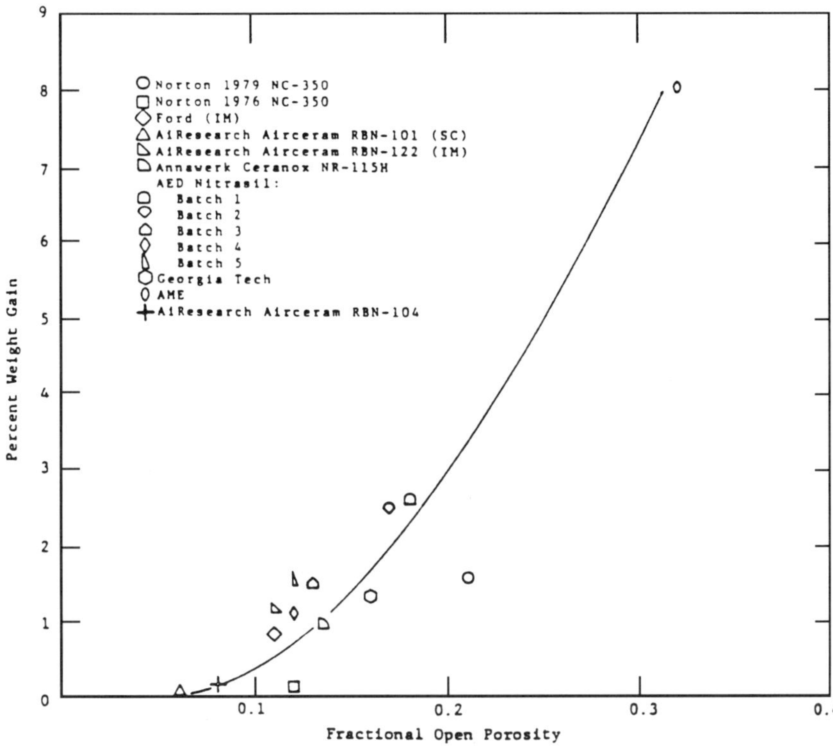

FIGURE 5.11 Weight gain–porosity relation for RS–Si_3N_4 materials exposed to static air for 1000 hr at 1400°C. (From Ref. 5.118; reprinted with permission of William Andrew, Inc.)

Sanders [5.119]. All these sintering aids formed pyrosilicate and cristobalite as reaction products. The lowest oxidation rate was produced by the addition of Y_2O_3, whereas the highest rates were produced by Sm_2O_3 and Ce_2O_3.

The weight change upon oxidation for Eq. (5.39) is the difference between the weight of 1 mol of Si_3N_4 and 3 mol of SiO_2, which Horton [5.120] represented by the mixed parabolic rate equation:

$$\Delta W^2 + A\Delta W = kt + c \tag{5.43}$$

where ΔW is the gain in weight per unit area, t is time, and A, k, and *c* are constants. Because linear kinetics is usually observed for the formation of very thin films of SiO_2, the first term in Eq. (5.43) would be negligibly small. Wang et al. [5.121] have shown that the formation of oxide layer thicknesses less than about 5 nm forming on silicon nitride powder containing 4 wt.% Y_2O_3 obey a linear rate law at 900, 950, and 1000°C. For the formation of very thick films, the second term becomes negligibly small. Oxidation rates in various atmospheres for powdered samples were in the following order: dry oxygen = humid air = 2× dry air. The oxidation product was amorphous at 1065°C and tridymite between 1125 and 1340°C. Franz and Langheinrich [5.122] reported the increased oxidation of amorphous CVD silicon nitride in wet oxygen over that in dry oxygen to be about four times at 1000°C for 10 hr. Horton pointed out that the exposed surface area for oxidation was actually the surface area of smaller particles that made up larger agglomerates, as solid particles of various mesh sizes all had the same BET surface area.

Reaction in Other Atmospheres. Even in environments other than pure oxygen or air, silicon nitride corrodes primarily through oxidation. An example of this is the work reported by de Jong et al. [5.123] in their studies of HP and RS materials at 1050°C in gas mixtures of H_2 plus 7% or 17% CO and 1300°C in a gas mixture of H_2 plus 1% CH_4 and 0.5% H_2O. Although the partial pressure of oxygen in these experiments was on the order of 10^{-20} atm, passive oxidation occurred at the lower temperature in H_2/CO, forming a surface layer of tridymite, as shown in Fig. 5.12. At the higher temperature in $H_2/CH_4 + H_2O$, active oxidation to SiO occurred, followed by the formation of SiC whiskers by the VLS mechanism, which deposited onto the Si_3N_4 surface as well as the internal parts of the furnace (see Fig. 5.13).

At 1200–1300°C in a gas mixture of $H_2S/H_2O/H_2$, Oliveira et al. [5.124] found that HP materials containing Y_2O_3 and Al_2O_3 (10–12 wt.% total) actively corroded through the forma-

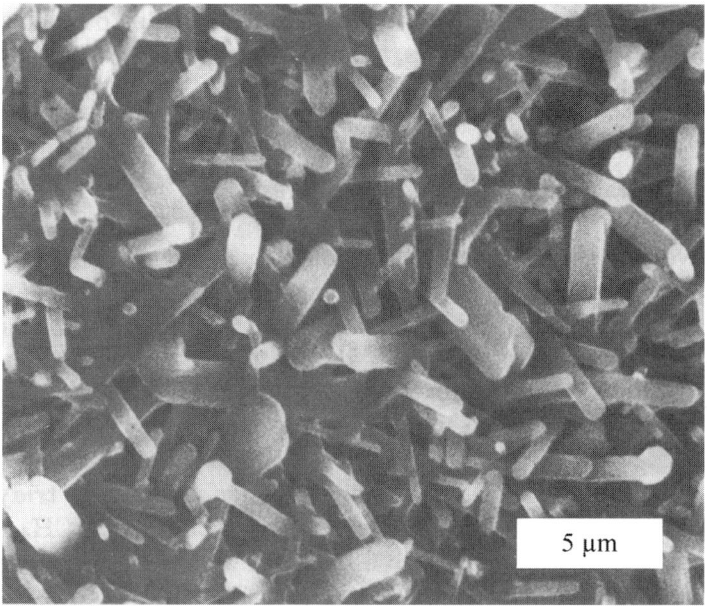

FIGURE 5.12 Tridymite surface formed on Si_3N_4 at 1050°C and 5×10^{-14} Pa oxygen for 1000 hr. Magnification = 3600×.

tion of gaseous SiO and SiS. The active corrosion of these materials depended upon the test temperature and the gas environment composition. In dry H_2S/H_2 at 1300°C, the outer nonprotective porous layer of the Si_3N_4 samples contained yttrium oxysulfide.

In their studies of the oxidation of CVD and HP materials, Kim and Moorehead [5.112] reported that the mechanism of oxidation changes at 1400°C, depending upon the partial pressure of water vapor present in H_2–H_2O mixtures. The Si_3N_4 in the CVD material decomposed to metallic silicon and nitrogen at very low partial pressures of water vapor (10^{-7} MPa). As the water vapor pressure increased to about 10^{-5} MPa, the following reaction occurred:

$$Si_3N_4 + 3H_2O_{(g)} \rightarrow 3SiO_{(g)} + 2N_{2(g)} + 3H_{2(g)} \qquad (5.44)$$

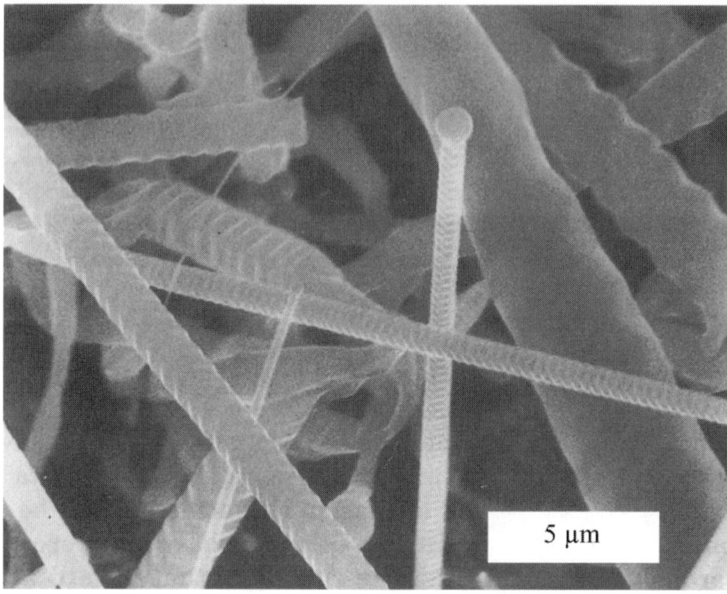

FIGURE 5.13 β-SiC whiskers deposited onto Si_3N_4 at 1300°C and 10^{-15} Pa oxygen for 100 hr. (Magnification = 4300×.)

with a resultant more severe weight loss. Less severe weight losses were noted at higher water vapor contents, presumably due to the formation of a discontinuous layer of SiO_2 and/or Si_2N_2O. Choi et al. [5.103], in contrast, reported that the oxidation of CVD Si_3N_4 was insensitive to the presence of H_2O in an inert atmosphere (He or Ar). Their studies in wet oxygen, however, indicated an increased oxidation rate over that of dry oxygen—exhibited by a decrease in the activation energy as the partial pressure of the water vapor increased from 2 to 80 kPa. This was thought to be due to a breaking of the Si–O–Si bonds in the silica structure as a result of OH solution. The various gaseous reaction products that may form (i.e., NO, NH_3, etc.) can alter the counterdiffusion kinetics thus modifying the inward diffusion of the oxidant and slightly changing the overall mechanism.

Kim and Moorehead [5.112] found a similar dependence upon water vapor pressure for the oxidation of HP-Si_3N_4 containing 6 wt.% Y_2O_3 and 1.5 wt.% Al_2O_3. In the low-pressure region, however, the magnitude of the weight loss for the HP material was about three times greater than that of the CVD material. This difference was attributed to the greater surface area for reaction in the HP material, because it had a much smaller grain size compared to the CVD material. At the higher water vapor pressures, the grain boundaries were not preferentially attacked and thus the two materials exhibited similar weight losses.

Other Nitrides

The crystalline solution series of materials of alumina dissolved into β-silicon nitride ($Si_{6-x}Al_xO_xN_{8-x}$) make up a truly interesting series of materials. The hopes were that these materials would yield properties that were the best of the two end members. One improvement over silicon nitride is the oxidation resistance with increasing amounts of alumina while maintaining the relatively low thermal expansion characteristics of the pure silicon nitride phase. Weight gain behaviors have been reported to be parabolic with mullite being the oxide that formed on the surface. Singhal and Lange [5.125] reported that mullite formed only in those compositions containing more than 20 wt.% alumina and that, above 40 wt.% alumina, additional unidentified phases occurred. Chartier et al. [5.126] prepared sialon crystalline solutions with $x = 0.4$. Because 14.05 wt.% yttria was added to the original mix, the final pressureless sintered samples contained β′-$Y_2Si_2O_7$ and a glassy phase as grain boundary phases. Oxidation was parabolic and very slow below 1380°C. Above this temperature, more rapid oxidation occurred with departures from parabolic behavior. A thin alumino-silicate film formed first, but as metal cation migration occurred (predominately yttrium) reaction with this film formed more complex silicates. This film was dense below 1400°C, and gradually became porous and nonprotective as the temperature was increased. Visual observation indicated a light gray zone under the surface scale that Chartier et al. reported to be attributable to selective oxidation of the grain boundary phase.

Wang et al. [5.127] investigated the oxidation in air of rare-earth aluminum oxynitrides with the ideal formula $LnAl_{12}O_{18}N$ (Ln = La, Ce, Pr, Nd, Sm, and Gd) at temperatures ranging from 700 to 1500°C. Noticeable oxidation started at about 700°C, and increased with temperature. The final reaction products depended upon the particular rare-earth, but progressed through several intermediate stages. At temperatures around 1000°C, the products were $LnAl_{11}O_{18}$ and α-alumina. $LnAlO_3$ also formed with or without the disappearance of $LnAl_{11}O_{18}$ at higher temperatures depending upon the rare-earth. In cerium-containing materials, CeO_2 formed at 900°C. At temperatures as low as 700°C, lattice parameter changes (increasing c/a ratio) were noted for the oxynitrides, which were attributed to the initiation of oxidation.

AlN is an important material in the electronic ceramics industry and is an example of when a small amount of oxidation is beneficial to the application. In this case, the formation of a thin (1–2 μm) protective coating of AlON is formed and is used to improve the adhesion of copper films. Suryanarayana [5.128] found the oxidation of AlN powders between 600 and 1000°C in flowing air to follow a linear rate law initially, and then a parabolic law as the oxide layer thickness became sufficient to require diffusion for further growth. In contrast, Abid et al. [5.129] found that the oxide layer that formed on polycrystalline AlN in air at 1200°C was α-Al_2O_3, whereas below 800°C, no oxidation was observed. Dutta et al. [5.130] reported that oxidation of sintered polycrystalline AlN between 20 and 200°C progressed from individual α-Al_2O_3 particles of 2–3 nm in size to a 50-nm-thick film after 150 hr at 200°C. They also commented that their data were consistent with the formation of an oxynitride layer, but believed α-Al_2O_3 to be the oxide formed at low temperature based on thermodynamic calculations. Others have shown that an oxynitride formed as an intermediate preceding alumina formation at high temperatures [5.131], and McCauley and Corbin [5.132] reported that a region of ALON stability occurred between Al_2O_3 and AlN at temperatures between 1800 and 2050°C in flowing nitrogen.

The oxidation of TaN to Ta_2O_5 was reported to commence at about 450°C by Montintin and Desmaison-Brut [5.133]. As the temperature is raised, the initially powdered reaction product densified; however, the high volume expansion of Ta_2O_5 generated stresses in the coating that caused failure and spalling at high stress regions. Between 590 and 770°C in oxygen, the kinetics of the reaction were characterized by a sigmoidal rate law associated with the formation of the non-protective Ta_2O_5.

Silicon Carbide

Oxidation. The oxidation of green hexagonal powdered SiC has been described by Ervin [5.134]. Ervin stated that oxidation at low oxygen pressures took place with the formation of SiO gas, while at atmospheric pressure under flowing air, SiO_2 formed. The rate-controlling step was thought to be the growth of an ordered lattice of SiO_2 by solid diffusion. The following reactions are representative of the oxidation of silicon carbide:

$$SiC + 3/2O_2 \rightarrow SiO_2 + CO \quad (5.45)$$

$$SiC + 3/2O_2 \rightarrow SiO + CO_2 \quad (5.46)$$

Jorgensen et al. [5.135] proposed that the rate-controlling step in the growth of the SiO_2 layer formed on powdered SiC may be the diffusion of either oxygen ions or silicon ions. They ruled out the diffusion of molecular O_2, CO_2, and CO based upon their experimentally determined activation energies being too large for molecular diffusion. Harris [5.136] studied the oxidation of crystals of 6H-α SiC and determined that the rate of oxidation on the (0001) carbon face was approximately seven times greater than that on the silicon face at 1060°C for 70 hr. The thin oxide layer on the (0001) silicon face grew according to a linear rate law at all temperatures, whereas the thick oxide on the carbon face initially grew with linear kinetics but then changed to parabolic when the thickness became greater than 250 nm. At high temperatures and/or long times during oxidation of powdered samples, the oxidation rate changed from parabolic to linear presumably because the growth of the linearly controlled face overtook that of the parabolically controlled face. This change in oxidation rate at high temperatures has been attributed to a change in the oxide layer from amorphous to crystalline by Ervin [5.134] and Jorgensen et al. [5.137], and suggested by Costello and Tressler [5.138].

The desorption of CO gas formed at the SiC/SiO_2 interface has been reported to be the rate-controlling step by Singhal [5.139]; however, Hinze et al. [5.140] and many others have reported that it is the inward diffusion of oxygen through the surface layer of SiO_2. Spear et al. [5.141] ruled out the diffusion of CO as rate-controlling based upon their experiments that exhibited a dependence of the oxidation rate upon the partial pressure of oxygen and the almost identical activation energies obtained for the oxidation of SiC and Si metal. Fergus and Worrell [5.142] have concluded that the various contradictions in reported kinetics were attributable to a change in the diffusing species from molecular to ionic oxygen at about 1400°C. This was based upon two observations: one being that the activation energy for the growth of amorphous silica on CVD SiC increased above 1400°C, and the other being that the activation energy for the growth of cristobalite increased, but at the higher temperature of 1600°C. Decreases in oxidation rates at low temperatures have been attributed to sufficiently long times to allow crystallization of the silica scale.

In an analysis of the various possible rate-controlling steps, Luthra [5.85] concluded that a mixed interface reaction/diffusion process was the limiting feature in the oxidation of SiC. This was based upon the following facts:

1. Oxidation rate is lower than for pure silicon
2. Presence of gas bubbles in the oxide layer

3. Oxidation rate of single crystals dependent upon crystallographic orientation
4. Higher activation energy than for pure silicon (although Spear et al. [5.141] reported similar energies)

Because all of the above, except the presence of gas bubbles, are consistent with interface reaction control and the fact that bubbles are present, a mixed controlled process was concluded. Luthra suggested that mixed control should yield a rate law more complex than the generally observed linear or parabolic laws.

For pure monolithic CVD SiC and Si_3N_4, Fox [5.143] reported oxidation rates for 100 hr at temperatures between 1200 and 1500°C in flowing dry oxygen to be similar. In silicon nitride, any additives present will affect the oxidation rate. In general, increased levels of additives or impurities result in higher oxidation rates. These higher oxidation rates are attributable to the migration of the additive to the oxidized layer, thus lowering the viscosity, which increases the diffusion of the oxidant to the SiC/SiO_2 interface. Fergus and Worrell [5.142] reported that 0.5 wt.% boron in sintered α-SiC did not, however, significantly affect the oxidation rate.

Understandably, the active oxidation of SiC has not been investigated quite as thoroughly as passive oxidation; however, it should be remembered that active oxidation to SiO gas can occur at any temperature if the oxygen partial pressure at the SiC surface falls below some critical value. Not all data reported in the literature agree. The variations reported for this transition are attributable to the differences in the SiC materials tested and in the experimental conditions used. The partial pressure of oxygen at the transition from passive-to-active oxidation decreases with an increase in the total gas flow through the system [5.127]. This is the result of a decreasing gaseous boundary layer thickness with increasing velocity. The total gas pressure of the system can also affect results as suggested by Narushima et al. [5.144], because molecular gas flow exists at low pressures and viscous gas flow exists at higher pressures, thus changing the gas diffusion phenomena. Because the rate-controlling mechanism in active oxidation is the oxygen diffusion through the gaseous boundary layer, the characteristics of the gaseous boundary layer play a major role in the oxidation. If experiments were conducted at very high flow rates and very low total pressures, as was the case for the work of Rosner and Allendorf [5.145], the rate-controlling step may no longer be oxygen diffusion through the gaseous boundary layer, but the kinetics of gas arrival and removal from the surface. The active-to-passive transition as determined by several investigators and compiled by Vaughn and Maahs [5.102] is shown in Fig. 5.14. The variations for the reported transition are attributable to differences in the gas flow rates of the tests and possibly differences in the SiC material tested.

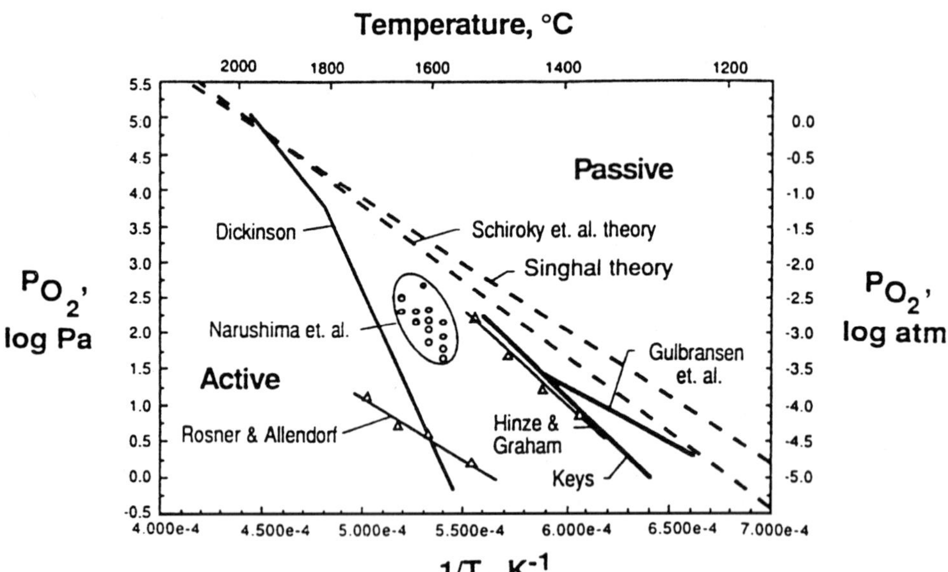

FIGURE 5.14 Literature data for active-to-passive oxidation transition for SiC. (From Ref. 5.102; reprinted with permission of The American Ceramic Society,www.ceramics.org. Copyright ©1990. All rights reserved.)

The oxidation of SiC fibers and whiskers is about as diverse as it is for other forms of SiC. Not only is the corrosive degradation of fibers and whiskers complexed by their chemistry (containing impurities of C, and SiO_2) and structure (containing more than one polymorph), but their surface area-to-volume ratio greatly enhances reaction rates when compared to an equal weight of some other form. SiC fibers manufactured from polycarbosilane polymer precursors generally contain excess carbon, silica, and some combined nitrogen. Jaskowiak and DiCarlo [5.146] reported the weight loss behavior of SiC fibers at temperatures ranging from 1000 to 2200°C at argon pressures of 0.1 and 138 MPa and under vacuum (10^{-9} MPa). Although the high external pressure delayed the onset of weight loss from about 1250 to 1550°C, active oxidation occurred through the formation of SiO. Wang et al. [5.147] measured the oxide layer thickness on SiC whiskers for the low temperature (600, 700, and 800°C) linear region at times less than 4 hr to be between 2 and 10 nm as determined by XPS analysis and X-ray photoelectron spectroscopy.

The wide variation of oxidation rates and activation energies reported in the literature is a result of one or a combination of many factors, including:

1. Decrease in reactive area with advancing oxidation (taken into account by some but not all)
2. Differences in materials studied (α, β, or amorphous)
3. Density and porosity variations
4. Variation and amount of preexistent surface oxide
5. Differences in oxide layer formed (crystalline, amorphous, or liquid)
6. Amounts and type of additives and impurities

Reaction in Other Atmospheres. McKee and Chatterji [5.22] reported no oxidation of SiC when exposed to gaseous environments of pure H_2, pure N_2, or H_2–10%H_2S at 900°C. No evidence of sulfide formation was found in the hydrogen–H_2S mixture. In a mixture of N_2–2%SO_2, which resulted in a partial pressure of oxygen of 10^{-10} atm, active oxidation was observed.

With the addition of 5% CH_4 to the mixture of N_2–2%SO_2, an initial (first hour) rapid weight loss was noted, presumably as a result of the formation of the volatile SiS.

Reaction of SiC in gas mixtures of 5% $H_2/H_2O/Ar$ when at 1300°C was predicted by Jacobson et al. [5.148] to fall within one of three regions: passive oxidation, active oxidation, or selective carbon removal depending upon the water content of the mixture. Gas phase diffusion (i.e., water transport to the SiC) was reported to be the rate-controlling step in the active oxidation region (oxygen partial pressures of 10^{-22} to 10^{-26} atm). In the carbon removal region (oxygen partial pressures less than 10^{-26} atm), iron impurities were found to react with the free silicon present to form iron silicides.

Maeda et al. [5.149] investigated the oxidation of several different SiC materials in flowing humid air containing 1–40 vol.% water vapor at a temperature of 1300°C for 100 hr. They found that water vapor greatly accelerated the oxidation of SiC, and that a linear relationship existed between percent water vapor and weight gain. Lu et al. [5.150] found that oxidation rates of SiC thin films were increased by 10–20 times in wet oxygen when compared to dry oxygen at temperatures of 950–1100°C. The active oxidation (i.e., weight loss) of SiC was reported to occur in 1 atm hydrogen containing water vapor at pressures of 10^{-6} to 10^{-3} MPa between 1400 and 1527°C by Readey [5.151]. At high water vapor pressures, a reaction product of SiO_2 was formed; however, active oxidation continued, because this SiO_2 was reduced to SiO by the hydrogen present. The reactions that took place can be represented by the following equations:

$$SiC_{(s)} + 3H_2O_{(g)} \rightarrow SiO_{2(s)} + CO_{(g)} + 3H_{2(g)} \tag{5.47}$$

$$SiO_{2(s)} + H_{2(g)} \rightarrow SiO_{(g)} + H_2O_{(g)} \tag{5.48}$$

In addition to water vapor, alkali vapors have been shown by Pareek and Shores [5.152] to enhance oxidation rates. They studied the oxidation of α-SiC in flowing gas mixtures of dry CO_2–O_2 (9:1 ratio) containing small quantities of K_2CO_3 and K vapors at 1300–1400°C for times of up to 42 hr. Water vapor was added in some tests; however, the vapor species in those cases was KOH. Pareek and Shores found, at low potassium levels, that the oxidation to SiO_2 followed a parabolic rate law; at higher potassium levels, the growth followed a linear law; and when low levels of water vapor were also present (i.e., KOH vapors), the growth kinetics were intermediate between parabolic and linear, indicative of a possible transition from one rate law to another. At moderate to high levels of potassium in the presence of water vapor, the kinetics of oxidation again followed a linear rate law. The increased oxidation in atmospheres containing potassium vapors was suggested to be a result of the enhanced mobility of the oxidant through the oxide layer containing dissolved potassium, although the reported activation energy of 225–463 kJ/mol was much higher than expected for oxygen diffusion through silica, which is about 115 kJ/mol. The scales were determined to be composed of cristobalite under most test conditions. At higher potassium levels and higher temperatures, the scale was sufficiently fluid to flow from the samples.

Federer [5.153] studied the effects of a vaporized solution of water containing 1 wt.% NaCl in air upon sintered α-SiC under a mechanical load at 1200°C. He reported that a molten reaction layer of sodium silicate formed causing premature failure under load within an average time of about 150 hr. The same material, when exposed to a 1200°C in air and the same loading conditions, could sustain the stress without failure for at least 1500 hr. In a similar test, Federer [5.154] exposed several types of SiC to a flowing atmosphere containing sodium sulfate and water vapor in air at 1200°C. In these tests, Federer reported that the reaction layer contained tridymite embedded in a sodium silicate liquid. Enhanced oxygen diffusion through this liquid allowed continued corrosion to take place. No discussion was given for the effects of SO_3 gas upon the corrosion as Jacobson and coworkers did [5.69,–5.72], other than to state that sodium sulfate vapor reacted with silica under low partial pressures of SO_3.

Park et al. [5.155] investigated the corrosion of sintered α-SiC in a gas mixture containing 2 vol.% chlorine and varying amounts of oxygen (0, 1, 2, and 4 vol.%) in argon at temperatures of 900, 1000, and 1100°C. They concluded that small amounts of oxygen were necessary to facilitate active corrosion by removing carbon as CO, thus allowing access to the silicon for formation of $SiCl_4$ (or $SiCl_3$). Volatile SiO may also form. At 1000°C, the rate of active corrosion increased as the amount of oxygen increased. Some amorphous SiO_2 began to form at 1000°C and 2 vol.% O_2, but it remained nonprotective even at 1100°C.

Other Carbides

Probably the next most important carbide after silicon carbide is tungsten carbide. The principal application of this material is in cemented carbide cutting tools. The carbides of titanium, tantalum, and niobium are used as alloying additions to WC. Addition of TiC to WC cutting tools causes the formation of a titanium oxide surface layer that greatly increases the tool's wear resistance. When WC oxidizes, it forms the volatile WO_3 oxide that offers no protection to wear.

Most of the oxidation studies conducted on TiC are rather dated; however, reasonable agreement exists among the various studies. Stewart and Cutler [5.156] found that the oxide layer that formed below 400°C was anatase and that above 600°C it was rutile. Single crystal studies indicated no difference in oxidation between the (100) and (110) faces at 1000°C. At low temperatures (752–800°C), the rate of oxidation exhibited a dependence upon the oxygen partial pressure to the 1/6 power, whereas at high temperatures, the dependence was to the 1/4 power. The actual mechanism of oxidation appeared to be mixed with a near-parabolic rate initially changing to a near-linear rate at longer times.

The use of ZrC at high temperatures (>450°C) has been limited due to excessive oxidation, although it possesses other excellent properties. The oxidation of powdered ZrC at low temperatures (between 380 and 550°C) was reported by Shimada and Ishii [5.157]. They reported that oxidation commenced at 300°C at all partial pressures of oxygen between 0.66 and 39.5 kPa, and that complete oxidation occurred at different temperatures depending on the oxygen pressure. Shimada and Ishii suggested that rapid initial oxidation occurred through the formation of an oxycarbide, $Zr(C_xO_{1-x})$. Following this initial oxidation, the mechanism changed around 470°C to one that formed cubic zirconia as the product of oxidation along with the generation of microcracks. Selected area electron diffraction around the edges of the ZrC grains oxidized below 470°C, exhibited the presence of cubic ZrO_2 nuclei that were not observable by XRD. At low temperatures, when the reaction was about 75% complete, carbon was found to be present as hexagonal diamond. Hexagonal diamond was also produced initially (reaction 40% complete) at higher temperatures.

Arun et al. [5.158] reported the following order of TiC > HfC > ZrC for the oxidation resistance for these three carbides at 1273 K. The oxidation of these materials is much greater when they are incorporated into hot pressed compositions of $TiC-ZrO_2$, $ZrC-ZrO_2$, and $HfC-HfO_2$. Arun et al. also reported a greater oxidation of TiC when incorporated into ZrO_2 as opposed to Al_2O_3.

Boron carbide is chemically very stable. It will dissociate in a vacuum above 2600°C into boron gas and solid carbon [5.159]. The oxidation of B_4C starts at about 600°C, forming a B_2O_3 film. Moisture in the air will lower this temperature to 250°C. Chlorine reacts with B_4C at 1000°C forming BCl_3 and graphite.

5.2.3 Borides

Several of the diborides are of considerable interest because of their high melting points and high strengths at elevated temperatures. Probably the one that has received the most attention is TiB_2; however, ZrB_2, HfB_2, NbB_2, and TaB_2 are also of interest. These are the most attractive because of their high stability compared to the other diborides. Like the carbides and nitrides, the diborides possess the undesirable characteristic of oxidation. The oxidation of the diborides generally forms B_2O_3 and a metallic oxide according to:

$$M^{4+}B_2 + 5/2O_2 \leftrightarrow M^{4+}O_2 + B_2O_3 \tag{5.49}$$

B_2O_3 readily vaporizes above 1100°C, and therefore applications at high temperatures result in porous reaction layers. At lower temperatures, where the B_2O_3 is molten ($T_m = 490°C$), a surface layer of glassy material is formed over an inner layer of metallic oxide [5.160].

When these diborides are used as particulate reinforcement for oxide matrices, various reactions may take place depending upon the oxide matrix. In a study of the oxidation of hot-pressed composites in air at 1650, 1850, and 2050°C, Vedula et al. [5.160] reported that in a zirconia matrix, the titania that formed by oxidation of the diboride went into solution into the zirconia and the B_2O_3 vaporized. In an yttria matrix tested in vacuum at 1600°C, they found significant reaction, but were unable to determine its exact nature because of the lack of published phase equilibria data. During hot pressing of a composite with an alumina matrix, the B_2O_3 that formed by oxidation of the diboride during heat-up reacted with the alumina forming a low melting liquid. Subsequent heating to 1600°C in vacuum caused reaction between the alumina and titania to form an intermediate aluminum titanate.

During sintering studies of titanium diboride, Walker and Saha [5.161] reported the following reactions:

$$TiB_2 + 2CO_2 \rightarrow TiC + B_2O_3 + CO \tag{5.50}$$

$$TiB_2 + 3CO \rightarrow TiC + B_2O_3 + 2C \tag{5.51}$$

In addition to these reactions, excess CO_2 or CO will oxidize the TiC formed to TiO_2. Davies and Phennah [5.162] have shown that TiB_2 reacts with CO_2 forming titanium borate, in addition to the TiO_2 and B_2O_3 formed.

Silicon hexaboride exhibits an oxidation resistance better than the above diborides as a result of the formation of a well-attached borosilicate film [5.159].

5.2.4 Silicides

The oxidation of $MoSi_2$ has been reported to occur by several mechanisms by Fitzer [5.163] depending upon the temperature and the oxygen partial pressure. Initially, only MoO_3 formed, but volatilized, allowing the formation of SiO_2. The partial pressure of oxygen at the interface $MoSi_2/SiO_2$ then decreased, allowing the oxidation of only silicon to continue. At very high temperatures (>1200°C) and low oxygen pressures (<10^{-6} atm), active oxidation occurred with the formation of volatile silicon monoxide, as long as the silicon content on the surface was sufficient. At low pressures of oxygen, selective oxidation of silicon occurred because of its greater affinity for oxygen than molybdenum. The selective oxidation of silicon led to the formation of a sublayer of Mo_5Si_3. At moderately high temperatures (around 1000°C) and high oxygen pressures (≈10^{-2} atm), the evaporation of the molybdenum oxides formed led to a protective SiO_2 layer. During the volatilization of the molybdenum oxides, the SiO_2 layer was very porous allowing rapid oxidation with temperature increase. At lower temperatures, where the molybdenum oxides did not volatilize but remained as solid oxide reaction products, a continuous silica layer could not form. This occurred at temperatures below 600°C, and is called *pesting*, which can lead to total destruction of the material. The actual amount of pesting that occurred, however, was dependent upon the microstructure [5.164]. Lin et al. [5.165] found that the oxidation of $MoSi_2$ in a combustion gas environment (i.e., natural gas) was very similar to that described above. Initially, SiO_2 and volatile MoO_3 formed until the silica layer was of sufficient thickness to diminish the H_2O/CO_2 activity (equivalent to a low partial pressure of oxygen) to a level where Mo_5Si_3 formed. The change in mechanisms took only minutes at temperatures of 1370 and 1600°C.

5.2.5 Superconductors

It is well known that $YBa_2Cu_3O_x$ (123) is unstable with respect to reaction with carbon dioxide or water vapor. This reaction is related to the large enthalpies of formation of barium carbonate and barium hydroxide, which are −64.4 and −35.4 kcal, respectively. Davison et al. [5.166] reported the formation of barium carbonate when 123 was held over water for 48 hr. Yan et al. [5.167] reported the formation of $Cu(OH)_2$ when 123 was exposed to 85°C and 85% relative humidity for 90 min. After exposing 123 to 80°C and 100% relative humidity for times ranging from 15 min to 24 hr, Fitch and Burdick [5.168] reported the formation of Y_2BaCuO_x, $BaCO_3$, CuO, $Cu(OH)_2$, and the possible formation of $Y(OH)_3$ and BaO. Fitch and Burdick, who noticed that corrosion was visibly present by a significant expansion of their samples, concluded that barium was leached first then reacted with atmospheric CO_2 to form $BaCO_3$ on the surface.

5.3 ATTACK BY SOLIDS

The stability of various materials to graphite is a good example of a solid–solid reaction. In this case, however, at least one of the products is a gas. The stability of a few selected refractory oxides in contact with graphite increases in the order TiO_2, Al_2O_3, ThO_2, MgO, $MgAl_2O_4$, SiO_2, and BeO, as reported by Klinger et al. [5.169].

5.3.1 Silica

Miller et al. [5.170] have shown that carbon reacts with SiO_2 to form the intermediate phase SiC, which then reacts with silica to form the gaseous phase SiO. The following equations were given to represent the reaction:

$$SiO_2 + 3C \rightarrow SiC + 2CO \tag{5.52}$$

$$2SiO_2 + SiC \rightarrow 3SiO + CO \tag{5.53}$$

They stated that these reactions were sufficiently rapid at 1000°C and, in the presence of iron, which acts as a catalyst for the reduction of silica by SiC, caused failure of silicate refractories in coal gasification atmospheres.

Probably one of more severe reactions of the past that has taken place in commercial glass furnaces is that between silica and alumina or alumina-containing refractories. When these two materials are in direct physical contact at high temperature, an interface of mullite forms. This reaction is accompanied by a substantial volume increase that tends to push the two original materials apart. Separation of silica and alumina by the more neutral material, zircon, has prevented this deleterious reaction in modern furnaces.

5.3.2 Magnesia

Magnesia vaporization is important in basic refractories where it migrates to form a region rich in magnesia by vaporization and condensation, and leaving behind a region of high porosity. The zone of high porosity causes a mechanically weak area that may crack or spall. Vaporization and condensation of magnesia can also occur in silicon nitride where it is used as a sintering aid (see discussion on page 203).

In pitch-containing high magnesia refractories, it has been found [5.171] that the carbon in the refractory can react with the magnesia to form magnesium gas according to the following equation:

$$C_{(s)} + MgO_{(s)} \rightarrow Mg_{(g)} + CO_{(g)} \tag{5.54}$$

This magnesium gas is then transported to the hot-face of the refractory where it can react with FeO in the slag forming iron metal liquid and a dense solid magnesia layer according to:

$$Mg_{(g)} + FeO_{(l)} \rightarrow Fe_{(l)} + MgO_{(s)} \tag{5.55}$$

5.3.3 Superconductors

The $YBa_2Cu_3O_x$ (123) superconductors, where $x = 6.5–7.0$, have been reported to exhibit reaction and/or decomposition when in contact with various materials. This has presented researchers with the problem of sample holders for the production of 123 materials. Williams and Chandhury [5.172] have conducted a thermodynamic study of the various materials that might react with 123. Based upon the heat of formation of CuO of −18.6 kcal/g atom and the following equations:

$$M + 2CuO \rightarrow MO_2 + 2Cu \tag{5.56}$$

or

$$M + 3CuO \rightarrow M_2O_3 + 3Cu \tag{5.57}$$

they reported that the nine elements Ru, Rh, Pd, Ag, Os, Ir, Pt, Au, and Hg should not react with CuO, and most likely would not react with 123. Murphy et al. [5.47] reported that 123 was nonreactive toward silver and to a lesser extent, gold.

One of the potential applications of superconductors is that of thin films on a semiconductor substrate; however, the most widely used semiconductor substrate material, silicon, reacts with 123. An examination of the various phase equilibria indicated that $BaSi_2O_5$ does not react with 123, because these two materials form a stable tie-line in the BaO–Y_2O_3–CuO–SiO_2 quaternary system. Thus this barium silicate could be used as a buffer layer for production purposes or during manufacture of thin films on semiconducting substrates.

Mikalsen et al. [5.173] reported that no reaction occurred between thin film superconductors in the Bi–Sr–Ca–Cu–O system and MgO substrates even after annealing at 850°C for 30 min. Thin films on Al_2O_3, however, reacted and became insulating and transparent. Abe et al. [5.174] and Ibara et al. [5.175] have reported that melts of $BiSrCaCu_2O_x$ or $BiPb_ySr CaCu_2O_x$ reacted with alumina crucibles contaminating their samples.

5.3.4 Attack by Metals

Because of the prevalence of platinum metal in various research and manufacturing operations, the reactions of various refractory oxides with platinum is of considerable importance. Ott and Raub [5.176] reported that platinum acts as a catalyst for the reduction of refractory oxides by hydrogen, carbon, CO, and organic vapors. These reactions can occur as low as 600°C and result because of the affinity of platinum for the metal of the oxide by forming intermetallic compounds and crystalline solutions.

Gotman and Gutmanas [5.177] reported that titanium powder reacted with silicon nitride forming a complex interfacial layer structure consistent with the Ti–N–Si and Ti–Si phase diagrams. A 1-hr heat treatment in an evacuated stainless steel bag at 1273 K yielded two morphologically different main layers. A fine-grained layer (1.5 μm thick) next to the silicon nitride was composed of mostly Ti_5Si_3. A very thin layer adjacent to the silicon nitride contained TiN with some dissolved silicon. As the silicon increased away from the silicon nitride, the phases changed into TiN plus Ti_5Si_3. The next layer was almost pure Ti_5Si_3 with some additional titanium adjacent to the next main layer. A coarse-grained layer came next, containing a mixture of what was thought to be Ti_3Si particles imbedded within Ti metal with a little dissolved silicon. Similar results were obtained when different times (20, 40, and 60 min) and temperatures (1173, 1223, and 1323 K) were used.

5.4 ADDITIONAL RELATED READING

Amoroso, G.G.; Fassina, V. *Stone Decay and Conservation*; Materials Science Monographs 11; Elsevier: Amsterdam, 1983; 453 pp.

Guthrie, G.D., Jr., Mossman, B.T., Eds.; *Health Effects of Mineral Dusts*, Reviews in Mineralogy; Min. Soc. Am.: Washington, DC, 1993; Vol. 28, 584 pp.

Lea, F.M. *The Chemistry of Cement and Concrete*; Edward Arnold Publishers: London, 1970.

Taylor, H.F.W. Mineralogy, microstructure, and mechanical properties of cements. Proc. Br. Ceram. Soc. 1979, *29*, 147–163.

5.5 EXERCISES, QUESTIONS, AND PROBLEMS

1. Determine whether or not molten aluminum will react with cordierite by calculating the free energy of reaction. If so what are the reaction products?
2. List the following oxides in the order of most thermodynamically stable to least stable: Al_2O_3, CaO, Fe_2O_3, MgO, and ZrO_2.
3. Discuss several cases where corrosion is beneficial.
4. Using the Al_2O_3–MgO–SiO_2 phase diagram, determine the interface, if any, that should form on an alumina crucible containing a silica melt at 1000, 1200, and 1400°C. What changes would one expect if a mullite or spinel crucible were substituted for the alumina? Which of three materials is the best at each temperature?
5. Discuss how forced convection (erosion) affects the reaction interface layer thickness.
6. Calculate the critical slag layer thickness for the passive-to-active oxidation of SiC in air at 1400°C using the following equation:

$$pO_2 = RTC^*/(1 + 2X_1/A)$$

Assume $C^* = 0.080\ mol/m^3$ and $A = 0.31\ \mu m$.
7. Does the tetrahedra packing density of the different polymorphs of silica affect their dissolution? If so, how?
8. Explain why the reported oxidation of silicon nitride and/or carbide by various investigators varies.
9. Describe the difference between active and passive oxidation of SiC.

REFERENCES

5.1. Sandhage, K.H.; Yurek, G.J. Indirect dissolution of $(Al,Cr)_2O_3$ in CaO–MgO–Al_2O_3–SiO_2 (CMAS) melts. J. Am. Ceram. Soc. 1991, *74* (8),1941–1954.

5.2. Oishi, Y.; Cooper, A.R., Jr.; Kingery, W.D. Dissolution in ceramic systems: III. Boundary layer concentration gradients. J. Am. Ceram. Soc. 1965, *48* (2), 88–95.

5.3. Bonar, J.A.; Kennedy, C.R.; Swaroop, R.B. Coal-ash slag attack and corrosion of refractories. Ceram. Bull. 1980, *59* (4), 473–478.

5.4. Sandhage, K.H.; Yurek, G.J. Indirect dissolution of sapphire into calcia–magnesia–alumina–silica melts: electron microprobe analysis of the dissolution process. J. Am. Ceram. Soc. 1990, *73* (12), 3643–3649.

5.5. Sandhage, K.H.; Yurek, G.J. Indirect dissolution of sapphire into silicate melts. J. Am. Ceram. Soc. 1988, *71* (6), 478–489.

5.6. McCauley, R.A. *unpublished data*, 1975.

5.7. Hilger, J.P.; Babel, D.; Prioul, N.; Fissolo, A. Corrosion of AZS and fireclay refractories in contact with lead glass. J. Am. Ceram. Soc. 1984, *64* (4), 213–220.

5.8. Matsushima, M.; Yadoomaru, S.; Mori, K.; Kawai, Y. A fundamental study on the dissolution rate of solid lime into liquid slag. Trans. Iron Steel Inst. Jpn. 1977, *17*, 442–449.

5.9. Bates, J.L. Heterogeneous dissolution of refractory oxides in molten calcium–aluminum silicate. J. Am. Ceram. Soc. 1987, *70* (3), C55–C57.

5.10. Umakoshi, M.; Mori, K.; Kawai, Y. Corrosion kinetics of refractory materials in molten CaO–FeO–SiO_2 slags. Kyushu Daigaku Kogaku Ih. 1980, *53*, 191–197.

5.11. Bonetti, G.; Toninato, T.; Bianchini, A.; Martini, P.L. Resistance of refractories to corrosion by lead-containing glasses. Proc. Br. Ceram. Soc. 1969, *14*, 29–40.

5.12. Clauss, H.; Salge, H. Electron micro-probe analysis of the dissolution behavior of fusion cast tank blocks. Glastech. Ber. 1974, *47* (7–8), 159–181.

5.13. Derobert, M. Microscope and x-ray diffraction identification of crystalline phases in refractories and their corrosion products in glass tanks. Bull. Soc. Fr. Ceram. 1975, *109*, 31–36.

5.14. Lakatos, T.; Simmingskold, B. The influence of constituents on the corrosion of pot clays by molten glass. Glass Technol. 1967, *8* (2), 43–47.

5.15. Lakatos, T.; Simmingskold, B. Influence of viscosity and chemical composition of glass on its corrosion of sintered alumina and silica-glass. Glastek. Tidskr. 1971, *26* (4), 58–68.

5.16. Orlova, R.G.; Moroz, Kh.I.; Naidenova, G.A. Kinetics of mullite dissolution in alkali and alkaline-earth melts (obtained during the firing of mullite porcelain). Neorg. Stekla Pokryt. Mater. 1979, (No. 4), 103–109 (Russ).

5.17. Thomas, E.A.; Brock, W.W. A post-mortem examination of zircon and bonded alumina–zirconia–silica paving. Proc. 10th International Congress on Glass, Ceram. Soc. Japan, No.2 Refractories and Furnaces, 2–9 to 2–19, 9 July 1974.

5.18. Chung, Y-D; Schlesinger, M.E. Interaction of $CaO-FeO-SiO_2$ slags with partially stabilized zirconia. J. Am. Ceram. Soc. 1994, *77* (3), 611–616.

5.19. Muan, A. Thermodynamics aspects of the application of ceramics/refractories in advanced energy technologies. In *Ceramics in Advanced Energy Technologies;* Krockel, H., Merz, M., Van der Biest, O., Eds.; D. Reidel Pub. Co.: Dordrecht 1984; 384–366.

5.20. Tsai, R.L.; Raj, R. Dissolution kinetics of β-Si_3N_4 in an Mg–Si–O–N glass. J. Am. Ceram. Soc. 1982, *65* (5), 270–274.

5.21. Ferber, M.K.; Ogle, J.; Tennery, V.J.; Henson, T. Characterization of corrosion mechanisms occurring in a sintered SiC exposed to basic coal slags. J. Am. Ceram. Soc. 1985, *68* (4), 191–197.

5.22. McKee, D.W.; Chatterji, D. Corrosion of silicon carbide in gases and alkaline melts. J. Am. Ceram. Soc. 1976, *59* (9–10), 441–444.

5.23. Deal, B.E.; Grove, A.S. General relationship for the thermal oxidation of silicon. J. Appl. Phys. 1965, *36* (12), 3770–3778.

5.24. Lea, F.M. *The Chemistry of Cement and Concrete*; Edward Arnold Publishers: London, 1970.

5.25. Taylor, H.F.W. Mineralogy, microstructure, and mechanical properties of cements. Proc. Br. Ceram. Soc. 1979, *29*, 147–163.

5.26. Jennings, H.M. Aqueous solubility relationships for two types of calcium silicate hydrate. J. Am. Ceram. Soc. 1986, *69* (8), 614–618.

5.27. McConnell, J.D.C. The hydration of larnite (β-Ca_2SiO_4) and bredigite ($\beta' Ca_2SiO_4$) and the properties of the resulting gelatinous mineral plombierite. Mineral. Mag. 1955, *30*, 672–680.

5.28a. *Scientific Basis for Nuclear Waste Management IV*, Materials Research Society Symposia Proceedings; Topp, S.V., Ed.; North-Holland: NY, 1982; Vol. 6.

5.28b. *Scientific Basis for Nuclear Waste Management V*, Materials Research Society Symposia Proceedings; Lutze, W., Ed.; MRS: Pittsburgh, PA, 1982; Vol. 11.

5.28c. *Scientific Basis for Nuclear Waste Management VI*, Materials Research Society Symposia Proceedings; Brookins, D.G., Ed.; MRS: Pittsburgh, PA, 1983; Vol. 15.

5.28d. *Scientific Basis for Nuclear Waste Management VII*, Materials Research Society Symposia Proceedings, McVay, G.L., Ed.; MRS: Pittsburgh, PA, 1984; Vol. 26.

5.28e. *Scientific Basis for Nuclear Waste Management VIII*, Materials Research Society Symposia Proceedings; Jantzen, C.M., Stone, J.A., Ewing, R.C., Eds.; MRS: Pittsburgh, PA, 1985; Vol. 44.

5.28f. *Scientific Basis for Nuclear Waste Management IX*, Materials Research Society Symposia Proceedings; Werme, L.O., Ed.; MRS: Pittsburgh, PA, 1986; Vol. 50.

5.28g. *Scientific Basis for Nuclear Waste Management X*, Materials Research Society Symposia Proceedings; Bates, J.K., W.B., Seefeldt, Eds.; MRS: Pittsburgh, PA, 1987; Vol. 84.

5.28h. *Scientific Basis for Nuclear Waste Management XI*, Materials Research Society Symposia Proceedings; Apted, M.J., Westerman, R.E., Eds.; MRS: Pittsburgh, PA, 1988; Vol. 112.

5.28i. *Scientific Basis for Nuclear Waste Management XII*, Materials Research Society Symposia Proceedings; Lutze, W., Ewing, R.C., Eds.; MRS: Pittsburgh, PA, 1989; Vol. 127 .

5.28j. *Scientific Basis for Nuclear Waste Management XIII*, Materials

Research Society Symposia Proceedings, Oversby, V.M., Brown, P.W., Eds.; MRS: Pittsburgh, PA, 1989; Vol. 176.

5.28k. *Scientific Basis for Nuclear Waste Management XIV*; Materials Research Society Symposia Proceedings; Abrajans, T., Jr., Johnson, L.H., Eds.; MRS: Pittsburgh, PA, 1991; Vol. 212.

5.29. *Advances in Ceramics Vol 8: Nuclear Waste Management*; Wicks, G.G., Ross, W.A., Eds.; Am. Ceram. Soc.: Columbus, OH, 1984; 746 pp.

5.30. *Advances in Ceramics Vol 20: Nuclear Waste Management II*. Clark, D.E., White, W.B., Machiels, A.J., Eds.; Am. Ceram. Soc.: Westerville, OH, 1986 (773 pp.).

5.31. *Ceramic Transactions Vol 9: Nuclear Waste Management III*; Mellinger, G.B., Ed.; Am. Ceram. Soc.: Westerville, OH, 1990; 595 pp.

5.32. *Ceramic Transactions Vol 23: Nuclear Waste Management IV*; Wicks, G.G., Bickford, D.F., Bunnell, L.R., Eds.; Am. Ceram. Soc.: Westerville, OH, 1991; 799 pp.

5.33. Sato, T.; Sato, S.; Okuwaki, A. Corrosion behavior of alumina ceramics in caustic alkaline solutions at high temperatures. J. Am. Ceram. Soc. 1991, *74* (12), 3081–3084.

5.34. Brady, P.V.; House, W.A. Surface-controlled dissolution and growth of minerals. In *Physics and Chemistry of Mineral Surfaces*; Brady, P.V., Ed.; CRC Press: New York, 1996; 225–305. Chap. 4.

5.35. Wilding, L.P.; Smeck, N.E.; Drees, L.R. Silica in soils: quartz, cristobalite, tridymite, and opal. In *Minerals in Soil Environments*; Dinauer, R.C., Ed.; Soil Sci. Soc. Am.: Madison, WI, 1977; 471–552. Chap. 14.

5.36. Schnitzer, M.; Kodama, H. Reactions of minerals with soil humic substances. In *Minerals in Soil Environments*; Dinauer, R.C., Ed.; Soil Sci. Soc. Am.: Madison, WI, 1977; 741–770. Chap. 21.

5.37. Graustein, W.C.; Cromack, K., Jr.; Sollins, P. Calcium oxalate: occurrence in soils and effect on nutrient and geochemical cycles. Science, New Series 1977, *198* (4323), 1252–1254.

5.38. Liang, D-T; Readey, D.W. Dissolution kinetics of crystalline and amorphous silica in hydrofluoric–hydrochloric acid mixtures. J. Am. Ceram. Soc. 1987, *70* (8), 570–577.

5.39. Guthrie, G.D., Jr. Biological effects of inhaled minerals. Am. Mineral. 1992, *77* (3/4), 225–243.

5.40. Hume, L.A.; Rimstidt, J.D. The biodurability of chrysotile asbestos. Am. Mineral. 1992, *77* (9/10), 1125–1128.

5.41. Webster, R.P.; Kukacka, L.E. Effects of acid deposition on Portland cement concrete. In *Materials Degradation Caused by Acid Rain*; Baboian, R., Ed.; ACS Symposium Series 318, Am. Chem. Soc.: Washington, DC, 1986; 239–249.

5.42. Amoroso, G.G.; Fassina, V. *Stone decay and conservation*. Materials Science Monographs 11; Elsevier: Amsterdam, 1983; 453 pp.

5.43. Kitano, Y. Polymorphic formation of calcium carbonite in thermal springs with an emphasis on the effect of temperature. Bull. Chem. Soc. Jpn. 1962, *35*,1980.

5.44. Kitano, Y. Behavior of various inorganic ions in the separation of calcium carbonate from a bicarbonate solution. Bull. Chem. Soc. Jpn. 1962, *35*, 1973.

5.45. Charola, A.E.; Lazzarini, L. Deterioration of brick masonry caused by acid rain. In *Materials Degradation Caused by Acid Rain*; Baboian, R., Ed.; ACS Symposium Series 318, Am. Chem. Soc.: Washington, DC, 1986; 250–258.

5.46. Yoshimura, M.; Hiuga, T.; Somiya, S. Dissolution and reaction of yttria-stabilized zirconia single crystals in hydrothermal solutions. J. Am. Ceram. Soc. 1986, *69* (7), 583–584.

5.47. Murphy, D.W.; Johnson, D.W., Jr.; Jin, S.; Howard, R.E. Processing techniques for the 93°K Superconductor $Ba_2YCu_3O_7$. Science August 19, 1988, *241*, 922–930.

5.48. Myhra, S.; Savage, D.; Atkinson, A.; Riviere, J.C. Surface modification of some titanate minerals subjected to hydrothermal chemical attack. Am. Mineral. 1984, *69* (9/10), 902–909.

5.49. Kastrissios, T.; Stephenson, M.; Turner, P.S. Hydrothermal dissolution of perovskite: implications for synroc formulation. J. Am. Ceram. Soc. 1987, *70* (7), C144–C146.

5.50. Buykx, W.J.; Hawkins, K.; Levins, D.M.; Mitamura, H.; Smart, R.St.C.; Stevens, G.T.; Watson, K.G.; Weedon, D.; White, T.J. Titanate ceramics for the immobilization of sodium-bearing high-level nuclear waste. J. Am. Ceram. Soc. 1988, *71* (8), 678–688.

5.51. Bright, E.; Readey, D.W. Dissolution kinetics of TiO_2 in HF–HCl solutions. J. Am. Ceram. Soc. 1987, *70* (12), 900–906.

5.52. Harris, L.A.; Cross, D.R.; Gerstner, M.E. Corrosion suppression on rutile anodes by high energy redox reactions. J. Electrochem. Soc. 1977, *124* (6), 839–844.

5.53. Horkans, J.; Shafer, M.W. Effect of orientation, composition, and electronic factors in the reduction of O_2 on single crystal electrodes of conducting oxides of molybdenum and tungsten. J. Electrochem. Soc. 1977, *124* (8), 1196–1202.

5.54. Horkans, J.; Shafer, M.W. An investigation of the electrochemistry of a series of metal dioxides with rutile-type structure: MoO_2, WO_2, ReO_2, RuO_2, OsO_2, and IrO_2. J. Electrochem. Soc. 1977, *124* (8), 1202–1207.

5.55. Clayton, J.C. In-pile and out-of-pile corrosion behavior of thoria- and urania-based nuclear fuels. Ceram. Eng. Sci. Proc. 1988, *9* (9–10).

5.56. Bowen, P.; Highfield, J.G.; Mocellin, A.; Ring, T.A. Degradation of aluminum nitride powder in an aqueous environment. J. Am. Ceram. Soc. 1990, *73* (3), 724–728.

5.57. Hirayama, H.; Kawakubo, T.; Goto, A.; Kaneko, T. Corrosion behavior of silicon carbide in 290°C water. J. Am. Ceram. Soc. 1989, *72* (11), 2049–2053.

5.58. Sato, T.; Tokunaga, Y.; Endo, T.; Shimada, M.; Komeya, K.; Komatsu, M.; Kameda, T. Corrosion of silicon nitride ceramics in aqueous hydrogen chloride solutions. J. Am. Ceram. Soc. 1988, *71* (12), 1074–1079.

5.59. Sato, T.; Tokunaga, Y.; Endo, T.; Shimada, M.; Komeya, K.; Komatsu, M.; Kameda, T. Corrosion of silicon nitride ceramics in aqueous HF solutions. J. Mater. Sci 1988, *23* (10), 3440–3446.

5.60. Seshadri, S.G.; Srinivasan, M. Liquid corrosion and high-temperature oxidation effects on silicon carbide/titanium diboride composites. J. Am. Ceram. Soc. 1988, *71* (2), C72–C74.

5.61. Grjotheim, K.; Holm, J.L.; Krohn, C.; Thonstad, J. Recent progress in the theory of aluminium electrolysis. In *Selected Topics in High Temperature Chemistry*; Forland, T., Grjotheim, K., Motzfeldt, K., Urnes, S., Eds.; Universitetsforlaget: Oslo, 1966; 151–178.

5.62. Siljan, O.-J.; Seltveit, A. Chemical reactions in refractory linings of alumina reduction cells. In *UNITECR '91 CONGRESS*, 2d Edition; German Refractories Association, Bonn Verlag Stahleisen mbH: Düsseldorf, 1991; 59–65.

5.63. Allaire, C. Refractory lining for alumina electrolytic cells. J. Am. Ceram. Soc. 1992, *75* (8), 2308–2311.

5.64. Lawson, M.G.; Kim, H.R.; Pettit, F.S.; Blachere, J.R. Hot corrosion of silica. J. Am. Ceram. Soc. 1990, *73* (4), 989–995.

5.65. Baumgartner, C.E. Metal oxide solubility in eutectic Li/K carbonate melts. J. Am. Ceram. Soc. 1984, *67* (7), 460–462.

5.66. Huseby, I.C.; Klug, F.J. Chemical compatibility of ceramics for directionally solidifying Ni-base eutectic alloys. Ceram. Bull. 1979, *58* (5), 527–535.

5.67. Borom, M.P.; Arendt, R.H.; Cook, N.C. Dissolution of oxides of Y, Al, Mg, and La by molten fluorides. Ceram. Bull. 1981, *60* (11), 1168–1174.

5.68. Ballman, A.A.; Laudise, R.A. Crystallization and solubility of zircon and phenacite in certain molten salts. J. Am. Ceram. Soc. 1965, *48* (3), 130–133.

5.69. Jacobson, N.S.; Smialek, J.L. Hot corrosion of sintered α-SiC at 1000°C. J. Am. Ceram. Soc. 1985, *68* (8), 432–439.

5.70. Smialek, J.L.; Jacobson, N.S. Mechanism of strength degradation for hot corrosion of α-SiC. J. Am. Ceram. Soc. 1986, *69* (10), 741–752.

5.71. Jacobson, N.S. Kinetics and mechanism of corrosion of SiC by molten salts. J. Am. Ceram. Soc. 1986, *69* (1), 74–82.

5.72. Fox, D.S.; Jacobson, N.S. Molten-salt corrosion of silicon nitride: I, sodium carbonate. J. Am. Ceram. Soc. 1988, *71* (2), 128–138.

5.73. Jacobson, N.S.; Fox, D.S. Molten-salt corrosion of silicon nitride: II. Sodium sulfate. J. Am. Ceram. Soc. 1988, *71* (2), 139–148.

5.74. Jacobson, N.S.; Stearns, C.A.; Smialek, J.L. Burner rig corrosion of SiC at 1000°C. Adv. Ceram. Mater. 1986, *1* (2), 154–161.

5.75. Sato, T.; Kubato, K.; Shimada, M. Corrosion kinetics and strength degradation of sintered α-silicon carbide in potassium sulfate melts. J. Am. Ceram. Soc. 1991, *74* (9), 2152–2155.

5.76. Cree, J.W.; Amateau, M.F. Mechanical behavior of SiC exposed to molten lithium and lithium salts. Ceram. Eng. Sci. Proc. 1987, *8* (7–8), 812–814.

5.77. Tressler, R.E.; Meiser, M.D.; Yonushonis, T. Molten salt corrosion of SiC and Si_3N_4 ceramics. J. Am. Ceram. Soc. 1976, *59* (5–6), 278–279.

5.78. Raeder, C.H.; Knorr, D.B. Stability of $YBa_2Cu_3O_{7-x}$ in molten chloride salts. J. Am. Ceram. Soc. 1990, *73* (8), 2407–2411.

5.79. Lee, B.J.; Lee, D.N. Calculation of phase diagrams for the $YO_{1.5}$–BaO–CuO_x system. J. Am. Ceram. Soc. 1989, *72* (2), 314–319.

5.80. Kim, S.M.; Lu, W.-K.; Nicholson, P.S.; Hamielec, A.E. Corrosion of aluminosilicate refractories in iron–manganese alloys. Ceram. Bull. 1974, *53* (7), 543–547.

5.81. Brondyke, K.J. Effect of molten aluminum on alumina–silica refractories. J. Am. Ceram. Soc. 1953, *36* (5), 171–174.

5.82. Siljan, O.-J.; Rian, G.; Pettersen, D.T.; Solheim, A.; Schøning, C. Refractories for molten aluminum contact: part I. Thermodynamics and kinetics. Refract. Appl. News 2002, *7* (6), 17–25.

5.83. Allaire, C.; Desclaux, P. Effect of alkalies and of a reducing atmosphere on the corrosion of refractories by molten aluminum. J. Am. Ceram. Soc. 1991, *74* (11), 2781–2785.

5.84. Lindsay, J.G.; Bakker, W.T.; Dewing, E.W. Chemical resistance of refractories to Al and Al–Mg alloys. J. Am. Ceram. Soc. 1964, *47* (2), 90–94.

5.85. Cornie, J.A.; Chiang, Y.-M.; Uhlmann, D.R.; Mortensen, A.; Collins, J.M. Processing of metal and ceramic matrix composites. Ceram. Bull. 1986, *65* (2), 293–304.

5.86. Wills, R.R.; Sekercioglu, I.; Niesz, D.E. The interaction of molten silicon with silicon aluminum oxynitrides. J. Am. Ceram. Soc. 1980, *63* (7–8), 401–403.

5.87. Nelson, H.G. A challenge to materials: advanced hypersonic flight hydrogen and high temperature materials. In *Proc. 1993 Conf. on Processing, Fabrication and Applications of Advanced Composites, Long Beach, CA*; Upadhya, K., Ed.; ASM: Ohio, Aug 9–11, 1993; 11–20.

5.88. de Jong, R.; McCauley, R.A.; Fordham, R.J.; Riley, F.L. High temperature corrosion of some silicon nitrides. *Proceedings of the European Materials Research Society Conference, Strasbourg, France*, Nov. 26–29, 1985.

5.89. Anderson, N.C. Basal plane cleavage cracking of synthetic sapphire arc lamp envelopes. J. Am. Ceram. Soc. 1979, *62* (1–2), 108–109.

5.90. van Hoek, J.A.M.; van Loo, F.J.J.; Metselaar, R. Corrosion of alumina by potassium vapor. J. Am. Ceram. Soc. 1992, *75* (1), 109–111.

5.91. Mayberry, M.L.; Boyer, W.H.; Martinek, C.A.; Neely, J.E. Effect of alternating oxidizing—reducing atmospheres on basic refractories. Presented at the Pacific Coast Regional Meeting of the Am. Ceram. Soc. Oct. 1970.

5.92. Wang, S.C.P.; Anghaie, S.; Collins, C. Reaction of uranium hexafluoride gas with alumina and zirconia at elevated temperatures. J. Am. Ceram. Soc. 1991, *74* (9), 2250–2257.

5.93. Muan, A. Reactions between iron oxides and alumina–silica refractories. J. Am. Ceram. Soc. 1992, *75* (6), 1319–1330.

5.94. McCauley, R.A. The effects of vanadium upon basic refractories. In *UNITECR '89 Proceedings*; Trostel, L.J., Jr., Ed.; Am. Ceram. Soc.: Westerville, OH, 1989; 858–863.

5.95. Parker, F.J.; McCauley, R.A. Investigation of the system CaO–MgO–V_2O_5: I. Phase equilibria. J. Am. Ceram. Soc. 1982, *65* (7), 349–351.

5.96. Parker, F.J.; McCauley, R.A. Investigation of the system CaO–MgO–V_2O_5: II. Crystalline solutions and crystal chemistry. J. Am. Ceram. Soc. 1982, *65* (9), 454–456.

5.97. McGarry, M.J.; McCauley, R.A. Subsolidus phase equilibria of the MgO–V_2O_5–SiO_2 system. J. Am. Ceram. Soc. 1992, *75* (10), 2874–2876.

5.98. Ready, D.W. High temperature gas corrosion of ceramic composites. Ceram. Eng. Sci. Proc. 1992, *13* (7–8), 301–318.

5.99. Lepistö, T.T.; Lintula, P.V.; Mäntylä, T.A. TZP-ceramics in humid conditions at 150°C. Ceram. Eng. Sci. Proc. 1988, *9* (9–10), 1517–1523.

5.100. Lepistö, T.T.; Mäntylä, T.A. A model for structural degradation of Y-TZP ceramics in humid atmosphere. Ceram. Eng. Sci. Proc. 1989, *10* (7–8), 658–667.

5.101. Singhal, S.C. Oxidation of silicon nitride and related materials. In *Nitrogen Ceramics*; Riley, R.L., Ed.; NATO Adv. Study Inst. Ser.:E, Appl. Sci., No. 23, Noordhoff: Leyden, 1977; 607–626.

5.102. Vaughn, W.L.; Maahs, H.G. Active-to-passive transition in the oxidation of silicon carbide and silicon nitride in air. J. Am. Ceram. Soc. 1990, *73* (6), 1540–1543.

5.103. Choi, D.J.; Fischbach, D.B.; Scott, W.D. Oxidation of chemically-vapor-deposited silicon nitride and single-crystal silicon. J. Am. Ceram. Soc. 1989, *72* (7), 1118–1123.

5.104. Du, H.; Tressler, R.E.; Spear, K.E. Thermodynamics of the Si–N–O system and kinetic modeling of oxidation of Si_3N_4. J. Electrochem. Soc. 1989, *136* (11), 3210–3215.

5.105. Luthra, K.L. Some new perspectives on oxidation of silicon carbide and silicon nitride. J. Am. Ceram. Soc. 1991, *74* (5), 1095–1103.

5.106. Luthra, K.L. A mixed-interface reaction/diffusion-controlled model for oxidation of Si_3N_4. J. Electrochem. Soc. 1991, *138* (10), 3001–3007.

5.107. Ogbuji, L.U.J.T. Role of Si_2N_2O in the passive oxidation of chemically-vapor-deposited Si_3N_4. J. Am. Ceram. Soc. 1992, *75* (11), 2995–3000.

5.108. de Jong, R. Incorporation of additives into silicon nitride by colloidal processing of metal organics in an aqueous medium. Univ. Microfilms Int. (Ann Arbor, MI), Order No. DA9123266. Diss. Abstr. Int. 1991, *B52* (3), 1660.

5.109. Joshi, P.N. Metal–organic surfactants as sintering aids for silicon nitride in an aqueous medium. MS thesis, Rutgers University, 1992.

5.110. Bergstrom, L.; Pugh, R.J. Interfacial characterization of silicon nitride powders. J. Am. Ceram. Soc. 1989, *72* (1), 103–109.

5.111. Tripp, W.C.; Graham, H.C. Oxidation of Si_3N_4 in the range 1300 to 1500°C. J. Am. Ceram. Soc. 1976, *59* (9–10), 399–403.

5.112. Kim, H.-E.; Moorehead, A.J. High-temperature gaseous corrosion of Si_3N_4 in H_2–H_2O and Ar–O_2 environments. J. Am. Ceram. Soc. 1990, *73* (10), 3007–3014.

5.113. Singhal, S.C. Thermodynamic analysis of the high-temperature stability of silicon nitride and silicon carbide. Ceramurg. Int. 1976, *2* (3), 123–130.

5.114. Cubicciotti, D.; Lau, K.H. Kinetics of oxidation of hot-pressed silicon nitride containing magnesia. J. Am. Ceram. Soc. 1978, *61* (11–12), 512–517.

5.115. Kiehle, A.J.; Heung, L.K.; Gielisse, P.J.; Rockett, T.J. Oxidation behavior of hot-pressed Si_3N_4. J. Am. Ceram. Soc. 1975, *58* (1–2), 17–20.

5.116. Plucknett, K.P.; Lewis, M.H. Microstructure and oxidation behavior of HIPed silicon nitride. Ceram. Eng. Sci. Proc. 1992, *13* (9–10), 991–999.

5.117. Gregory, O.J.; Richman, M.H. Thermal oxidation of sputter-coated reaction-bonded silicon nitride. J. Am. Ceram. Soc. 1984, *67* (5), 335–340.

5.118. Larsen, D.C., Adams, J.W., Johnson, L.R., Teotia, A.P.S., Hill, L.G., Eds.; *Ceramic Materials Advanced Heat Engines*; Noyes Publications: Park Ridge, NJ, 1985; 221.

5.119. Mieskowski, D.M.; Sanders, W.A. Oxidation of silicon nitride sintered with rare-earth oxide additions. J. Am. Ceram. Soc. 1985, *68* (7), C160–C163.

5.120. Horton, R.M. Oxidation kinetics of powdered silicon nitride. J. Am. Ceram. Soc. 1969, *52* (3), 121–124.

5.121. Wang, P.S.; Hsu, S.M.; Malghan, S.G.; Wittberg, T.N. Surface oxidation kinetics of Si_3N_4–4%Y_2O_3 powders studied by Bremsstrahlung-excited auger spectroscopy. J. Mater. Sci. 1991, *26*, 3249–3252.

5.122. Franz, I.; Langheinrich, W. Formation of silicon dioxide from silicon nitride. In *Reactivity of Solids: Proc. 7th International Symp. on Reactivity of Solids*; Anderson, J.S., Roberts, M.W. Stone, F.S., Eds.; Hall and Chapman: London, 1972; 303–314.

5.123. de Jong, R.; McCauley, R.A.; Tambuyser, P. Growth of twinned β-silicon carbide whiskers by the vapor–liquid–solid process. J. Am. Ceram. Soc. 1987, *70* (11), C338–C341.

5.124. Oliveira, F.C.; Edwards, R.A.H.; Fordham, R.J.; Riley, F.L. High temperature corrosion of Si_3N_4–Y_2O_3–Al_2O_3 ceramics in $H_2S/H_2O/H_2$ reducing environments. In *High Temperature Corrosion of Technical Ceramics*; Fordham, R.J., Ed.; Elsevier Applied Science: London, 1990; 53–68.

5.125. Singhal, S.C.; Lange, F.F. Oxidation behavior of sialons. J. Am. Ceram. Soc. 1977, *60* (3–4), 190–191.

5.126. Chartier, T.; Besson, J.L; Goursat, P. Microstructure, oxidation and creep behavior of a β'-sialon ceramic. Int. J. High Technol. Ceram. 1986, *2* (1), 33–45.

5.127. Wang, X.H.; Lejus, A.-M.; Vivien, D. Oxidation behavior of lanthanide aluminum oxynitrides with magnetoplumbite-like structure. J. Am. Ceram. Soc. 1990, *73* (3), 770–774.

5.128. Suryanạrayana, D. Oxidation kinetics of aluminum nitride. J. Am. Ceram. Soc. 1990, *73* (4), 1108–1110.

5.129. Abid, A.; Bensalem, R.; Sealy, B.J. The thermal stability of AlN. J. Mater. Sci. 1986, *21*, 1301–1304.

5.130. Dutta, I.; Mitra, S.; Rabenberg, L. Oxidation of sintered aluminum nitride at near-ambient temperatures. J. Am. Ceram. Soc. 1992, *75* (11), 3149–3153.

5.131. Katnani, A.D.; Papathomas, K.I. Kinetics and initial stages of oxidation of aluminum nitride: thermogravimetric analysis and x-ray photoelectron spectroscopy study. J. Vac. Sci. Technol. 1987, *A5*, 1335.

5.132. McCauley, J.W.; Corbin, N.D. Phase relations and reaction sintering of transparent cubic aluminum oxynitride spinel (ALON). J. Am. Ceram. Soc. 1979, *62* (9–10), 476–479.

5.133. Montintin, J.; Desmaison-Brut, M. Oxidation behavior of hot-isostatic-pressed tantalum nitride. In *High Temperature Corrosion of Technical Ceramics;* Fordham, R.J., Ed.; Elsevier Applied Science: London, 1990; 121–130.

5.134. Ervin, G. Oxidation behavior of silicon carbide. J. Am. Ceram. Soc. 1958, *41* (9), 347–352.

5.135. Jorgensen, P.J.; Wadsworth, M.E.; Cutler, I.B. Effects of water vapor on oxidation of silicon carbide. J. Am. Ceram. Soc. 1961, *44* (6), 258–260.

5.136. Harris, R.C. Oxidation of 6H-α silicon carbide platelets. J. Am. Ceram. Soc. 1975, *58* (1–2), 7–9.

5.137. Jorgensen, P.J.; Wadsworth, M.E.; Cutler, I.B. Oxidation of silicon carbide. J. Am. Ceram. Soc. 1959, *42* (12), 613–616.

5.138. Costello, J.A.; Tressler, R.E. Oxidation kinetics of hot-pressed and sintered α-SiC. J. Am. Ceram. Soc. 1981, *64* (6), 327–331.

5.139. Singhal, S.C. Oxidation of silicon-based structural ceramics. In *Properties of High Temperature Alloys with Emphasis on Environmental Effects*; Foroulis, A.Z., Pettit, F.S., Eds.; Electrochem. Soc. Inc.: Princeton, NJ, 1977; 697–712.

5.140. Hinze, J.W.; Tripp, W.C.; Graham, H.C. The high temperature oxidation of hot-pressed silicon carbide. In *Mass Transport Phenomena in Ceramics*; Cooper, A.R., Heuer, A.H., Eds.; Plenum Press: New York, 1975.

5.141. Spear, K.E.; Tressler, R.E.; Zheng, Z.; Du, H. Oxidation of silicon carbide single crystals and CVD silicon nitride. In *Ceramic*

Transactions, Vol 10: Corrosion and Corrosive Degradation of Ceramics; Tressler, R.E., McNallan, M., Eds.; Am. Ceram. Soc.: Westerville, OH, 1990; 1–18.

5.142. Fergus, J.W.; Worrell, W.L. The oxidation of chemically vapor deposited silicon carbide. In *Ceramic Transactions, Vol 10: Corrosion and Corrosive Degradation of Ceramics*; Tressler, R.E., McNallan, M., Eds.; Am. Ceram. Soc.: Westerville, OH, 1990; 43–51.

5.143. Fox, D.S. Oxidation kinetics of CVD silicon carbide and silicon nitride. Ceram. Eng. Sci. Proc. 1992, *13* (7–8), 836–843.

5.144. Narushima, T.; Goto, T.; Iguchi, Y.; Hirai, T. High-temperature active oxidation of chemically vapor-deposited silicon carbide in an $Ar-O_2$ atmosphere. J. Am. Ceram. Soc. 1991, *74* (10), 2583–2586.

5.145. Rosner, D.E.; Allendorf, H.D. High temperature kinetics of the oxidation and nitridation of pyrolytic silicon carbide in dissociated gases. J. Phys. Chem. 1970, *74* (9), 1829–1839.

5.146. Jaskowiak, M.H.; DiCarlo, J.A. Pressure effects on the thermal stability of silicon carbide fibers. J. Am. Ceram. Soc. 1989, *72* (2), 192–197.

5.147. Wang, P.S.; Hsu, S.M.; Wittberg, T.N. Oxidation kinetics of silicon carbide whiskers studied by x-ray photoelectron spectroscopy. J. Mater. Sci. 1991, *26*, 1655–1658.

5.148. Jacobson, N.S.; Eckel, A.J.; Misra, A.K.; Humphrey, D.L. Reactions of SiC with $H_2/H_2O/Ar$ mixtures at 1300°C. J. Am. Ceram. Soc. 1990, *73* (8), 2330–2332.

5.149. Maeda, M.; Nakamura, K.; Yamada, M. Oxidation resistance evaluation of silicon carbide ceramics with various additives. J. Am. Ceram. Soc. 1989, *72* (3), 512–514.

5.150. Lu, W.-J.; Steckl, A.J.; Choe, T.P.; Katz, W. Thermal oxidation of sputtered silicon carbide thin films. J. Electrochem. Soc. 1984, *131* (8), 1907–1914.

5.151. Readey, D.W. Gaseous corrosion of ceramics. In *Ceramic Transactions, Vol 10: Corrosion and Corrosive Degradation of*

Ceramics; Tressler, R.E., McNallan, M., Eds.; Am. Ceram. Soc.: Westerville, OH, 1990; 53–80.

5.152. Pareek, V.; Shores, D.A. Oxidation of silicon carbide in environments containing potassium salt vapor. J. Am. Ceram. Soc. 1991, *74* (3), 556–563.

5.153. Federer, J.I. Stress–corrosion of SiC in an oxidizing atmosphere containing NaCl. Adv. Ceram. Mater. 1988, *3* (3), 293–295.

5.154. Federer, J.I. Corrosion of SiC ceramics by Na_2SO_4. Adv. Ceram. Mater. 1988, *3* (1), 56–61.

5.155. Park, D.S.; McNallan, M.J.; Park, C.; Liang, W.W. Active corrosion of sintered α-silicon carbide in oxygen–chlorine gases at elevated temperatures. J. Am. Ceram. Soc. 1990, *73* (5), 1323–1329.

5.156. Stewart, R.W.; Cutler, I.B. Effect of temperature and oxygen partial pressure on the oxidation of titanium carbide. J. Am. Ceram. Soc. 1967, *50* (4), 176–181.

5.157. Shimada, S.; Ishii, T. Oxidation kinetics of zirconium carbide at relatively low temperatures. J. Am. Ceram. Soc. 1990, *73* (10), 2804–2808.

5.158. Arun, R.; Subramanian, M.; Mehrotra, G.M. Oxidation behavior of TiC, ZrC, and HfC dispersed in oxide matrices. In *Ceramic Transactions Vol 10: Corrosion and Corrosive Degradation of Ceramics*; Tressler, E., McNallan, M., Eds.; Am. Ceram. Soc.: Westerville, OH, 1990; 211–223.

5.159. Markarenko, G.N. Borides of the IVb group. In *Boron and Refractory Borides*; Matkovich, V.I., Ed.; Springer-Verlag: New York, 1977; 310–330. Chap. 7.

5.160. Vedula, K.; Abada, A.; Williams, W.S. Potential for diboride reinforcement of oxide matrix composites. In *Materials Research Society Symposium Proceedings, Vol. 125: Materials Stability and Environmental Degradation*; Barkatt, A., Verink, E.D., Jr., Smith, L.R., Eds.; Mater. Res. Soc.: Pittsburgh, PA, 1988; 61–69.

5.161. Walker, J.K.; Saha, C.K. Formation of a surface carbide layer during sintering of titanium diboride. J. Am. Ceram. Soc. 1988, *71* (4), C207–C209.

5.162. Davies, M.W.; Phennah, P.J. Reactions of boron carbide and other boron compounds with carbon dioxide. J. Appl. Chem. 1959, *9* (4), 213.

5.163. Fitzer, E. Oxidation of molybdenum disilicide. In *Ceramic Transactions Vol 10: Corrosion and Corrosive Degradation of Ceramics*; Tressler, E., McNallan, M., Eds.; Am. Ceram. Soc.: Westerville, OH, 1990; 19–41.

5.164. Berztiss, D.A.; Cerchiara, R.R.; Gulbransen, E.A.; Pettit, F.S.; Meier, G.H. Oxidation of $MoSi_2$ and comparison with other silicide materials. Mater. Sci. Eng. 1992, *A155*, 165–181.

5.165. Lin, W.; Hsu, J.-Y.; Speyer, R.F. Stability of molybdenum disilicide in combustion gas environments. J. Am. Ceram. Soc. 1994, *77* (5), 1162–1168.

5.166. Davison, S.; Smith, K.; Zhang, Y.-C.; Liu, J.-H.; Kershow, R.; Dwight, K.; Rieger, P.H.; Wold, A. Chemical problems associated with the preparation and characterization of superconducting oxides containing copper. In *Chemistry of High-Temperature Superconductors*; Nelson, D.L., Whittingham, M.S., George, T.F., Eds. Am. Chem. Soc.: Washington, D.C., 1987; 65–78. Chap. 7.

5.167. Yan, M.F.; Barns, R.L.; O'Bryan, H.M., Jr.; Gallagher, P.K.; Sherwood, R.C.; Jin, S. Water interaction with the superconducting $YBa_2Cu_3O_7$ phase. Appl. Phys. Lett. 1987, *51* (7), 532–534.

5.168. Fitch, L.D.; Burdick, V.L. Water corrosion of $YBa_2Cu_3O_{7-x}$ superconductors. J. Am. Ceram. Soc. 1989, *72* (10), 2020–2023.

5.169. Klinger, N.; Strauss, E.L.; Komarek, K.L. Reaction between silica and graphite. J. Am. Ceram. Soc. 1966, *49* (7), 369–375.

5.170. Miller, P.D.; Lee, J.G.; Cutler, I.B. The reduction of silica with carbon and silicon carbide. J. Am. Ceram. Soc. 1979, *62* (3–4), 147–149.

5.171. Martin, D.M.; Hunter, O. Jr. Polymer bonded refractories. In *The Bond in Refractories*. Tenth Annual Symposium on Refractories; Staley, W., Jr., Givan, G., Eds.; Am. Ceram. Soc. Westerville, OH, 5 Apr 1974.

5.172. Williams, R.S.; Chandhury, S. Chemical compatibility of high-temperature superconductors with other materials. In *Chemistry of High-Temperature Superconductors II*; Nelson, D.L., George, T.F., Eds.; Am. Chem. Soc.: Washington, DC, 1988; 291–302. Chap. 22.

5.173. Mikalsen, D.J.; Roy, R.A.; Yee, D.S.; Shivashankar, S.A.; Cuomo, J.J. Superconducting thin films of the Bi–Sr–Ca–Cu–O system prepared by multilayer metal deposition. J. Mater. Res. 1988, *3*, (4), 613–618.

5.174. Abe, Y.; Hosono, H.; Hosoe, M.; Iwase, J.; Kubo, Y. Superconducting glass–ceramic rods in $BiCaSrCu_2O_x$ prepared by crystallization under a temperature gradient. Appl. Phys. Lett. 1988, *53* (14), 1341–1342.

5.175. Ibara, Y.; Nasu, H.; Imura, T.; Osaka, Y. Preparation and crystallization process of the high-T_c superconducting phase in Bi, Pb–Sr–Ca–Cu–O glass–ceramics. Jpn. J. Appl. Phys. 1989, *28* (1), L37–L40.

5.176. Ott, D.; Raub, C.J. The affinity of the platinum metals for refractory oxides. Platin. Met. Rev. 1976, *20* (3), 79–85.

5.177. Gutman, I.; Gutmanas, E.Y. Interaction of Si_3N_4 with titanium powder. J. Mater. Sci. Lett. 1990, *9*, 813–815.

5.178. Yamaguchi, A. Reactions between alkaline vapors and refractories for glass tank furnace. In *10th International Congress on Glass, No. 2 Refractories and Furnaces*. Ceram. Soc. Japan; 9 July, 1974; 2-1–2-8.

5.179. Mukerji, J.; Biswas, S.K. Synthesis, properties, and oxidation of alumina–titanium nitride composites. J. Am. Ceram. Soc. 1990, *73* (1), 142–145.

5.180. Borom, M.P.; Brun, M.K.; Szala, L.E. Kinetics of oxidation of carbide and silicide dispersed phases in oxide matrices. Adv. Ceram. Mater. 1988, *3* (5), 491–497.

6

Corrosion of Specific Glassy Materials

Perhaps the preceding millennia have not had a Glass Age because it is still to come.

HUBERT SCHROEDER

6.1 INTRODUCTION

The corrosion of glassy materials is predominantly through the action of aqueous media. The attack by gases quite often is that of water vapor or some solution after various species condense and dissolve in the water. Therefore this chapter will be devoted mostly to aqueous attack.

In general, very high silica ($>96\%$ SiO_2), aluminosilicate, and borosilicate compositions have excellent corrosion resistance to a variety of environments. Silicate glasses, in general, are less resistant to alkali solution than they are to acid solution. A list of about 30 glass compositions with their resistance to weathering, water, and acid has been published by Hutchins and Harrington [6.1], and is shown in Tables 6.1 and 6.2. The dissolution rate vs. pH for several composition types is depicted in Figure 6.1.

6.2 SILICATE GLASSES

Generally, silicate glass corrosion processes are typified by diffusion-controlled alkali ion exchange for H^+ or H_3O^+, followed by matrix dissolution as the solution pH drifts toward higher values. This concept was perhaps first reported in 1958 by Wang and Tooley [6.2]. The initial exchange reaction produces a transformed gel-like surface layer. This surface layer may contain various crystalline phases depending upon the overall glass composition and solution pH. Diffusion through this layer becomes the rate-controlling step. This layer is formed through the process of network hydrolysis and condensation of network bonds releasing alkali, a process that is very similar to the second, essentially simultaneous, step of network dissolution. Thus the dissolution of silicate glasses is dependent upon the test conditions of time, temperature, pH, and the sample composition (i.e., structure). Although many references are made to the effects of glass composition upon dissolution, the actual correlation is with glass structure not composition. This is so because composition determines structure. An example of this was indicated by Brady and House [6.3]. They determined that glasses that were silica-rich and highly polymerized dissolved more slowly than those containing large amounts of other cations. The key structural factor is that highly polymerized glasses dissolved more slowly.

The deterioration of a glass surface by atmospheric conditions, commonly called *weathering*, is very similar to that described above. If droplets of water remain on the glass surface, ion exchange can take place with a subsequent increase in the pH. As the volume of the droplets is normally small

TABLE 6.1 Properties of Commercial Glasses

Glass code[a]	Type	Forms usually available[b]	Weathering	Water	Acid
0010	potash–soda–lead	T	2	2	2
0080	soda–lime	BMT	3	2	2
0120	potash–soda–lead	TM	2	2	2
1720	aluminosilicate	BT	1	1	3
1723	aluminosilicate	BT	1	1	3
1990	potash–soda–lead		3	3	4
2475	soda–zinc	T	3	2	2
3320	borosilicate		1	1	2
6720	soda–zinc	P		1	2
6750	soda–barium	BPR		2	2
6810	soda–zinc	BPR		1	2
7040	borosilicate	BT	3	3	4
7050	borosilicate	T	3	3	4
7052	borosilicate	BMPT	2	2	4
7056	borosilicate	BTP	2	2	4
7070	borosilicate	BMPT	2	2	2
7250	borosilicate	P	1	2	2
7570	high lead		1	1	4
7720	borosilicate	BPT	2	2	2
7740	borosilicate	BPSTU	1	1	1
7760	borosilicate	BP	2	2	2
7900	96% silica	BPTUMF	1	1	1
7913	96% silica	BPRSTF	1	1	1
7940	fused silica	UF	1	1	1
8160	potash–soda–lead	PT	2	2	3
8161	potash–lead	PT	2	1	4
8363	high lead	LC	3	1	4
8871	potash–lead		2	1	4
9010	potash–soda–barium	P	2	2	2
9700	borosilicate	TU	1	1	2
9741	borosilicate	BTU	3	3	4

The ratings listed are: 1 = high resistance, 2 = occasionally troublesome; and 3 = careful consideration for use necessary. (Reprinted from Ref 6.1, Copyright © 1966 by John Wiley & Sons, Inc. This material is used with permission from John Wiley & Sons, Inc.)
[a] The 4-digit glass codes, e.g., 0010, refer to Corning Glass Works glasses.
[b] B = blown glass; P = pressed ware; S = plate glass; M = sintered slip cast ware (multiform); R = rolled sheet; T = tubing and cane; LC = large castings; F = fibers; U = panels.

TABLE 6.2 Corrosion of Glass, Applies only to Durable Compositions

Type of reagent	Temperature	Degree of attack	Remarks
Water	Boil	Negligible	No absorption or swelling
	100–260°C	0.001–0.01 mg/cm^2, in 6 hr	Depends on glass
Seawater, 5% sea salt	Boil	0.03–0.08 mg/cm^2, 24 hr	1 yr in ocean, no visible effect
Acids			
HF	All	Severe	Not recommended
21% H_3PO_4	100°C	0.005 mg/cm^2, 24 hr	Glass satisfactory except
85% H_3PO_4	100°C	0.014 mg/cm^2, 24 hr	at high concentrations or raw acid with fluorides
Other inorganic	Boil	Negligible	
Organic	Boil	Negligible	
Bases			
Strong, 5% NaOH	80°C	0.3 mg/cm^2, 6 hr	
6.9% KOH	80°C	0.2 mg/cm^2, 6 hr	
Weak, 3% NH_4OH	80°C	0.33 mg/cm^2, 100 hr	
Halogens	to 150°C	Negligible	Dry fluorine questionable
Metal salts			
Acid	to 150°C	Negligible	

Neutral	to 150°C	Negligible	
Basic 0.5 N Na_2CO_3	100°C	0.18 mg/cm^2, 6 hr	
5% Na_2CO_3	150°C	10 mg/cm^2, 6 hr	
Inorganic nonmetallic halides	to 150°C	Negligible	Fluorides excepted
Sulfur dioxide	to 150°C	Negligible	Slight bloom may appear
Ammonia (dry)	to 150°C	Negligible	See bases for NH_4OH
Oxidizing chemicals	to 150°C	Negligible	
Reducing chemicals	to 150°C	Negligible	
Hydrocarbons	to 150°C	Negligible	Includes chlorinated compounds
Amines	to 150°C	Negligible	Those with pronounced basic reaction questioned
Polyhydroxyl aliphatics	to 150°C	Negligible	
Mercaptans	to 150°C	Negligible	
Oils and fats	to 150°C	Negligible	

A weight loss of 1 mg/cm^2 is equivalent to a depth loss of 0.01 mm/(specific gravity of glass) for those cases where the attack is not selective. (From Ref 6.1, Copyright © 1966 by John Wiley & Sons, Inc. This material is used by permission of John Wiley & Sons, Inc.)

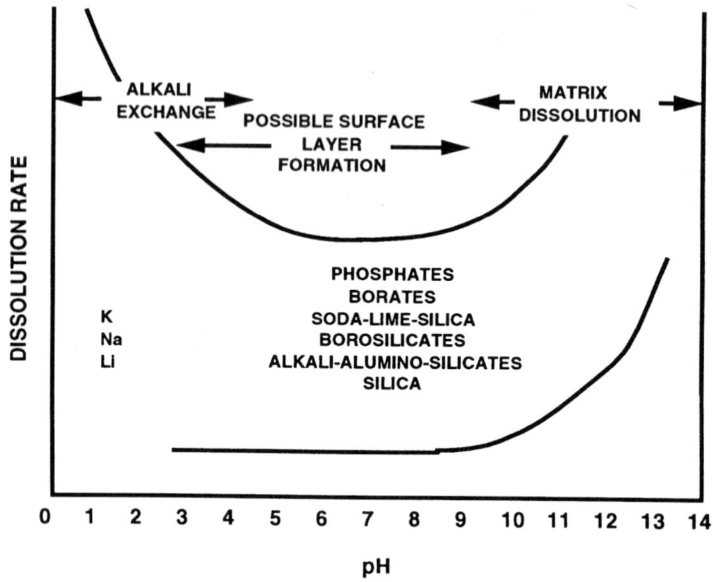

FIGURE 6.1 Dissolution rate vs. pH.

compared to the surface area they contact, drastic increases in pH can occur causing severe etching of the surface. The rough surface formed can then collect additional solution causing further deterioration. In some cases, the alkali-rich droplets can react with atmospheric gases forming deposits of sodium and calcium carbonate [6.4]. Chen and Park [6.5] have identified crystals on the surface of a Na-disilicate glass exposed to room temperature air as $Na_3H(CO_3)_2 \cdot 2H_2O$. These deposits can act as a barrier to further weathering; however, they detract from the visual aesthetics of the glass. Whether deakalization or matrix dissolution is the predominant mode of attack depends upon the volume and flow of water in contact with the surface.

Manufacturers of flat glass have, for many years, known of the beneficial effects of SO_2 gas treatment in increasing the weatherability of their products. This treatment, prior to the glass being annealed, allows the sodium in the surface layers to react with the SO_2, forming sodium sulfate. The sulfate deposit is then washed off prior to inspection and packing. The first step in weathering is then diminished because of the low alkali content of the surface.

According to Charles [6.6], the corrosion of an alkali-silicate glass by water proceeds through three steps. These are:

1. H^+ from the water penetrates the glass structure. This H^+ replaces an alkali ion, which goes into solution. A nonbridging oxygen is attached to the H^+,
2. the OH^- produced in the water destroys the Si–O–Si bonds, forming nonbridging oxygens, and
3. the nonbridging oxygens react with an H_2O molecule, forming another nonbridging oxygen—H^+ bond and another OH^- ion. This OH^- repeats step 2. The silicic acid thus formed is soluble in water under the correct conditions of pH, temperature, ion concentration, and time.

It is questionable as to whether the first step described above involves the penetration of a proton or a hydronium, H_3O^+ ion. There is evidence that supports the exchange of hydronium for alkalies [6.7]. In addition, the dissolution of silicate minerals, which is very similar to silicate glasses, has been reported to take place by exchange of hydronium ions for alkalies [see Ref. 2.36 in Chapter 2].

The development of films on the glass surface has been described by Sanders and Hench [6.8]. They showed that a 33 mol% Li_2O glass corroded more slowly than a 31 mol% Na_2O glass by 2 orders of magnitude. This difference was caused by the formation of a film on the Li_2O glass with a high silica content. Scratching the glass surface produced an unusually high release of silica. The nonbridging oxygen–H^+ groups may form surface films or go into solution. The thickness of this film and its adherence greatly affected the corrosion rate. In Na_2O–SiO_2 glasses, Schmidt [6.9] found that films formed only on glasses containing more than 80 mol% SiO_2 at 100°C for 1 hr.

Several workers have investigated the concentration profiles of glass surfaces after leaching by water and attempted to explain the variations observed. Boksay et al. [6.10] postulated a theory that fit the profiles observed in K_2O-SiO_2 glass, but did not explain the profiles in Na_2O-SiO_2 glass, presumably due to a concentration-dependent diffusion coefficient. Doremus [6.11] developed a theory that included a concentration-dependent diffusion coefficient to explain the profiles in Li_2O-SiO_2 glass; however, his theory still did not fit the observations for sodium determined by Boksay et al. [6.12]. Das [6.13] attributed the differences in the profiles between the sodium and potassium glasses as being a result of a difference in the structure of the leached layer caused by the relative difference in size between the H_3O^+ and the Na^+ ions and the similarity in size between H_3O^+ and K^+ ions. In general, the dissolution rate (i.e., dealkalization) decreased as the ion radius of the alkali decreased.

Douglas and coworkers [6.14–6.17] found that alkali removal was a linear function of the square root of time in alkali-silicate glass attacked by water. At longer times, the alkali removal was linear with time. Silica leached from alkali-silicate glasses decreased as the amount of silica in the glass increased, unlike that of the alkalies. Wood and Blachere [6.18] investigated a $65SiO_2-10K_2O-25PbO$ (mol%) glass and did not find a square root of time dependence for removal of K or Pb but found a dependence that was linear with time. This behavior was also reported by Eppler and Schweikert [6.19] and by Douglas and coworkers. Wood and Blachere proposed that an initial square root of time dependence occurred but that the corrosion rate was so great that it was missed experimentally.

The pH of the extracting solution is also very important as found by Douglas and El-Shamy [6.17]. They found that above $pH = 9$, the leaching rate of alkalies decreased with increasing pH, whereas below $pH = 9$ the leaching rate was independent of pH. A somewhat different relationship was found for the leaching rate of silica—above $pH = 9$ the rate increased with increasing pH, whereas below $pH = 9$ the amount of silica extracted was close to the detection limits of the apparatus. Two reactions were identified: one where alkalies passed into solution as a result of ion exchange with protons from the solution and one where silica passed into solution as a consequence of the breaking of siloxane bonds by attack from hydroxyl groups from the solution. Thus removal of silica was favored by an increase in hydroxyl ion activity (i.e., increased pH), which was accompanied by a reduction in proton activity and thus a reduction in alkali extraction.

The dependence of dissolution upon pH can be seen by an examination of Eq. (2.16) in Chapter 2 for the dissolution of minerals. Similarly, glasses in contact with aqueous solutions can be represented by the following ion exchange reaction:

$$MSiO_2(glass) + nH^+(aq) \leftrightarrow H_2SiO_3 + M^{n+}(aq) \tag{6.1}$$

which has as the equilibrium constant:

$$k = \frac{a_{H_2SiO_3} a_{M^{n+}}}{a_{MSiO_3} a_{H^+}} \tag{6.2}$$

Expressing this in logarithm form then gives:

$$\log a_{H_2SiO_3} = \log k - \log a_{M^{n+}} - npH \tag{6.3}$$

Thus it should be obvious that the exchange reaction of a proton for the leachable ionic species in the glass is dependent upon the pH of the solution and also the leached ion activity in the solution.

Das [6.20] has shown that substitutions of Al_2O_3 or ZrO_2 for SiO_2 in sodium silicate glasses shifted the pH at which increased dissolution occurred to higher values, creating glasses that were more durable and less sensitive to pH changes. Paul [6.21] has also reported the beneficial effects of alumina and zirconia upon durability.

Manufacturers of soda–lime–silicate glasses have known for a long time that the addition of lime to sodium silicate glass increased its durability. Paul [6.21] reported that substitutions of up to 10 mol% CaO for Na_2O rapidly decreased the leaching of Na_2O. Above about 10 mol% substitution, the leaching of Na_2O remained constant. With the larger amounts of CaO devitrification problems during manufacture occurred, requiring the substitution of MgO for some of the CaO. According to Paul [6.21] calcium-containing glasses should exhibit good durability up to about pH = 10.9. He also indicated that replacement of CaO by ZnO extended this durability limit to about pH = 13, although these compositions were attacked in acid solutions at pH < 5.5.

The effects of MgO, CaO, SrO, and BaO upon leaching of Na_2O at 60 and 98°C in distilled water were reported by Paul [6.21]. At higher temperature, the durability decreased with increasing ionic size, whereas at the lower temperature, the durability was relatively the same for all four alkaline earths. This was attributed to the restricted movement at the lower temperature for the larger ions.

Expanding upon the ideas originally proposed by Paul and coworkers [6.22–6.24], Jantzen and coworkers [6.25–6.27] have shown that network or matrix dissolution was proportional to the summation of the free energy of hydration of all the glass components as given by the equation:

$$\Delta G^{\circ} = ART \log L \tag{6.4}$$

where A is the proportionality constant and L is a normalized loss by leaching in mass per unit area. Jantzen [6.28] has shown that high-silica glasses exhibited weak corrosion in acidic-to-neutral solutions and that low-silica glasses exhibited active corrosion at pH from <2 to 3. Between pH 2 and 10 in an oxidizing solution, hydrolysis occurred through nucleophilic attack with the formation of surface layers by reprecipitation or chemisorption of metal hydroxides from solution. In reducing solutions, surface layers tended to be silicates that exhibited weak corrosion or were even immune. In alkaline solutions at pH greater than about 10, both low- and high-silica glasses exhibited active corrosion with low-silica glasses having a potential for surface layer formation.

Ernsberger [6.29] has described the attack of silica or silicate glasses by aqueous hydrofluoric acid in detail and related it to the structure of silica glasses. The silicon–oxygen tetrahedra are exposed at the surface in a random arrangement of four possible orientations. Protons from the water solution will bond with the exposed oxygens, forming a surface layer of hydroxyl groups. The hydroxyl groups can be replaced by fluoride ions in aqueous hydrofluoric solutions. Thus the silicon atoms may be bonded to an OH^- or and F^- ion. The replacement of the exposed oxygens of the tetrahedron by $2F^-$ causes a deficiency in the silicon atom coordination, which is six with respect to fluorine. This causes the additional bonding of fluoride ions, with a particular preference for bifluoride. Thus the four fluoride ions near the surface provide an additional four-coordinated site for the silicon. A shift of the silicon to form SiF_4 can take place by a small amount of thermal energy. The ready availability of additional fluoride ions will then cause the $(SiF_6)^{2-}$ ion to form. This mechanism is supported by data that show a maximum in corrosion rate with bifluoride ion concentration. Although giving a slightly different description of the possible reactions, Liang and Readey [6.30] reported that the dissolution of fused silica varied with HF concentration and was controlled by a surface reaction rather than diffusion through the liquid.

The solubility in nitric acid has been reported by Elmer and Nordberg [6.31] to be a function of acid concentration; however, the rate decreased with increasing concentration (from 0.8 to 7.0 N), just the opposite as that found in HF. In concentrations greater than 3 N, saturation was reached in about 24 hr. At 0.1 N, the rate was considerably lower than the other concentrations, not reaching saturation even after 96 hr.

White et al. [6.32] found that for Na_2O-SiO_2 (33/67% composition) and Li_2O-SiO_2 glass compositions, environments that caused surface corrosion also caused enhanced crack growth. The environments studied were distilled water, hydrazine, formamide, acetonitrite, and methyl alcohol. White et al. found that acetonitrite was noncorrosive and that water was the most effective in leaching alkali, while hydrazine was the most effective in leaching silica. Formamide was only mildly effective in leaching alkali. The mechanism of corrosion for water, formamide, and hydrazine was reported to be alkali ion exchange with H^+ or H_3O^+.

The durability of gel-derived 20 mol% Na_2O–80 mol% SiO_2 glass subjected to various temperatures in deionized water was studied by Hench et al. [6.33]. They concluded that both lower soda contents (compared to a 33 mol% Na_2O glass tested in a previous study [6.34] and higher densities improved the durability.

The effect of dissolved water in soda–lime glass upon the rate of dissolution in water was related to the influence of absolute humidity at the time of forming and annealing by Bacon and Calcamuggio [6.35]. Very high resistance was obtained by use of very dry air. Similar results were obtained by Wu [6.36] on a soda–silica glass containing K_2O, Al_2O_3, and ZnO with dissolved water contents between 4 and 8 wt.%. Wu, however, reported leach rates independent of water contents at concentrations less than 4 wt.%. Tomozawa et al. [6.37] concluded that many Si–O bonds in the glass are possibly hydrolyzed by the dissolved water content, thus eliminating some steps during the dissolution of the glass in water and increasing the rate of attack.

Little information seems to have been published in the area of molten salt attack on glasses. The dissolution of several glass compositions was reported by Bartholomew and Kozlowski [6.38] to be extensive and nonuniform in molten hydroxides. Samples attacked by sodium hydroxide exhibited an opaque and frosted surface, whereas those attacked by potassium hydroxide were transparent. Bartholomew and Kozlowski used the mechanism proposed by Budd [6.39] to interpret the attack shown in their studies. Considering the hydroxide ion as basic, a vigorous reaction should take place with an acidic glass. This was confirmed experimentally by testing glasses of different chemistries.

Loehman [6.40] reported no trends in leaching with nitrogen content for several Y–Al–Si–O–N glasses, although two of his compositions exhibited lower weight losses by at least a factor of 2 than fused silica when tested in distilled water at

95°C for 350 hr. In their study of soda–lime–silicate glasses, Frischat and Sebastian [6.41] reported that a 1.1 wt.% addition of nitrogen considerably increased the leach resistance to 60°C water for 49 hr. The release of sodium was 55% less and calcium 46% less for the nitrogen-containing glass. An additional indication of the greater resistance of the nitrogen-containing glass was the change in pH of the leaching solution with time. Starting with a solution pH of 6, the solution pH drifted to 9 for the nitrogen-free glass after 7 hr, but reached 9 for the nitrogen-containing glass after only 25 hr. The improved leach resistance of this glass was attributed to a greater packing density for the nitrogen-containing glass.

White and Day [6.42] reported no detectable weight loss of a $1\times1\times0.2$ cm rare-earth aluminosilicate (REAS) glass sample before 6 weeks in 100 mL of distilled water (pH = 7) or saline (pH = 7.4) at 37, 50, or 70°C. Dissolution rates of $\leq3\times10^{-9}$ g/cm^2 min were determined after 6 weeks. In a comparison study of fused silica, a Corning glass (CGW-1723™*) and yttria aluminosilicate (YAS), Oda and Yoshio [6.43] showed that YAS was significantly more durable than fused silica in saturated steam at 300°C and 8.6 MPa. The dissolution mechanism is very important for applications in the human body; however, it is very difficult to determine whether these glasses exhibit congruent or incongruent dissolution. Surface analyses of microspheres and bulk glasses indicated that the mechanism was congruent [6.42]. Using inductively coupled plasma and atomic adsorption spectroscopy it has been determined that the yttrium release from YAS microspheres in distilled water or saline at 37 or 50°C was below detectable limits [6.44].

In the manufacture of flat glass by the float process,† a cooperative diffusion process takes place where tin diffuses into

* CGW-1723™ is a clear aluminosilicate glass.

† The float process for the manufacture of flat glass involves floating molten glass onto molten tin in a chamber, called the float bath, containing a reducing atmosphere.

the glass and the constituents of the glass diffuse into the tin. The reaction zone in the glass is about 25 μm thick. Many investigators have studied the tin oxide gradient of float glass and have reported a rather complex behavior [6.45–6.54]. Stannous tin is dominant at the near surface. A typical hump occurs in the tin profile at between 5 and 10 μm where stannic (or oxidized tin) is predominant. This hump has been attributed to the additional tin from the ion exchange with calcium by Franz [6.55]. Investigation of extremely thin layers of glass has indicated tin oxide contents as high as 36% at the surface [6.52] (see Fig. 6.2). The amount of the tin contained within the glass surface and the depth to which it penetrates is dependent upon the exposure time and temperature (which relates to glass production tonnage and thickness), and the amount and type of impurities (especially sulfur) contained in the tin. Thin glass travels through the bath faster than thick glass and therefore has less time for the various reactions to take place.

At the hot end of the bath, iron oxide in the glass will migrate toward the bottom surface where it is reduced (by reaction with either stannous oxide or hydrogen) to iron metal and dissolves into the tin. At the cooler end of the bath, this tin/iron alloy will oxidize (oxygen coming from air ingress) forming both iron oxide and tin oxide. Iron has a greater potential to oxidize than tin and therefore acts as a scavenger for oxygen.

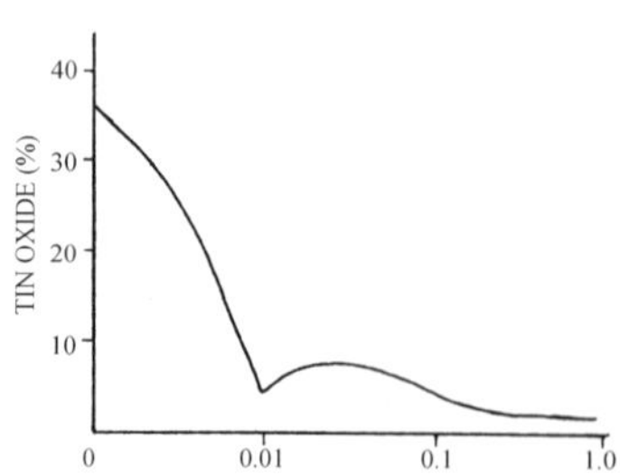

FIGURE 6.2 Tin oxide penetration into bottom surface of float glass. (From Ref. 6.52.)

Being essentially insoluble in tin these two oxides will enter into the glass either by diffusion or by exchange for calcium oxide [6.55]. Calcium oxide is also insoluble in molten tin and will therefore form a deposit on the bottom surface of the glass. The deposit can be washed from the bottom surface of the glass by a vinegar solution.* Thus iron that enters the tin at the hot end of the bath will reenter the glass at the cold end, setting up an equilibrium concentration of iron in the tin. This equilibrium can be altered if the glass composition is changed from one of high iron content to one of lower iron content (or vice versa).

Although the interaction layer thickness is quite small, the presence of tin in the surface of the glass ribbon causes some secondary fabrication problems. Many fabrication methods require that the flat glass piece be bent. This is carried out by reheating the glass on a metal frame and allowing the glass to sag to the desired shape. This reheating process can provide additional oxidation of the tin (from stannous to stannic oxide) in the bottom surface. This oxidation is accompanied by an expansion of the tin-rich layer causing a microwrinkled surface. This wrinkled surface becomes visible as a faint iridescent haze—known as the defect *bloom* [6.56]. This phenomenon can also occur when glass is reheated for tempering.

6.3 BOROSILICATE GLASSES

The durability of borosilicate glasses has been extensively investigated by the nuclear waste glass community. No attempt will be made here to review all the literature related to nuclear waste glasses; however, the article by Jantzen [6.28] described quite well the use of Pourbaix diagrams in predicting the dissolution of nuclear waste glasses. Jantzen has performed

* The deposit of calcium oxide reacts with atmospheric carbon dioxide forming calcium carbonate on the glass surface that is insoluble in water and must be washed off with a vinegar solution.

a very thorough job in explaining the interrelationship of pH, Eh, activity, free energy of hydration, and glass dissolution. It was shown that solution Eh had an effect upon network dissolution that was 20 times less than that of pH. But when redox-sensitive elements were leached from the glass, the solution Eh could have a much larger effect. Jantzen also concluded that less durable glasses had a more negative free energy of hydration and thus released more silicon and boron into solution. Higher boron release over that of silicon was attributed to the greater solution activity of vitreous boria compared to that of vitreous silica at any given *p*H. Refs. 5.28–5.32 listed at the end of the previous chapter are a good source of information for the reader interested in the aqueous attack upon borosilicate glasses and nuclear waste materials in general.

In borosilicate glasses requiring a heat treatment step after initial melting and cooling to produce phase separation, a surface layer is formed by selective evaporation of Na_2O and B_2O_3. These surface layers have been observed by several workers. This silica-rich surface layer can influence the subsequent leaching process that would be needed to produce Vycor™*-type glass [6.57]. If the hydrated surface layer were removed before heat treatment, the silica-rich layer would be almost entirely eliminated.

The leaching rate in 3 N HCl solution for borosilicates glasses with an interconnected microstructure was shown by Takamori and Tomozawa [6.58] to be dependent upon the composition of the soluble phase. The composition and size of this interconnected microstructure was also dependent upon the temperature and time of the phase separation heat treatment process. Taylor et al. [6.59] have shown that phase separated low soda borosilicate glasses form a less durable Na_2O plus B_2O_3-rich phase dispersed within a more durable

* Vycor™ is manufactured by Corning, Inc.

silica-rich phase. The overall durability in distilled deionized water was strongly dependent upon the soda content and was best for a composition containing about 3 mol% Na_2O. The durability was also dependent upon the SiO_2/B_2O_3 ratio, with the higher silica content glasses being more durable. In a study of soda borosilicate glasses, Kinoshita et al. [6.60] related the effects of the Si/B ratio to the dissolution rates. At low Si/B ratios, the glasses dissolved congruently at rapid constant rates at a pH = 2 in HCl/glycine solutions. Higher Si/B ratios caused the selective leaching of sodium and boron leaving behind a silica-rich layer that caused the dissolution rate to decrease with time.

In a study closely related to borosilicate glasses, El-Hadi et al. [6.61] investigated the addition of soda to B_2O_3 and the effect upon durability, which is generally very poor for borate glasses. Increased durability toward both acids and bases was related to the change in coordination of the boron from three to four as the alkali level was increased. Alkali borate glasses also increased in density as the alkali content was increased, suggesting that the change in coordination caused a more compact, more difficult to leach, structure. Addition of various divalent metal oxides to a lithium borate glass also increased the durability in the order: CdO > ZnO > PbO > SrO > BaO. Tait and Jensen [6.62] found an order-of-magnitude increase in durability (in deionized water) of a sodium borosilicate glass containing 8.5 mol% ZnO. CaO and Al_2O_3 also increased the durability.

The attack by various acids was studied by Katayama et al. [6.63], who determined that the corrosion of a barium borosilicate glass decreased in the order acetic, citric, nitric, tartaric, and oxalic acid, all at a pH of 4 at 50°C. The mechanism of attack by orthophosphoric acid was shown to vary with temperature by Walters [6.64]. The considerable degradation above 175°C was attributed to acid dehydration. At higher temperatures, the acid condensed and reacted with the glass forming a protective layer of SiP_2O_7. The formation of this barrier layer formed sufficient stresses to produce strength loss and caused mechanical failure.

Metcalfe and Schmitz [6.65] studied the stress corrosion of E-glass (borosilicate) fibers in moist ambient atmospheres and proposed that ion exchange of alkali by hydrogen ions led to the development of surface tensile stresses that could be sufficient to cause failure.

The effect of dissolved water content upon the resistance of borosilicate glasses to acid vapor attack (over boiling 20% HCl) was investigated by Priest and Levy [6.66]. Increasing water contents correlated with increasing corrosion resistance.

The use of borosilicate foamed glass blocks to line the outlet ducts of coal burning power plants was reported by Koch and Syrett [6.67] to perform better than silicate cement gunite, as well as nickel-based or titanium alloys in an 18-month test. This was attributed to the high concentration of aluminum in the outlet flue gas that formed soluble complexes with fluorine that are not detrimental to borosilicate glass.

Fast ion conduction glasses, such as lithium–borate and lithium–chloroborate glasses, were studied by Velez et al. [6.68] to determine their resistance to molten lithium at temperatures between 180 and 250°C. They found that those compositions with a minimum B_2O_3 content resulted in the best resistance to attack.

Recently, Conzone et al. [6.69] reported the development of borate glasses for use in treatment of rheumatoid arthritis, as these glasses are potentially more reactive with physiological liquids. Borate glasses containing only alkali ions dissolved uniformly (i.e., congruently) in simulated physiological liquids at temperatures ranging from 22 to 75°C. When the borate glasses contain other cations (such as Ca, Mg, Fe, Dy, Ho, Sm, and Y) in amounts ranging from 2 to 30 wt.% dissolution was nonuniform (i.e., incongruent) with the formation of new compounds. Day [6.70] gave an example of Dy_2O_3-containing borate solid glass microspheres that reacted to form hollow spheres, shells of concentric layers, or microspheres filled with homogeneous gel-like material depending upon the Dy_2O_3 content. The dissolution mechanism involved the selective leaching of lithium and boron allowing the rare earth (i.e.,

Dy) to react and form an insoluble phosphate.* When calcium-containing borate glasses were reacted a semicrystalline or gel calcium phosphate formed that had a composition very similar to hydroxyapatite. Although early work by Hench and colleagues has indicated the need for the formation of a silica gel surface layer for silicate glasses to be bioactive, the work of Day and colleagues has indicated that a silica gel is not always necessary for bioactivity.

6.4 LEAD-CONTAINING GLASSES

Yoon [6.71] found that lead release was a linear function of pH when testing lead-containing glasses in contact with various beverages. Low pH beverages such as orange juice or colas, leached lead more slowly than did neutral pH beverages such as milk. This dependence upon pH was also reported by Das and Douglas [6.16] and by Pohlman [6.72]. Later, Yoon [6.73] reported that if the ratio of moles of lead plus moles of alkali per moles of silica were kept below 0.7, release in 1 hr was minimized. If this ratio was exceeded, lead release increased linearly with increasing PbO content. Lehman et al. [6.74] reported a slightly higher threshold for more complex compositions containing cations of Ca^{2+} and Al^{3+} or B^{3+}, in addition to the base $Na_2O-PbO-SiO_2$ composition. The lead release in these complex compositions was not linear but increased upward with increased moles of modifiers. Lehman et al. related the mechanism of release or corrosion to the concentration of nonbridging oxygens. A threshold concentration was necessary for easy diffusion of the modifier cations. This threshold was reported to be where the number of nonbridging oxygens per mole of glass-forming cations equaled 1.4.

Krajewski and Ravaglioli [6.75] correlated the release of Pb^{2+} by acid attack to the site coordination of the network

*Phosphorus is from a phosphate-buffered saline simulated physiological liquid.

modifiers. The presence of cations with cubic coordination produced increased Pb^{2+} release, whereas cations with antiprismatic coordination produced a decreased Pb^{2+} release.

In general, it has been determined that mixed alkalies lower the release of lead by attack from acetic acid below that of a single alkali–PbO–silicate glass; lead release increased with increasing ionic radius of the alkaline earths; however, combinations of two or more alkaline earths exhibited lower lead release; Al_2O_3 and ZrO_2 both lowered the lead release; and B_2O_3 increased the lead release. Thinner glaze coatings on clay-based ceramic bodies decreased lead release because of interaction of the glaze and the body, providing higher concentration of Al_2O_3 and SiO_2 at the glaze surface [6.76].

Haghjoo and McCauley [6.77] found that small substitutions (0.05–0.15 mol%) of ZrO_2 and TiO_2 to a lead bisilicate glass lowered the solubility of lead ion in 0.25% HCl by an order of magnitude. Additions of Al_2O_3 had a lesser effect, while additions of CaO had essentially no effect.

The mechanism of release or corrosion for these glasses containing lead is similar to those proposed by Charles [6.6] for alkali-silicate glasses. The rate of this reaction depends upon the concentration gradient between the bulk glass and the acid solution and the diffusion coefficient through the reacted layer. In general, maximum durability can be related to compact, strongly bonded glass structures, which in turn exhibit low thermal expansion coefficients and high softening points [6.78].

6.5 PHOSPHORUS-CONTAINING GLASSES

The study of phosphate glass corrosion has shown that the glass structure plays a very important role in the rate of dissolution. Phosphate glasses are characterized by chains of PO_4 tetrahedra. As the modifier (alkalies or alkaline earths) content of these glasses is increased, there is increased cross linking between the chains. When very little cross linkage exists, corrosion is high. When the amount of cross linkage is high, corrosion is low.

Similar phenomena should exist for other glass-forming cations that form chain structures (B^{3+} and V^{5+}).

During the study of aqueous attack of soda–lime–silica glasses containing P_2O_5, Clark et al. [6.79] found that a double reaction layer was formed, consisting of a silica-rich region next to the glass and a Ca–P-rich reaction next to the water solution. This Ca–P film eventually crystallized into an apatite structure and provided a good mechanism to bond the glass to bone in implant applications. In order for these compositions to be highly active toward aqueous media, the bioactive glass composition must contain less than 60 mol% SiO_2, a high content of Na_2O and CaO, and a high CaO/P_2O_5 ratio [6.80]. When the SiO_2 content was greater than 60 mol%, the hydroxyapatite reaction layer did not form within 2–4 weeks. For a glass to be beneficial as an implant, the reactions leading to the formation of the $CaO-P_2O_5$-rich surface film must occur within minutes of implantation. The dependency of bioactivity upon the structure of the glass is thus a very important concern in the development of these materials. When the silica content exceeds 60 mol%, the glass structure changes from one of two-dimensional sheets containing chains of polyhedra to a three-dimensional network common to the high silica glasses. The two-dimensional structure being a more open structure allows more rapid ion exchange and thus faster hydroxyapatite film formation.

Potassium phosphate glasses containing various oxide additions were tested for water solubility by Minami and Mackenzie [6.81], with Al_2O_3 and WO_3 additions yielding the greatest improvement. In alkali phosphate glasses containing Al_2O_3 or WO_3, the durability increased as the ionic radius of the alkali cation decreased, a trend that was common in most glasses.

Reis et al. [6.82] investigated the durability of zinc–iron phosphate glasses in distilled water at 90°C for up to 32 days. They found the durability to be 100 times better than window glass and the dissolution rate to decrease with increasing iron content. Excellent durability of glasses containing more than 30 mol% Fe_2O_3 was related to the presence of the Fe–O–P bond.

According to Hench [6.83] in his discussion of bioactive glasses, the dissolution kinetics are a function of the following variables:

1. Composition
2. Particle size
3. Pore size distribution, average size, and volume %
4. Surface area
5. Thermal stabilization temperature
6. Chemical stabilization temperature

The alumina content of bioactive glasses is very important in controlling the durability of the glass surface. The bioactivity, although dependent upon the bulk composition of the glass, decreased beyond acceptable levels once the alumina content increased above 1.0–1.5 wt% [6.84]. This same phenomenon was present for glass compositions containing cations such as Ta_2O_5 except higher levels were tolerable (1.5–3.0 wt.%).

Avent et al. [6.85] studied the dissolution of Na–Ca–phosphate glasses containing small amounts of silver in an attempt to develop biocompatible controlled release glasses for applications in medical equipment such as urological catheters. It has been known for a long time that traces of silver have bactericidal properties. With that in mind, Avent et al. investigated the dissolution of several glass compositions in distilled water and two different simulated urine solutions at 25 and 35°C. They found that silver release was dependent upon the Na/Ca ratio of the glass and that silver release was double in simulated urine compared to distilled water. They concluded that these glass compositions dissolved by destruction of the links between polyphosphate species with the dominant polyphosphate specie being *cyclo*–hexophosphate.

6.6 FLUORIDE GLASSES

The corrosion of fluoride glasses has become rather important recently because of their potential application as optical components because of their excellent IR transmission properties

[6.86] and their application as membranes in fluoride-ion-selective electrodes [6.87]. The corrosion of these glasses is generally characterized by a double interfacial layer, an inner portion of hydrated species and an outer nonprotective layer of crystalline precipitates, generally ZrF_4 [6.88], except when highly soluble compounds are present [6.87,6.89]. The reaction:

$$F^-(\text{glass}) + OH^-(\text{aq}) \leftrightarrow F^-(\text{aq}) + OH^-(\text{glass}) \qquad (6.5)$$

reported by Ravaine and Perera [6.87] depicts the exchange reaction that forms this interfacial hydrated layer.

Simmons and Simmons [6.89] studied the corrosion of fluorozirconate glasses in water (pH = 5.6). A direct correlation was found between the solubility of the modifier additive and the glass durability. Those additives with the greatest water solubility (AlF_3, NaF, LiF, and PbF_2) were determined to cause the greatest solubility of the glasses. ZrF_4, BaF_2, and LaF_3 exhibited lower solubilities. The corrosion behavior of all the glasses was controlled by the Zr and Ba contents and the pH drift of the solution. The other modifier additives had only a limited effect upon corrosion. The order of leach rate for ZBL glass was $Zr > Ba \gg La$. The order when Al was added changed to $Al > Zr \gg Ba > La$, and when Li was added changed to $Li > Al > Zr > Ba \gg La$. When Na replaced Li, the Al leach rate was lower than the Na, and the others remained the same. The addition of Pb had the greatest effect by not exhibiting the marked decrease in the leach rate with time for the various components.

The major difference between fluorozirconate and silicate glasses was the drift in pH during the corrosion process. The fluorozirconate exhibited a solution pH drift toward acidic values. The equilibrium solution pH for a ZrBaLaAlLi–fluoride glass was found to be 2.6. Additional studies upon crystalline forms of the various additives indicated that the main cause of the drop in pH was the hydrolysis of ZrF_4 forming the complex species:

$$[ZrF_x(OH)_y]^{+4-x-y}[H_2O]_n$$

It is interesting that these glasses exhibited minimal corrosion from atmospheric moisture, even when exposed to 100% RH at 80°C for up to one week. Gbogi et al. [6.90] reported similar results for a ZBL glass exposed to ambient conditions for 30 days, and Robinson and Drexhage [6.91] reported no corrosion for ThF_4-containing fluoride glasses up to 200°C.

The time dependency of leaching rates varied with the composition of the heavy metal fluoride additive [6.87]. Compositions containing Zr, Ba, and Th; U, Ba, and Mn; and Sc, Ba, and Y displayed a continuous decrease in corrosion rate with time. Those containing Th, Ba, Mn, and Yb or Th, Ba, Zn, and Yb displayed a minimum. Those containing Pb, K, Ga, Cd, Y, and Al displayed a plateau. Ravaine and Perera also reported a direct relationship between fluoride ion conductivity and corrosion rate. Only the Sc, Ba, and Y composition did not form the outer layer of crystalline precipitates.

Thorium-based glasses containing Zn–Ba–Y–Th, Zn–Ba–Yb–Th, or Zn–Ba–Yb–Th–Na have been reported to be 50–100 times more resistant to dissolution than the corresponding zirconium-based glasses [6.92].

6.7 CHALCOGENIDE – HALIDE GLASSES

Lin and Ho [6.93] studied the chemical durability of As–S–I glasses exposed to neutral, acidic, and basic solutions. These glasses exhibited excellent resistance to neutral and acidic (pH 2–8) solutions; however, in basic solutions they formed thioarsenites or thioarsenates:

$$As_2S_3 + 3NaOH \rightarrow Na_3AsS_3 + As(OH)_3 \tag{6.6}$$

$$2As(OH)_3 \rightarrow As_2O_3 + 3H_2O \tag{6.7}$$

or:

$$8As_2S_5 + 30NaOH \rightarrow 10Na_3AsS_4 + 3As_2O_5 + 15H_2O \tag{6.8}$$

As pH increased from 10 to 14, the rate of attack increased about 400 times. Higher iodine contents lowered the durability.

For a given iodine content, increased arsenic contents also lowered durability. Plots of weight loss vs. the square root of time were linear, indicative of a diffusion-controlled process. The rate of attack on alkaline solutions increased linearly with temperature. Lin and Ho concluded that the low solubility of these glasses was consistent with the fact that the As–S bond is highly covalent in nature.

6.8 ADDITIONAL RELATED READING

Clark, D.E.; Zoitos, B.K.; Eds. *Corrosion of Glass, Ceramics and Ceramic Superconductors*; Noyes Publications: Park Ridge, NJ, 1992.

Paul, A. *Chemistry of Glasses*; Chapman and Hall: New York, 1982; 293 pp.

6.9 EXERCISES, QUESTIONS, AND PROBLEMS

1. Discuss how pH affects dissolution of silicate glasses including the different mechanisms at low and high pH.
2. Discuss how glass structural variations relate to dissolution and how this is related to composition.
3. What structural factor and what pH relates to the minimum dissolution rate?
4. Describe the surface area/volume ratio of the attacking fluid effects upon dissolution rate.
5. How does a surface treatment of SO_2 gas diminish dissolution rates?
6. Why do Al_2O_3 and/or ZrO_2 substitutions for SiO_2 increase durability?
7. How does the Si/B ratio affect dissolution in borosilicate glasses?
8. Why is the number of nonbridging oxygens important to dissolution?
9. Explain how softening points and/or thermal expansion coefficients may relate to dissolution.

REFERENCES

6.1. Hutchins, J.R. III; Harrington, R.V. Glass. In *Encyclopedia of Chemical Technology*; 2nd Ed; Wiley: New York, 1966; Vol. 10. 572.

6.2. Wang, F.F-Y.; Tooley, F.V. Influence of reaction products on reaction between water and soda–lime–silica glass. J. Am. Ceram. Soc. 1958, *41* (12), 521–524.

6.3. Brady, P.V.; House, W.A. Surface-controlled dissolution and growth of minerals. In *Physics and Chemistry of Mineral Surfaces,* Brady, P.V., Ed.; CRC Press: New York, 1996; 225–305. Chap. 4.

6.4. Simpson, H.E. Study of surface structure of glass as related to its durability. J. Am. Ceram. Soc. 1958, *41* (2), 43–49.

6.5. Chen, H.; Park, J.W. Atmospheric reaction at the surface of sodium disilicate glass. Phys. Chem. Glasses 1981, *22* (2), 39–42.

6.6. Charles, R.J. Static fatigue of glass: I. J. Appl. Phys. 1958, *29* (11), 1549–1553.

6.7. Scholze, V.H.; Helmreich, D.; Bakardjiev. Investigation of the behavior of soda–lime–silica glasses in dilute acids (Gr). Glass Technol. Ber. 1975, *48* (12), 237–247.

6.8. Sanders, D.M.; Hench, L.L. Mechanisms of glass corrosion. J. Am. Ceram. Soc. 1973, *56* (7), 373–377.

6.9. Schmidt, Yu.A. *Structure of Glass*; Consultants Bureau: New York, 1958; Vol. 1. Translated from the Russian.

6.10. Boksay, Z.; Bouquet, G.; Dobos, S. Diffusion processes in the surface layer of glass. Phys. Chem. Glasses 1967, *8* (4), 140–144.

6.11. Doremus, R.H. Interdiffusion of hydrogen and alkali ions in a glass surface. J. Non-Cryst. Solids 1975, *19*, 137–144.

6.12. Boksay, Z.; Bouquet, G.; Dobos, S. The kinetics of the formation of leached layers on glass surfaces. Phys. Chem. Glasses 1968, *9* (2), 69–71.

6.13. Das, C.R. Reaction of dehydrated surface of partially leached glass with water. J. Am. Ceram. Soc. 1979, *62* (7–8), 398–402.

6.14. Rana, M.A.; Douglas, R.W. The reaction between glass and water:

Part 1. Experimental methods and observations. Phys. Chem. Glasses 1961, *2* (6), 179–195.

6.15. Rana, M.A.; Douglas, R.W. The reaction between glass and water. Part 2. Discussion of results. Phys. Chem. Glasses 1961, *2* (6), 196–205.

6.16. Das, C.R.; Douglas, R.W. Studies on the reaction between water and glass: Part 3. Phys. Chem. Glasses 1967, *8* (5), 178–184.

6.17. Douglas, R.W.; El-Shamy, T.M.M. Reaction of glass with aqueous solutions. J. Am. Ceram. Soc. 1967, *50* (1), 1–8.

6.18. Wood, S.; Blachere, J.R. Corrosion of lead glasses in acid media: I. Leaching kinetics. J. Am. Ceram. Soc. 1978, *61* (7–8), 287–292.

6.19. Eppler, R.A.; Schweikert, W.F. Interaction of dilute acetic acid with lead-containing vitreous surfaces. Ceram. Bull. 1976, *55* (3), 277–280.

6.20. Das, C.R. Chemical durability of sodium silicate glasses containing Al_2O_3 and ZrO_2. J. Am. Ceram. Soc. 1981, *64* (4), 188–193.

6.21. Paul, A. Chemical durability of glass. In *Chemistry of Glasses*; Chapman and Hall: New York, 1982; 108–147. Chap. 4.

6.22. Paul, A. Chemical durability of glasses. J. Mater. Sci. 1977, *12* (11), 2246–2268.

6.23. Paul, A.; Youssefi, A. Alkaline durability of some silicate glasses containing CaO, FeO, and MnO. J. Mater. Sci. 1978, *13* (1), 97–107.

6.24. Newton, R.G.; Paul, A. A new approach to predicting the durability of glasses from their chemical composition. Glass Technol. 1980, *21* (6), 307–309.

6.25. Jantzen, C.M. Thermodynamic approach to glass corrosion. In *Corrosion of Glass, Ceramics and Ceramic Superconductors*; Clark, D.E., Zoitos, B.K., Eds.; Noyes Publications: Park Ridge, NJ, 1992; 153–215. Chap. 6.

6.26. Plodinec, M.J.; Jantzen, C.M.; Wicks, G.G. Thermodynamic approach to prediction of the stability of proposed radwaste glasses. In *Advances in Ceramics, Vol. 8: Nuclear Waste Management*; Wicks, G.G., Ross, W.A., Eds.; Am. Ceram. Soc.: Columbus, OH, 1984; 491–495.

6.27. Jantzen, C.M.; Plodinec, M.J. Thermodynamic model of natural, medieval and nuclear waste glass durability. J. Non-Cryst. Solids 1984, *67*, 207–223.

6.28. Jantzen, C.M. Nuclear waste glass durability: I. Predicting environmental response from thermodynamic (Pourbaix) diagrams. J. Am. Ceram. Soc. 1992, *75* (9), 2433–2448.

6.29. Ernsberger, F.M. Structural effects in the chemical reactivity of silica and silicates. J. Phys. Chem. Solids 1960, *13* (3–4), 347–351.

6.30. Liang, D-T.; Readey, D.W. Dissolution kinetics of crystalline and amorphous silica in hydrofluoric–hydrochloric acid mixtures. J. Am. Ceram. Soc. 1987, *70* (8), 570–577.

6.31. Elmer, T.H.; Nordberg, M.E. Solubility of silica in nitric acid solutions. J. Am. Ceram. Soc. 1958, *41* (12), 517–520.

6.32. White, G.S.; Greenspan, D.C.; Freiman, S.W. Corrosion and crack growth in 33% Na_2O–67%SiO_2 and 33% Li_2O–67% SiO_2 glasses. J. Am. Ceram. Soc. 1986, *69* (1), 38–44.

6.33. Hench, L.L.; Wallace, S.; Wang, S.; Prassas, M. The processing and environmental behavior of a 20-mol% Na_2O–80 mol% SiO_2 (20 N) gel-glass. Ceram. Eng. Sci. Proc. 1983, *4* (9–10), 732–739.

6.34. Hench, L.L.; Prassas, M.; Phalippou, J. Surface behavior of gel-derived glasses. Ceram. Eng. Sci. Proc. 1982, *3* (9–10), 477–483.

6.35. Bacon, F.R.; Calcamuggio, G.L. Effect of heat treatment in moist and dry atmospheres on chemical durability of soda–lime glass bottles. Am. Ceram. Soc. Bull. 1967, *46* (9), 850–855.

6.36. Wu, C.K. Nature of incorporated water in hydrated silicate glasses. J. Am. Ceram. Soc. 1980, *63* (7–8), 453–457.

6.37. Tomozawa, M.; Erwin, C.Y.; Takata, M.; Watson, E.B. Effect of water content on the chemical durability of $Na_2O{\cdot}3SiO_2$ glass. J. Am. Ceram. Soc. 1982, *65* (4), 182–183.

6.38. Bartholomew, R.F.; Kozlowski, T.R. Alkali attack of glass surfaces by molten salts. J. Am. Ceram. Soc. 1967, *50* (2), 108–111.

6.39. Budd, S.M. Mechanism of chemical reaction between silicate glasses and attacking agents: I. Electrophilic and nucleophilic mechanism of attack. Phys. Chem. Glasses 1961, *2* (4), 111–114.

6.40. Loehman, R.E. Preparation and properties of yttrium–silicon–aluminum oxynitride glasses. J. Am. Ceram. Soc. 1979, *62* (9–10), 491–494.

6.41. Frischat, G.H.; Sebastian, K. Leach resistance of nitrogen-containing Na_2O–CaO–SiO_2 glasses. J. Am. Ceram. Soc. 1985, *68* (11), C305–C307.

6.42. White, J.E.; Day, D.E. Rare earth aluminosilicate glasses for in vivo radiation delivery. In *Rare Elements in Glasses*; Shelby, J.E., Ed.; Key Engineering Materials, Trans Tech Pub.: Switzerland, 1994; Vol. 94–95, 181–208.

6.43. Oda, K.; Yoshio, T. Properties of Y_2O_3–Al_2O_3–SiO_2 glasses as a model system of grain boundary phase of Si_3N_4 ceramics: Part 2. Leaching characteristics. J. Ceram. Soc. Jpn. 1991, *99,* (11), 1150–1152.

6.44. Erbe, E.M.; Day, D.E. In *Proceedings: Science and Tech of New Glasses*; Sakka, S., Soga, N., Eds.; Ceram. Soc. Japan: Tokyo, 1991; 105.

6.45. Baitinger, W.E.; French, P.W.; Swarts, E.L. Characterization of tin in the bottom surface of float glass by ellipsometry and EPS. J. Non-Cryst. Solids 1980, *38–39*, 749–754.

6.46. Colombin, L.; Charlier, H.; Jelli, A., et al. Penetration of tin in the bottom surface of float glass—a synthesis. J. Non-Cryst. Solids 1980, *38–39* (Pt II), 551–556.

6.47. Kondrashov, V.; Bezlyudnaya, V.; Zverev, Y. Effect of regimes and methods of glass forming on the tin content in float glass. Glass Ceram. 1998, *55* (5–6), 174–175.

6.48. Meisel, W. Depth profile of tin in float glass—A CEMS study. Glastech. Ber. Glass Sci. Technol. 1998, *72* (9), 291.

6.49. Paulson, T.E.; Spear, K.E.; Pantano, C.G. Thermodynamic analysis of the tin penetration profile in high-iron float glass. Proc. ICG, Am. Cer. Soc. 1998.

6.50. Sieger, J.S. Chemical characteristics of float glass surfaces. J. Non-Cryst. Solids 1975, *19*, 213–220.

6.51. Sieger, J.S.; Parsons, J.M. Tin penetration profiles in float glass.

Am. Cer. Soc. Bull. 1974, *53* (4), 350; Abstract of a presentation at the 76th Annual Meeting of the Am. Cer. Soc.

6.52. Swift, H.R. How surface chemistry affects float glass properties. Glass Ind. 1984, *65* (5), 27–30.

6.53. Wang, T-J. Penetration of tin in the surface of float glass. Glass Technol. 1997, *38* (3), 104–106.

6.54. Williams, K.; Johnson, C.; Greengrass, J.; Tilley, B.; Gelder, D.; Johnson, J. Tin oxidation state, depth profiles of Sn^{2+} and Sn^{4+}, and oxygen diffusivity in float glass by Mossbauer spectroscopy. J. Non-Cryst. Solids 1997, *211* (1–2), 164–172.

6.55. Franz, H. Surface chemistry of commercial glasses. Ceram. Eng. Sci. Proc. 1995, *16* (2), 221–227.

6.56. Mishra, A.; Pecoraro, G.A.; Paulson, T.E.; Pantano, C.G. Glass–tin interactions during the float glass forming process. Ceram. Trans. 1998, *82*, 205–217.

6.57. Hood, H.P.; Nordberg, M.E.; Method of Treating Borosilicate Glasses. U.S. Patent 2,215,039, September 17, 1940.

6.58. Takamori, T.; Tomozawa, M. HCl leaching rate and microstructure of phase-separated borosilicate glasses. J. Am. Ceram. Soc 1978, *61* (11–12), 509–512.

6.59. Taylor, P.; Ashmore, S.D.; Owen, D.G. Chemical durability of some sodium borosilicate glasses improved by phase separation. J. Am. Ceram. Soc. 1987, *70* (5), 333–338.

6.60. Kinoshita, M.; Harada, M.; Sato, Y.; Hariguchi, Y. Percolation phenomenon for dissolution of sodium borosilicate glasses in aqueous solutions. J. Am. Ceram. Soc. 1991, *74* (4), 783–787.

6.61. El-Hadi, Z.A.; Gammal, M.; Ezz-El-Din, F.M.; Moustaffa, F.A. Action of aqueous media on some alkali-borate glasses. Cent. Glass Ceram. Res. Inst. Bull 1985, *32* (1–2), 15–19.

6.62. Tait, J.C.; Jensen, C.D. Effect of Zn(II) ion adsorption on the durability of sodium borosilicate glasses. J. Non-Cryst. Solids 1982, *49* (1–3), 363–377.

6.63. Katayama, J.; Fukuzuka, M.; Kawamoto, Y. Corrosion of heavy crown glass by organic acid solutions. Yogyo Kyokai Shi 1978, *86* (5), 230–237.

6.64. Walters, H.V. Corrosion of a borosilicate glass by orthophosphoric acid. J. Am. Ceram. Soc 1983, *66* (8), 572–574.

6.65. Metcalfe, A.G.; Schmitz, G.K. Mechanism of stress corrosion in E glass filaments. Glass Technol. 1972, *13* (1), 5–16.

6.66. Priest, D.K.; Levy, A.S. Effect of water content on corrosion of borosilicate glass. J. Am. Ceram. Soc. 1960, *43* (7), 356–358.

6.67. Koch, G.H.; Syrett, B.C. Progress in EPRI research on materials for flue gas desulphurization systems. In *Dewpoint Corrosion*; Holmes, D.R., Ed.; Ellis Horwood Ltd: Chichester, UK, 1985; 105–124.

6.68. Velez, M.H.; Tuller, H.L.; Uhlmann, D.R. Chemical durability of lithium borate glasses. J. Non-Cryst. Solids 1982, *49* (1–3), 351–362.

6.69. Conzone, C.D.; Brown, R.F.; Day, D.E.; Ehrhardt, G.J. In vitro and in vivo dissolution behavior of a dysprosium lithium borate glass designed for the radiation synovectomy treatment of rheumatoid arthritis. J. Biomed. Mater. Res. 2002, *60* (2), 260–268.

6.70. Day, D.E. Reactions of bioactive borate glasses with physiological liquids. Glass Res 2002–2003, *12* (1–2), 21–22.

6.71. Yoon, S.C. Lead release from glasses in contact with beverages; M.S. thesis, Rutgers University, New Brunswick, NJ, 1971.

6.72. Pohlman, H.J. Corrosion of lead-containing glazes by water and aqueous solutions. Glastech. Ber. 1974, *47* (12), 271–276.

6.73. Yoon, S.C. Mechanism for lead release from simple glasses, Univ. Microfilms Int. (Ann Arbor, Mich.) Order No. 73-27,997; Diss. Abstr. Int. 1973, *B34* (6) 2599.

6.74. Lehman, R.L.; Yoon, S.C.; McLaren, M.G.; Smyth, H.T. Mechanism of modifier release from lead-containing glasses in acid solution. Ceram. Bull. 1978, *57* (9), 802–805.

6.75. Krajewski, A.; Ravaglioli, A. Lead-ion stability in vitreous systems. J. Am. Ceram. Soc 1982, *65* (5), 265–269.

6.76. *Lead Glazes for Dinnerware*, International Lead Zinc Research Organization Manual, Ceramics I, International Lead Zinc Research Organization and Lead Industries Association, New York, 1974.

6.77. Haghjoo, M.; McCauley, R.A. Solubility of lead from ternary and quaternary silicate frits. Ceram. Bull. 1983, *62* (11), 1256–1258.

6.78. Moore, H. The structure of glazes. Trans. Br. Ceram. Soc. 1956, *55*, 589–600.

6.79. Clark, A.E.; Pantano, C.G.; Hench, L.L. Spectroscopic analysis of bioglass corrosion films. J. Am. Ceram. Soc. 1976, *59* (1–2), 37–39.

6.80. Hench, L.L. Surface modification of bioactive glasses and ceramics. In *Corrosion of Glass Ceramics and Ceramic Superconductors*; Clark, D.E., Zoitos, B.K., Eds.; Noyes Publications: Park Ridge, NJ, 1992; 298–314.

6.81. Minami, T.; Mackenzie, J.D. Thermal expansion and chemical durability of phosphate glasses. J. Am. Ceram. Soc. 1977, *60* (5–6), 232–235.

6.82. Reis, S.T.; Karabulut, M.; Day, D.E. Chemical durability and structure of zinc–iron phosphate glasses. J. Non-Cryst. Solids 2001, *292* (1–3), 150–157.

6.83. Hench, L.L. Bioactive glasses help heal, repair and build human tissue. Glass Res. 2002–2003, *12* (1–2), 18.

6.84. Hench, L.L.; Wilson, J. Introduction. In *An Introduction to Bioceramics*; Advanced Series in Ceramics; World Scientific Publishing Co. Ltd: Singapore, 1993; Vol. 1, 1–24.

6.85. Avent, A.G.; Carpenter, C.N.; Smith, J.D.; Healy, D.M.; Gilchrist, T. The dissolution of silver–sodium–calcium–phosphate glasses for the control of urinary tract infections. J. Non-Cryst. Solids 2003, *328*, 31–39.

6.86. Murch, G.E., Ed.; *Materials Science Forum, Halide Glasses I and II,* Proceedings of the 3rd International Symposium on Halide Glasses, Rennes, France, Trans Tech Publications: Aedermannsdorf, Switzerland, 1985.

6.87. Ravaine, D.; Perera, G. Corrosion studies of various heavy-metal fluoride glasses in liquid water: application to fluoride-ion-selective electrode. J. Am. Ceram. Soc. 1986, *69* (12), 852–857.

6.88. Doremus, R.H.; Bansal, N.P.; Bradner, T.; Murphy, D. Zirconium fluoride glass: surface crystals formed by reaction with water. J. Mater. Sci. Lett. 1984, *3* (6), 484–488.

6.89. Simmons, C.J.; Simmons, J.H. Chemical durability of fluoride glasses: I. Reaction of fluorozirconate glasses with water. J. Am. Ceram. Soc. 1986, *69* (9), 661–669.

6.90. Gbogi, E.O.; Chung, K.H.; Moynihan, C.T.; Drexhage, M.G. Surface and bulk -OH infrared absorption in ZrF_4- and HfF_4-based glasses. J. Am. Ceram. Soc. 1981, *64* (3), C51–C53.

6.91. Robinson, M.; Drexhage, M.G. A phenomenological comparison of some heavy metal fluoride glasses in water environments. Mater. Res. Bull. 1983, *18*, 1101–1112.

6.92. Simmons, C.J.; Azali, S.; Simmons, J.H. Chemical Durability Studies of Heavy Metal Fluoride Glasses. Extended Abstract # 47, 2nd International Symp. on Halide Glasses, Troy, NY; 1983.

6.93. Lin, F.C.; Ho, S.-M. Chemical durability of arsenic–sulfur–iodine glasses. J. Am. Ceram. Soc. 1963, *46* (1), 24–28.

7

Corrosion of Composites Materials

The whole is most always better than the sum of the parts.

ANONYMOUS

7.1 INTRODUCTION

Although the term composite historically meant any product made from a combination of two (or more) materials, the modern meaning is less broad in scope. In general, a composite is manufactured in an attempt to obtain the best properties of two materials or at least to capture a specific property of each material that is potentially better in the composite. It is also possible for the composite to have a particular property that neither component exhibited individually. According to Holmes and Just [7.1], a true composite is where distinct materials are combined in a nonrandom manner to produce overall structural characteristics superior to those of the individual components. Although, in a very broad sense, products such as glazed ceramic tile, enameled metal, and ceramic coated metal (e.g., thermal barrier coatings) could be considered composites, they will not be considered as such here. Only those materials where a substantial intermixing of the different materials exists on a microscopic scale will be considered composites.

The concept of composite materials is not a new idea and is definitely not limited to ceramics. Nature has provided us with several excellent examples of composite materials. Wood is a composite of cellulose fibers contained in a matrix of lignin. Bone, another example, is composed of the protein collagen and the mineral apatite. In all these materials, the result is a product that is lighter and stronger than either of the components individually. Because of this, they can be used in more severe environments, e.g., space exploration. A list of the more desirable properties of a composite is given in Table 7.1. In a very broad sense, all engineering materials are composites of one kind or another.

The matrix and the reinforcement, quite often fibrous, provide two different functions. The reinforcement is most often a discontinuous phase whether it be a fibrous material or a particulate material. It is important that the reinforcement be discontinuous, especially if it is a ceramic, so that cracks will

TABLE 7.1 Desirable Properties of Composites

1. Very strong in tension
2. Very lightweight
3. Very stiff
4. High strength-to-weight ratio
5. High stiffness-to-weight ratio
6. Good corrosion resistance
7. Low thermal expansion
8. High abrasion resistance
9. High melting point

not be able to propagate through it. The matrix must not damage the reinforcement and it must transmit any stresses to the reinforcement. Thus the adhesion of the matrix to the reinforcement is of prime importance for mechanical integrity and is the region of greatest importance related to corrosion. Since it is necessary to have weak interfaces to maximize toughness (i.e., resistance to crack propagation), the development of optimum fiber/matrix interfaces is quite difficult. To obtain these optimum characteristics, it is sometimes required to coat the reinforcement fibers with various materials to obtain the proper debonding, sliding, and/or reaction characteristics. Fibers that do not debond do not enhance toughening and lead only to increased brittle fracture of the composite [7.2–7.7].

A recent development in composites is that of a nanosized second phase or reinforcement material. The second phase particles are generally less than 300 nm and are present in amounts equal to 1–30 vol.%. These new composites unfortunately have been called *nanocomposites*.

Before going into the specifics of corrosion of composite materials, a few words must be said about those materials that have been called *cermets*. Historically, the term cermet was derived to cover those materials composed of cobalt-bonded tungsten carbide and used as cutting tools. Since cermets contain both ceramics and metals, some confusion has existed in the literature as to an exact meaning. The term, however, has been used to cover a broad list of materials. It appears that the ceramic community confines cermets to essentially cutting tool materials, whatever the matrix or reinforcement, whereas the metals community confines cermets to only those materials with a metal matrix. Since the broader concept of composites includes those materials called cermets, only the term composite will be used in the discussion below.

The actual corrosion of composite materials quite often begins with reaction of the reinforcement material and especially with any interface material (called the *interphase*) used to coat the reinforcement for debonding. One property that exacerbates this is a mismatch in thermal expansion coefficients between the reinforcement and the matrix, leading to microcracks. These microcracks allow the ingress of corrosive gases (e.g., oxygen). Courtright [7.8] has given the value of 10^{-12} g O_2/cm sec for the limit of oxygen ingress that causes nonoxide fiber deterioration. Microcracks are also quite often a product of sample preparation techniques, and thus great care must be used in cutting and grinding/polishing samples for testing. If the composite is cut or machined, any exposed fiber reinforcement will be susceptible to attack by the environment. Because of this inherent problem, protective coatings are often applied to the exterior surfaces. Actually, the whole corrosion process of composite materials is not unlike that of other polyphase ceramic materials where the grain boundary phase is the first to corrode. A complete understanding of all the phases that make up the microstructure of the composite must also be known for an accurate interpretation of any corrosion. For example, Munson and Jenkins [7.9] reported that their samples were actually attacked internally by molten metal from a small amount of free aluminum present as a residue during the manufacture of Dimox™* (a melt-infiltrated alumina). Actually, a large amount of the literature on composites is concerned with an evaluation of the internal reactions that take place among the various reinforcement, interphase, and matrix materials. The time-dependent loss of strength due to the corrosive nature of moist environments at room temperature is a major concern for composites containing glass or glass-ceramics as either the matrix or the reinforcement [7.10]. As temperatures are increased, the concern shifts toward oxidation problems associated with nonoxide materials. See the discussions in Chap. 5, Sec. 5.2.2, Nitrides and Carbides, and Chapter 8, Properties and Corrosion, for more details of oxidation and its effects upon the properties of nonoxides.

---

*DIMOX™ (directed metal oxidation) is the name given to composites manufactured by a process developed by Lanxide Corp., Newark, DE in 1986.

TABLE 7.2 Acronyms Used in the Discussion of Composites

BMC—brittle matrix composite
C/C—carbon–carbon composite
CCC—carbon–carbon composite
CFCC—continuous fiber ceramic composite
CFCMC—continuous fiber ceramic matrix composite
CFRC—continuous fiber reinforced ceramic
CFRCMC—continuous fiber reinforced ceramic matrix composite
CFRP—carbon fiber reinforced plastic, polymer, or polyester
CMC—ceramic matrix composite
CPC—ceramic particulate composite
FRCMC—fiber reinforced ceramic matrix composite
FRP—fiber reinforced plastic, polymer, or polyester
GCMFC—glass-ceramic matrix fiber composite
GFRP—glass fiber reinforced plastic
GRC—glass fiber reinforced cement
GRP—glass reinforced plastic, polymer, or polyester
IMC—intermetallic matrix composite
MMC—metal matrix composite
PMC—polymer matrix composite
RCC—reinforced carbon–carbon composite
SFRC—short fiber reinforced ceramic
TMC—titanium metal matrix composite
WRA—whisker reinforced alumina

With the advancement of the development of composites, there is an increasing number of acronyms with which one must contend. To aid the reader, a list is given in Table 7.2 of the most common acronyms.

7.2 REINFORCEMENT

7.2.1 Fibers

Various types of materials have been used as the fibrous reinforcement. These include various glasses, metals, oxides, nitrides, and carbides either in the amorphous or crystalline state. The surface chemistry and morphology of fibers is very important in determining their adherence to the matrix. Fiber internal structure and morphology determines the mechanical strength. A tremendous amount of literature is available that discusses the degradation of mechanical properties as temperatures are increased in various atmospheres; however, there is very little interpretation of any corrosion mechanisms that may be involved. Although many composites are classified as continuous-fiber-reinforced, some composites contain fibers that are actually not continuous but of a high aspect ratio (i.e., length-to-width). The actual matrix material will determine the aspect ratio required to obtain a certain set of properties. Thus the term "high aspect ratio" is a relative term.

Boron fibers can generally be heated in air to temperatures of about 500°C without major strength deterioration. Above 500°C, the oxide that formed at lower temperatures becomes fluid increasing the oxidation rate and drastically reducing the strength [7.11]. Galasso [7.11] discussed the benefits of coating boron fibers with either SiC or by nitriding the surface. The SiC coating was more protective than the nitride with strength retention even after 1000 hr at 600°C in air. Boron carbide (B_4C) is stable to 1090°C in an oxidizing atmosphere, whereas boron nitride is stable to only 850°C.

Carbon or graphite fibers have been used since the early 1970s as reinforcement for composites. Strength loss due to oxidation occurs at temperatures above 500°C in air. An interesting structural feature of carbon fibers is that they have a relatively large negative axial thermal expansion coefficient.

Glass fibers generally are used as reinforcement for composites that are to be used at low temperatures (i.e., <500°C) due to the softening of glasses at elevated temperatures. These composites are generally of the polymer matrix type and are used for marine or at least moist environments. It is well known that glass is attacked by moist environments with the specific mechanism dependent upon the pH (see Chap. 6). It has been shown by Metcalfe and Schmitz [7.12] that borosilicate glass fibers when exposed to moist ambient environments developed surface tensile stresses caused by exchange of alkali for hydrogen sufficient to cause failure.

A large portion of the CMC today contains SiC fiber reinforcement. This is mainly due to the excellent properties of SiC—low reactivity to many matrix materials, its strength at elevated temperatures, and its oxidation resistance. It is this latter property (i.e., oxidation resistance) that generally causes deterioration in these materials. SiC will oxidize readily when heated to temperatures greater than 1000°C. As discussed in Chap. 5, Silicon Carbide, page 223, at low partial pressures of oxygen, active corrosion takes place with the formation of gaseous products of CO and SiO. At higher partial pressures, passive oxidation occurs with the formation of CO and SiO_2 that may be protective if cracks do not form. The formation of cracks is dependent upon the heat treatment and whether the oxide layer is crystalline or amorphous. These reactions generally result in the decrease of fiber strength. Nicalon™ fiber*, being formed by the pyrolysis of organometallics, actually contains some remnant oxygen (~9%) and carbon (~11%) that will affect the subsequent oxidation of the fiber. Two different grades of Nicalon™ fiber have been examined by various investigators [7.13–7.15]. Clark et al. [7.13] reported these fibers to exhibit weight losses of 13% and 33% after being treated in argon at 1400°C. Both grades of fiber gained weight (on the order of 2–3%) when treated in flowing wet air at 1000°C, 1200°C, and 1400°C. As-received Nicalon™ fibers have protective sizing (i.e., polyvinyl acetate) on their surfaces. When heated in air, this sizing will burn off at temperatures between 250°C and 500°C. At temperatures above about 1250°C, the SiC_xO_y amorphous phase contained in these fibers decomposed to SiO and CO [7.16].

* Nicalon™, Nippon Carbon Co., Tokyo, Japan.

Titanium nitride (TiN) resists attack from iron or nickel aluminides better than does SiC and thus is a better reinforcement for these metal alloy matrix composites [7.17].

7.2.2 Fiber Coatings or Interphases

Protective coatings (also called *interphases*) such as graphite or BN, in addition to providing proper debonding and pullout [7.18,7.19], are used to provide some degree of oxidation resistance [7.20,7.21] for fibers such as SiC. Bender et al. [7.21] concluded that the BN protects the SiC fiber from the matrix since BN will not react with SiO_2, which is generally present on the surface of the fibers. Boron nitride-coated mullite, carbon, and SiC fibers were tested in a mullite matrix with varying degrees of success by Singh and Brun [7.22].

Boron nitride-coated SiC fibers have shown a slight improvement over carbon-coated fibers with an increase of about 100–200°C in composite embrittlement (see Sec. 7.3 for a discussion concerning embrittlement) temperatures [7.23]. Since some matrices are grown in situ, techniques to coat fibers become problematic. A combination coating of BN and SiC was developed by Fareed et al. [7.24] to eliminate the undesirable reaction of molten aluminum in contact with Nicalon™ fibers forming alumina and aluminum carbide during the directed metal oxidation method (at 900–1000°C) of forming an alumina matrix. When used alone as a coating, BN oxidation inhibited complete oxidation of the aluminum. In combination with SiC, however, Fareed et al. believed that any oxidation of BN led to the formation of boria glass that acted as a sealant to any microcracks, thus minimizing oxygen ingress and protection of the composite. The SiC outer coating protected the BN inner coating during growth of the matrix. Ogbuji [7.25] reported that the BN first oxidized to B_2O_3, which then dissolved some of the SiC fiber and matrix forming a borosilicate liquid. If any moisture were present, the boria may be volatilized by hydrolysis releasing $B(OH)_4$ gas. This reaction resulted in a silica residue that cemented the fibers together embrittling the composite.

In an effort to find an interphase or coating for alumina and mullite fibers, Cooper and Hall [7.26] developed a synthetic fluorophlogopite*, based upon their geochemical approach, that when reacted with alumina formed an intermediate spinel phase that was stable after heating to 1200°C in air for 150 hr. Thus by coating alumina fibers with spinel and then using the fluorophlogopite as an interphase, an alumina matrix composite proved successful. Above about 1280°C, the alumina reaction with fluorophlogopite produced forsterite, leucite, and spinel along with the volatile fluorides SiF_4, AlF_3, and KF [see Eq. (7.1) below], making the spinel-coated alumina fiber/fluorophlogopite laminate unstable at those high temperatures. Reactions between mullite and fluorophlogopite formed cordierite in addition to the phases mentioned above. This was not successful as a mullite fiber composite since the cordierite allowed potassium diffusion from the fluorophlogopite continually deteriorating the mullite. In the alumina fiber case, the spinel coating acted as a barrier to potassium diffusion.

$$\begin{aligned} 4KMg_3AlSi_3O_{10}F_2 + 2Al_2O_3 &\rightarrow 5Mg_2SiO_4 \\ &+ 3KAlSi_2O_6 + 2MgAl_2O_4 + AlF_3 + SiF_4 + KF \end{aligned} \tag{7.1}$$

Cooper and Hall reported that reaction (7.1) occurred at temperatures above 1230°C in flowing dry argon, although thermodynamic calculations indicated that the reaction proceeded only after the temperature reached 1279°C. This was attributed to the partial pressures of the gaseous phases not summing to 1 atm during the experiment in flowing argon.

7.2.3 Particulates

For a discussion of the various mechanisms involved in toughening composites when particulates are used as the reinforce-

*Cooper and Hall use the term fluorophlogopite interchangeably with the terms mica, fluoromica, and fluorophyllosilicate, which may cause some confusion unless the reader is well versed in mineralogy.

ment, one should read the article by Rice [7.27]. Particulate reinforcement can be oxides, carbides, nitrides, borides, or even metals. Some of the common materials used are listed below:

$Al_2O_3, ZrO_2, SiC, TiC, BN, TiB_2, NbB_2, Co.$

The corrosion of most of these materials has been discussed in Chap. 5 (Corrosion of Specific Crystalline Materials) and will not be repeated here.

Actually, most of the ceramic products manufactured today could be considered ceramic particulate reinforced ceramic matrix composites. For example, one product not generally considered a composite by the ceramics community is MgO:carbon refractories. This product contains a substantial quantity of carbon particles in an effort to improve the corrosion resistance of the MgO in molten metal applications. In addition to the obvious oxidation of carbon to monoxide and/or dioxide, carbon will also react with MgO at temperatures above 1400°C forming magnesium vapor and carbon monoxide.

7.3 CERAMIC MATRIX COMPOSITES

Ceramic matrix composites are characterized by a high modulus of elasticity, excellent high temperature and corrosion resistance, but generally poor crack propagation resistance. The composite systems that have probably received the most attention are those of SiC or carbon fiber reinforced SiC. The major problem with these materials is one of oxidation of either the carbon or the SiC. Below about 600°C, oxidation is generally not a problem. Above 1000°C, the oxidation of both carbon and SiC are rapid. It is the temperature range of 600–1000°C that is the most difficult where the carbon oxidation can be rapid but the SiC is relatively inert [7.28]. Any microcracks or pores can allow ingress of oxygen for continued oxidation of the carbon.

Stress corrosion cracking (discussed in "Introduction" in Chap. 8) of the matrix is a major problem in CMC. Another problem is that of *oxidation embrittlement* that can manifest itself through the oxidation of the interface between the fiber reinforcement and the matrix subsequently causing a strong bond between the two leading to embrittlement. Embrittlement may not be noticed when samples are tested in flexure due to load redistribution, thus requiring that samples be tested in tension [7.29].

7.3.1 Oxide–Matrix Composites

Al_2O_3–Matrix Composites

A BN/SiC-coated Nicalon™ fiber/Al_2O_3 composite has been reported by Heredia et al. [7.29] to become embrittled when heated to temperatures between 650°C and 850°C. The tensile strength after heat treatment for 24 hr was reduced by one-half. Only the outer regions of the ~3-mm-thick samples were fully embrittled.

A 21 vol.% SiC in alumina composite was reported by Borom et al. [7.30] to form a reaction zone upon oxidation at 1530°C for 150 hr that contained mullite and an amorphous aluminosilicate phase containing bubbles from the formation of CO. The SiO_2 formed by the oxidation of the SiC reacted with the alumina matrix to form the mullite. It is important that the formation of silica in the outer layer is sufficient for complete conversion of the alumina to mullite [7.31]. Insufficient silica causes a rigid scale that delaminates. Too much silica forms a scale containing mullite and silica on an alumina substrate that may also delaminate due to expansion mismatch during thermal cycling. A matrix of mullite works much better than alumina since the scale is more compatible with the substrate, both containing mullite, and thus forms a protective layer. Luthra [7.32] reported that the products of reaction of SiC with alumina should be mullite and alumina when the SiC content is below 24.4 vol.% and silica and mullite when it is greater than 24.4 vol.%. In practice, this limit will vary due to mullite forming over a range of compositions.

A TiN/Al_2O_3 composite was reported by Mukerji and Biswas [7.33] to exhibit linear oxidation kinetics above 820°C after a short (<120 min) parabolic induction period. The change from parabolic to linear kinetics was reported to be due to the difference in specific volumes between TiN and TiO_2 that caused an expansion of the oxidized layer forming cracks, which allowed oxidation to continue. The rutile that formed above 820°C was reported to grow epitaxially with a preferential growth direction of [211] and [101]. At 820°C and 710°C, this oriented growth was not present. Tampieri and Bellosi [7.34] reported this oriented growth to occur in the [221] and [101] directions and only above 900°C. Contrary to Mukerji and Biswas, Tampieri and Bellosi reported parabolic growth between 900°C and 1100°C for times up to 1200 min. These differences must be attributed to differences in starting materials and experimental conditions since the authors did not report any specific reasons that one may assign to the variation in results.

TiN decomposes to form titanium oxides and aluminum titanates at temperatures in excess of 1550°C [7.35]. TiN will also react with alumina to form titanium oxides and aluminum titanates at temperatures as low as 1450–1500°C. Therefore during processing by hot pressing, the temperature must be kept below 1500°C and the pressure must be high.

The oxidation at 1500°C of TiC-containing (25 vol.%) alumina matrix composite has been reported to form Al_2TiO_5 as the reaction product by Borom et al. [7.31]. Approximately a 30-vol.% expansion accompanied this reaction that caused delamination of the oxide reaction product layer.

In a previous study, Borom et al. [7.30] reported the oxidation at 1520°C of $MoSi_2$ (10 vol.%) dispersed within a matrix of alumina to form a reaction layer of mullite and volatile MoO_3 that completely escaped. It was suggested that this reaction layer contained an interconnected network of porosity through which the MoO_3 escaped, although no evidence of such porosity was given. Linear growth kinetics was reported for the formation of this nonprotective layer of mullite. A unique-appearing periodic change in density (porosity) was developed at about 200-μm intervals within the mullite reaction layer along with a slight bulging of the layer, both of which were reported to be due to volume changes during reaction and thermal expansion mismatch among the phases present during cooling.

Several investigators have shown that by adding particulate nickel to alumina, a composite with improved mechanical properties can be obtained. Volume percentages ranging from 5% to 35% Ni have been studied. Although the oxidation resistance of nickel was quite good, compared to other metals, its resistance was decreased when added to alumina. Wang et al. [7.36] studied the oxidation of Ni-toughened alumina from 1000°C to 1300°C for up to 1000 hr with nickel contents ranging from 5 to 15 vol.%. They found that oxygen diffused into the composite and nickel diffused out to the surface forming NiO first and then a dense $NiAl_2O_4$ spinel on the surface.

Although the reinforcement is the problem area most of the time, the alumina matrix can also deteriorate under some conditions. One of those is in an environment of hydrogen. The upper limit of usage has been estimated to be between 1200°C and 1300°C by Nelson [7.37].

Other Oxide–Matrix Composites

Glass and Glass-Ceramics (Alkali and Alkaline-Earth Aluminosilicates). Probably the first glass matrix composites were those reinforced with graphite fibers [7.38]. Due to the oxidation problems with graphite, other fiber reinforcements were developed, predominantly Al_2O_3 and SiC. Due to the inherent softening of glasses, glass matrix composites are relegated to temperatures that are generally less than 1000°C. To improve upon this temperature limitation, researchers are now investigating glass-ceramics, which are formed in the glassy state and then subsequently crystallized to obtain maximum temperature stability.

Nicalon™ SiC fibers (i.e., generally any polymer-derived fibers of the Si–C–O and Si–C–N–O types) have been reported by several investigators to react with various alkali and alkaline-earth aluminosilicate glass matrices (LAS* [7.39], CAS* [7.40,7.41], and BaMAS* [7.42]) during fabrication to form carbon at the fiber/matrix interface. Although the oxygen contained within the as-fabricated SiC fibers can react to form SiO, SiO_2, C, CO, or CO_2 depending upon which of the following equations:

$$SiC + 3/2\ O_2 \rightarrow SiO + CO_2, \tag{7.2}$$

$$SiC + 3/2\ O_2 \rightarrow SiO_2 + CO, \tag{7.3}$$

$$SiC + O_2 \rightarrow SiO_2 + C, \tag{7.4}$$

$$SiC + 2CO \rightarrow SiO_2 + 3C, \text{ or} \tag{7.5}$$

$$C + 1/2\ O_2 \leftrightarrow CO \tag{7.6}$$

is operative, it is also possible for the reaction to be caused by oxygen and/or CO that is dissolved in the glass or glass-ceramic matrix material [7.43]. Pantano et al. [7.44] developed a model, based upon stoichiometric SiC that explained many experimental observations when Nicalon™ fiber was used, showing that it was the effective pO_2 of the glass that was the driving force for the reaction. Long time exposure in air eventually oxidized the carbon and the SiC fibers resulting in deterioration of the composite.

Glass-ceramics being materials composed of several phases are considered composites by many investigators and by all definitions they should be. These are materials that are formed as a glass and then either heated to an intermediate temperature or held at some intermediate temperature upon cooling to crystallize the glass either completely or partially. These are composites where the reinforcement is generally particulate and crystalline and the matrix is the remaining glassy phase. Therefore any corrosion that takes place attacks the matrix glassy phase first and generally more severely [7.45].

A lithia-aluminosilicate glass-ceramic containing a small percentage of Nb_2O_5 was reported by Prewo et al. [7.46] to react

---

* LAS = lithia-aluminosilicate, CAS = calcia-aluminosilicate, and BaMAS = baria-magnesia-aluminosilicate.

with the SiC reinforcement fibers forming particles of NbC on the surface of the fibers. A very thin carbon-rich layer formed between the SiC fibers and the NbC that contributed to the excellent toughness and crack deflection of these composites.

The reactions between Nicalon™ fiber and Solaramic™* glass (a baria and chromia silicate) were examined by Mendelson [7.47]. He was concerned that reaction between the two might cause sufficient fiber/matrix bonding to degrade the mechanical properties. Based upon the assumption that glass constituents of less than ~5 wt.% would not degrade the fibers or the composite, Mendelson calculated the free energy of reaction at 1350°C and 1 atm pressure among the remaining constituents and SiC. His calculations indicated that the formation of Cr_3C_2, $CrSi_2$, SiO_2, and BaC_2 was probable and that SiO formation was not. Upon heat treatment of composite samples at 1150°C and 1350°C, Mendelson found that Cr_3C_2, $CrSi_2$, and SiO_2 did indeed form and that barium diffused into the fibers at the higher temperature causing embrittlement and degraded strengths. Mendelson attributed this incompatibility to the excess carbon and oxygen in the fibers and to the Cr- and Ba-containing glass matrix. The diffusion of barium into Nicalon™ fibers was reported also by Herron and Risbud [7.48,7.49]. Their matrix glass was a BaSiAlON. Diffusion of the barium extended into the fibers to a depth of ~15 μm. Aluminum was also found to diffuse into the fibers but to a lesser depth (~ 2–3 μm). Herron and Risbud offered no explanation as to the effect of this diffusion of Ba and Al upon the mechanical properties of the composites.

Heredia et al. [7.29] reported the embrittlement of a carbon/boron-coated Nicalon™ fiber/magnesium aluminosilicate composite when heat-treated to temperatures between 500°C and 750°C. Similar results were reported by Wetherhold and Zawada [7.50] for a Nicalon™ fiber/alkaline-earth aluminosilicate resulting from the oxidation at 650°C in air of the carbon interface and subsequent embrittlement of the compo-

* Solaramic™, Arcilla Research, Epen, the Netherlands.

site. This embrittlement was less severe if heat treatments were performed at a higher temperature (800°C or 850°C for 15 min). This was attributed to the higher temperature causing the formation of a glassy phase, presumably amorphous SiO_2 that flowed and sealed the matrix inhibiting oxygen ingress, as reported in an earlier study by Bischoff et al. [7.51]. Wetherhold and Zawada suggested that a short high-temperature heat treatment could protect these composites against embrittlement when exposed to lower temperatures.

The reaction of various small quantities of materials contained within a composite must not go unexamined, especially when reinforced with SiC. Pannhorst et al. [7.52] found that a composite containing a small amount of TiO_2 formed TiC during fabrication. This reaction was sufficiently severe to degrade the SiC fibers causing the composite to exhibit bend strengths of about 100 MPa.

Mullite, Spinel, Titania, and Zirconia. The oxidation at temperatures between 1310°C and 1525°C of a 30-vol.% SiC in mullite composite was reported by Borom et al. [7.31] to obey parabolic kinetics and form a reaction layer of mullite and an amorphous aluminosilicate phase containing bubbles from CO evolution. Similar results were obtained by Luthra and Park [7.53] and Hermes and Kerans [7.54]. By changing the matrix from mullite to a strontium-aluminosilicate ($SrO{\cdot}Al_2O_3{\cdot}2SiO_2$) phase, Borom et al. also showed that the presence of alkaline earth cations increased the oxidation rates by 1 to 2 orders of magnitude, presumably due to the formation of nonbridging oxygens in the silicate glass that allowed much higher transport rates.

In a composite containing 18.5 vol.% SiC in a matrix of mullite (40 vol.%), alumina (26 vol.%), zirconia (12 vol.%), and spinel (3.5 vol.%), Baudin and Moya [7.55] reported passive oxidation at 1200°C, 1300°C, and 1400°C in air. No weight changes were reported for 800°C and 900°C and minimal changes were noted at 1000°C and 1100°C. The oxidized layer contained cordierite along with mullite, zirconia, and alumina at 1200°C and 1300°C. At 1400°C, mullite and zircon were detected along with a viscous amorphous phase. The silica oxidation product apparently reacted with the free alumina and zirconia present to form additional mullite and zircon.

Hermes and Kerans [7.54] found that the magnesium from a spinel matrix composite containing 30 vol.% SiC heated to 1250°C in air diffused to the surface faster than the aluminum or silicon forming an outer layer of MgO over a dense intermediate layer of spinel. This is an example of the demixing of a mixed oxide (i.e., spinel) along an oxygen chemical potential gradient. Since diffusion of magnesium and aluminum is much greater than oxygen, the metals diffuse from low partial pressures of oxygen to high partial pressures. A third innermost layer was composed of porous cordierite. None of the scale layers contained SiC. At a temperature of 1450°C, the nonprotective scale was essentially one porous layer composed of cordierite and small grains of spinel. Panda and Seydel [7.56] found that a spinel (prepared from hydrated magnesium nitrate and aluminum hydroxide) matrix reacted with SiC fibers (Versite-grade single crystal fibers) to form a composition they assigned to sapphirine. They concluded that the SiC fibers oxidized during the calcining of the composite forming surface silica that reacted with the spinel to form the sapphirine. Although the identification of sapphirine was not conclusive (undetected by XRD), Panda and Seydel based their result upon the chemistry obtained from energy dispersive spectroscopy (EDS). If one were to examine the $MgO–Al_2O_3–SiO_2$ phase diagram, it should be apparent that, if the spinel were stoichiometric, cordierite should have formed. A reasonable amount of spinel nonstoichiometry (i.e., Al_2O_3-rich) would be required for any sapphirine to form. Thus the reaction product that Panda and Seydel thought to be sapphirine was most likely cordierite*.

* Both these phases contain the same elements but in different ratios (sapphirine $M_4A_5S_2$ and cordierite $M_2A_2S_5$) and in small quantities could easily be confused with one another by EDS analysis.

Molybdenum disilicide is the most oxidation-resistant of all the silicides. Borom et al. [7.30] reported the oxidation of $MoSi_2$ (8 vol.%) dispersed within a matrix of mullite at 1500°C for 6 hr. At the low partial pressure of oxygen near the original surface, the silicon from the $MoSi_2$ was selectively oxidized, similar to that reported by Fitzer [7.57], leaving behind a region of metallic Mo and silica dispersed within the mullite matrix. The addition of silica to this region increased the optical transparency that was very noticeable with examination by optical microscopy. As one proceeded toward the surface with increasing oxygen pressure, the molybdenum was oxidized first to MoO_2 and then to MoO_3. The additional silica that formed was incorporated into the matrix by dissolution and diffusion in the liquid state. Since the MoO_3 that formed was volatile, it mechanically forced this aluminosilicate liquid toward the surface. Mullite was present throughout all the various zones; however, the crystal size and quantity changed due to the other reactions taking place. Thus the oxidation of the $MoSi_2$-mullite composite initially exhibited a weight gain but then shifted to one of weight loss.

The flexural behavior of alkali-resistant high zirconia glass fiber reinforced cement composites was evaluated by Bentur et al. [7.58] after exposure to water at 20°C and 50°C for times up to 2 years. The principal mode of degradation was not the etching of the glass fiber surfaces (indicated by their smooth surfaces) but by the growth of hydration products between the glass filaments. Each glass fiber strand was composed of about 200 individual filaments. Initially, crack stresses and MOR of the unreinforced matrix changed very little with exposure. In general, the MOR of the composites degraded considerably approaching the MOR of the matrix. After 6 months, considerable differences existed among the several types of fibers investigated. One exhibited embrittlement, one degraded to 50% of its original toughness, and one degraded very little. These differences were attributed by Bentur et al. to the differences in growth of hydration products, predominantly $CaO{\cdot}H_2O$, between the glass filaments. Within the first year of exposure, chemical attack of the glass fibers did not appear to play a role in the degradation of the mechanical properties. Even after 2 years, it was minimal. The degree of hydration product growth and its density was directly related to the degree of embrittlement. This embrittlement was attributed to an increase in pull-out bond strength due to the growth of $CaO \cdot H_2O$. In a composite where no hydration products formed, ductile fracture occurred as evidenced by fiber pull-out.

Ready [7.59] suggested that the water pressures developed during the reaction of hydrogen with particulate $NiTiO_3$ in a matrix of titania at temperatures between 700°C and 1000°C were sufficient to cause microstructural degradation (i.e., grain boundary cracks). Thermodynamic data indicated that pressures as high as 6 MPa could be developed. The proposed mechanism involved the diffusion of hydrogen through TiO_2 grains, reduction of $NiTiO_3$ producing Ni metal and H_2O gas at the $TiO_2/NiTiO_3$ interface, and subsequent grain boundary separation. The separation of the grain boundaries allowed additional hydrogen ingress. According to Ready, the location of pores (i.e., water vapor bubbles) at only the matrix/particulate interface suggested that the reduction of the $NiTiO_3$ was controlled by oxygen diffusion out of the TiO_2 grains toward the interface.

Arun et al. [7.60] reported the following order of TiC > HfC > ZrC for the oxidation resistance of these three carbides at 1273 K. The oxidation of these materials was much greater when they were incorporated into hot-pressed compositions of $TiC–ZrO_2$, $ZrC–ZrO_2$, and $HfC–HfO_2$. Arun et al. also reported a greater oxidation of TiC when incorporated into ZrO_2 as opposed to Al_2O_3.

7.3.2 Nonoxide–Matrix Composites

Si_3N_4 Matrix Composites

A Si_3N_4 composite containing 30 wt.% ZrO_2 (also containing 3 mol% Y_2O_3), when oxidized at 1200°C, exhibited decomposition of the zirconia grains as reported by Falk and Rundgren [7.61].

The oxidation proceeded by first forming faceted cavities close to the zirconia grain boundaries due to release of nitrogen dissolved in the zirconia. Prolonged oxidation formed silica-rich films upon the pore walls. Hot pressing at 1800°C apparently formed zirconia containing a variation in the amount of yttria that led to the formation of some monoclinic zirconia after oxidation for 20 min at 1200°C. At shorter times, only cubic and tetragonal zirconia were detected. Cristobalite formed in the oxide scale after 2 hr of oxidation. Short-term oxidation was suggested as a means to enhance mechanical properties; however, long-term oxidation resulted in disintegration of the composite.

SiC–Matrix Composites

Oxidation. As discussed in Sec. 7.2.2 on fiber coatings, an interphase material is either deposited onto the fibers before composite fabrication or the interphase is formed in situ. If the interphase is carbon, the composite must receive an exterior surface protective coating. This is the case for SiC fiber/SiC matrix composites. Once the carbon interphase has been oxidized leaving behind an annular cavity surrounding the fibers, continued oxidation fills the cavity with silica. The amount of silica present is dependent upon the proximity of the reaction site to the location of oxygen ingress. The time/temperature schedule required for complete filling of the cavities with silica is also dependent upon the interphase layer thickness. Filipuzzi et al. [7.16] reported that a time of 10 hr in flowing oxygen at temperatures between 900°C and 1300°C was required to consume completely a 1-μm-thick carbon interphase in a 13 × 3 × 3 mm sample. Filipuzzi et al. reported that composites with thin interphase layers (on the order of 0.1 μm) resulted in microcracking due to the volume increase associated with the SiC to SiO_2 conversion. Microcracking was not observed at high temperatures (i.e., 1200°C) presumably due to stress release through the lower viscosity silica glass nor was it observed in composites with a thick interphase presumably due to the cavity not being filled with silica.

In a graphite-coated Nicalon™ fiber/SiC composite tested at 600 and 950°C in air, Lin and Becher [7.62] found that lifetimes were more dependent upon open porosity (15–25%) than upon parameters such as graphite coating thickness or fiber layout design. Increases in performance were obtained by the use of boron-containing oxidation inhibitors. This was attributed to the oxidation of the boron forming a glass that sealed cracks in the matrix thus minimizing the ingress of oxidation. The oxidation of graphite was the predominant mode of deterioration at low temperatures, but oxidation of SiC occurred at temperatures of 425°C. Verrilli et al. [7.63] found similar results in their investigations of graphite-coated Nicalon™ fiber/SiC composites tested at 500–1300°C. Oxidation of the interfacial graphite occurred first and then oxidation of the SiC fibers occurred evidenced by the formation of surface pits and radius reduction (most severe between 700 and 800°C). Other investigators have reported the degradation of carbon-coated Nicalon™ fiber/SiC [7.29] resulting from the oxidation of the fibers at intermediate temperatures (600–800°C).

The graphite oxidizes to CO and CO_2 [reactions (7.7) and (7.8) below]. Then additional oxygen reacts with the SiC forming free Si, which then continues to react to form SiO_2, filling the space originally occupied by the graphite [reactions (7.9) and (7.10) below]. These reactions are all temperature- and oxygen partial pressure-dependent as discussed in Chaps. 2 and 5. This causes embrittlement and loss of toughness.

$$C + 1/2\ O_2 \rightarrow CO \tag{7.7}$$

$$CO + 1/2\ O_2 \rightarrow CO_2 \tag{7.8}$$

$$SiC + 1/2\ O_2 \rightarrow Si + CO \tag{7.9}$$

$$Si + 1/2\ O_2 \rightarrow SiO + 1/2\ O_2 \rightarrow SiO_2 \tag{7.10}$$

The latest preference for SiC/SiC composite is one with fibers of improved microstructure and chemistry called Sylramic™* incorporated into a matrix of melt-infiltrated SiC matrix. In addition, the interphase material of choice has become BN, although its oxidation is essentially the same as carbon. Ogbuji [7.25] attributed the problems with BN interphase to one involving a thin film of carbon that formed under the BN either from carbon-rich fibers or sizing char that oxidized first exposing the BN interphase. PVA sizing and carbon-free fibers, although not completely removing this problem, at least tremendously decreased the severe pesting† that did occur.

Moisture Attack. In the study of alumina composites reinforced with SiC whiskers, Kim and Moorhead [7.64] found that the room-temperature flexural strength after exposure to H_2/H_2O at 1300°C and 1400°C was significantly affected by pH_2O. Reductions in strength were observed when active oxidation of the SiC occurred at $pH_2O < 2 \times 10^{-5}$ MPa. Kim and Moorhead also reported that long-term exposures greater than 10 hr resulted in no additional loss in strength. At higher water vapor pressures, reductions in strength were less severe due to the formation of an aluminosilicate glass and mullite upon the surface of the sample. For exposures at 1400°C for 10 hr above $pH_2O = 5 \times 10^{-4}$ MPa, strength increases were observed due to the healing of cracks caused by glass formation at the sample surface.

Even at high temperatures, moisture may attack silica containing materials in the same fashion it does silicate glasses at ambient conditions. This is of concern for those materials like silicon nitride and carbide that form a protective layer of silica on their surfaces at high temperatures. Once the protective layer is broken, oxidation of the underlying material may take place. The protective layer does not even need to be broken for

* Sylramic™, Dow Corning Corp., Midland, MI.

† Pesting was originally used to describe the formation of a powder-like deposit on the surface of metallic silicides during oxidation; however, it is now used to describe a similar phenomenon on any material.

continued oxidation in moist environments. According to Williams [7.65], the diffusion of oxygen through silica is an order of magnitude greater when moisture is present. This same enhanced diffusion of oxygen in moist environments was noted by Singhal [7.66] and Narushima et al. [7.67].

Other Gases. Hydrogen can react with SiC-forming silicon or with carbon-forming methane [7.37]. This reaction is negligible below 1100°C in essentially dry environments (containing ≤100 ppm of moisture). With increasing amounts of moisture, this temperature limit increases reaching about 1300°C at moisture contents as high as 10%. The dissociation of molecular hydrogen occurs at temperatures above 1100°C or at lower temperatures by heterogeneous surface reactions. Once dissociated, hydrogen can become extremely reactive. The dissociation of hydrogen by heterogeneous surface reactions is much easier on metals than on carbon or ceramics. Hallum and Herbell [7.68] reported a weight loss at 1000°C, grain boundary corrosion at 1100°C, and both grain and grain boundary corrosion at 1300°C for samples of SiC exposed to pure hydrogen. The effects of weight loss and corrosion were noted at times as low as 50 hr. After 500 hr at 1100°C and 1300°C, the room temperature MOR decreased by one-third.

Carbon–Carbon Composites

Carbon–carbon (i.e., carbon fiber reinforcement and carbon matrix) composites are probably the only materials that possess a combination of high strength/weight ratio, very low thermal expansion, excellent thermal shock resistance, and strength retention over a wide temperature range. This combination of properties makes them highly desirable in the aerospace industry. The major drawback for widespread use of C/C composites, however, is their poor oxidation resistance above 500°C. Only through the use of oxygen barrier coatings can C/C composites be useful at elevated temperatures in oxidative environments. Silicon nitride applied by CVD has proven to work well as an oxygen barrier for applications in rapid thermal cycles up to 1800°C [7.69]. For less rapid cycling to lower temperatures (<1500°C) and thermal soaking at temperatures between 600°C and 1000°C, multilayer coatings containing boron with CVD Si_3N_4 or SiC overlays have been tested. Although the boron can form a low-temperature liquid that will seal microcracks, its use at high temperatures may be limited due to volatility and the high fluidity of the glass formed. Labruquere et al. [7.70] reported that silicon-rich coatings on the carbon fibers enhanced their oxidation resistance. Although the coating did not stop oxidation, it oxidized at a slower rate than the carbon fibers and was confined to the region between the coating and the fiber and between the coating and the matrix. After testing silicon-rich and boron-rich Si–B–C coatings, Labruquere et al. concluded that to protect the carbon/carbon composite efficiently, the following parameters were required:

1. The coating must have a Pilling and Bedworth* coefficient greater than 1.
2. The coating must oxidize at a faster rate than the carbon.
3. The coating must have a minimum thickness (~100 nm).
4. The coating must not react with the carbon fibers.
5. The coating must form a stable oxide.

In another study, Labruquere et al. [7.71] found that wet air (500 L/hr of 3 vol.% H_2O) increased the oxidation rate by a factor of 3.

Cawley et al. [7.28] reported that the crystallographic orientation of well-crystallized carbon determined the rate of oxidation. Thomas [7.72] has shown that oxidation parallel to the *c*-axis was lower by 30 times compared to the basal plane and that the rate exhibited about 10% anisotropy in the basal plane at 800°C.

The reaction of carbon with hydrogen forms methane at low temperatures. Other products begin to form as the temper-

---

*The Pilling and Bedworth coefficient is defined as the ratio of the volume of silica formed to the volume of the ceramic consumed. See N.G. Pilling and R.E. Bedworth, J. Int. Met., *29*, 529 (1923).

ature is raised above 1400°C. The severity of the reaction is dependent upon the amount of atomic hydrogen present. Molecular hydrogen is relatively inert.

Other Nonoxide Matrix Composites

The ternary compound Ti_3SiC_2 has been reported by Li et al. [7.73] to have excellent properties. When reinforced with SiC (in situ formed), a weight gain of 7.9 mg/cm^2 was reported for heat treatment in air at 1200°C for 21 hr. At all temperatures between 1000°C and 1500°C, TiO_2 formed on the surface. SiO_2 was present on the surface only at 1000°C. At higher temperatures, the outer surface layer was porous allowing oxygen diffusion to the interior, which allowed the formation of silica along with TiO_2 in the interior layer. The porous outer layer caused the weight gain at 1300°C to be approximately 2.5 times greater than at 1200°C.

Nicalon™ fiber reinforced TiB_2 matrix composites have been tested for use in the production of aluminum. Exposure to molten aluminum for 24 hr caused heavy attack of the TiB_2 and delamination of the composite. This deterioration was attributed by Lowden et al. [7.74] to retain chlorine that caused premature dissolution of TiB_2 into aluminum. The retained chlorine was from the $TiCl_4$ precursor material and an infiltration temperature of 900°C. When an infiltration temperature of 1200°C was used, the TiB_2 was unaffected even after 10 weeks of exposure to molten aluminum. These longer exposure times, however, resulted in attack of the reinforcement fibers.

Various ceramic matrix materials have been manufactured from the pyrolysis of polymers, called *preceramic polymer precursors*. Although these materials are supposedly nonoxides, one must be careful as to the actual production route and source. According to French [7.23], the polysilazane materials are moisture-sensitive and therefore yield a ceramic high in oxygen. Decomposition reactions of the type:

$$\text{polysilazane} - \text{containing moisture} \rightarrow SiO_g + CO_g \qquad (7.11)$$

may occur. French commented that reaction (7.11) was not appreciable below 1400°C.

One of the early investigations into improving the oxidation of carbides was that by Accountius et al. [7.75]. They attempted to form glassy protective layers on TiC, B_4C, and SiC.

7.4 METAL MATRIX COMPOSITES

Aluminum is probably the most common metal used for the production of metal matrix composites. This is due mostly to its low density, excellent mechanical properties, and low melting point that aids in fabrication. The addition of zinc creates an alloy with improved strength, wear resistance, and corrosion resistance. Since aluminum has a relatively low melting point, other metals have been used as the matrix when increased service temperatures are required. See Table 7.3 for a list of some of the metals used and their service temperature limits.

The stress corrosion cracking in a 3.5% NaCl solution of a pure aluminum matrix composite reinforced by alumina borate whiskers ($Al_{18}B_4O_{33}$) was studied by Hu et al. [7.76]. Crack propagation was predominantly along the whisker/matrix interface. This was attributed by Hu et al. to dissolution of the matrix surrounding the whiskers as evidenced by pits that formed at the crack tips.

A SiC-coated graphite fiber/Al alloy matrix composite was developed to overcome the reaction of graphite with aluminum

TABLE 7.3 Matrix Metals and Their Upper Limits of Temperature Service

Metal	Upper service temperature (°C)
Aluminum	400
Titanium alloys	650
Superalloys (Ni)	980–1100
Refractory metals	1100–1650

above 500°C forming Al_4C_3 [7.77]. This phase formed as hexagonal plates with subsequent degradation of fiber strength. In addition, Al_4C_3 is hygroscopic causing further deterioration. A silicon enrichment on the outer surface of the fibers inhibited the formation of Al_4C_3; however, it allowed improvements of only short duration.

Although there is not much on corrosion, the book by Taya and Arsenault [7.78] contains a lot of information about the properties and behavior of MMC.

Galvanic corrosion (see Chap 2, page 25 for a discussion on galvanic corrosion) between the matrix metal and the reinforcement fibers can occur, especially when in contact with aqueous solutions. This has been reported by Trzaskoma [7.79] for magnesium alloy matrix–graphite fiber composites.

The seawater corrosion of SiC/Al was found to be more resistant than graphite/Al by Aylor and Kain [7.80]. This was attributed to a lack of a galvanic driving force between the SiC and the aluminum matrix, although both composites exhibited similar mechanisms of corrosion—essentially pitting of the metal matrix around the reinforcement material.

The reactions of hydrogen and SiC reinforced titanium-based matrix composites have been described by Nelson [7.37]. Since these reactions are ones involving various metallic phases rather than the SiC fiber or the interface, the details will not be given here. However, as reported by Nelson, the solubility of hydrogen in metals was problematic and, in some cases, formed metal hydrides (generally at higher temperatures) that led to mechanical degradation.

The stability of TiN toward reaction with various metals such as iron and nickel aluminides has been reported to be better than SiC by Nolan et al. [7.81]. Thus TiN would be the better choice as the reinforcement for these metal matrix composites.

Since the intermetallic Ni- and Ti-aluminides have service temperatures limited to about 1200°C, molybdenum disilicide has been investigated for applications where greater temperatures may be reached. One of the major drawbacks of $MoSi_2$ is its oxidation resistance. Cook et al. [7.82] investigated the incorporation of 30 vol.% TiB_2, ZrB_2 HfB_2, and SiC as a reinforcement in hopes of developing a composite of greater oxidation resistance than the base $MoSi_2$. Specimen were exposed to isothermal testing at 800°C, 1200°C, 1400°C, and 1500°C for 24 hr in air, in addition to a thermal cycle consisting of 55 min at 1200°C or 1500°C and then 5-min ambient cooling with subsequent reheating. All the boride-containing materials exhibited a greater deterioration than the silicon carbide-containing composite, although none exhibited a greater oxidation resistance than the base $MoSi_2$. See Sec. 5.2.3 on borides for a discussion of the oxidation of these materials.

Although not generally thought of as metal matrix composites, a relatively new class of materials called *fibrous monolithic ceramics* [7.83] actually may contain a metal as the matrix that surrounds cells of a fibrous polycrystalline ceramic. One example of such a material investigated by Baskaran et al. [7.84] contained fibrous polycrystalline alumina cells surrounded by nickel. The nickel cell boundary thickness varied from 1 to about 15 μm. Oxidation at 1200°C for 10 hr initially formed NiO that subsequently reacted with the alumina forming $NiAl_2O_4$. The formation of the aluminate was thought to provide protection toward additional oxidation.

7.5 POLYMER MATRIX COMPOSITES

Two publications by ASTM discuss the environmental effects upon polymeric composites [7.85,7.86]. The largest amount of composites produced is probably of this type reinforced with glass fibers, called glass-reinforced plastics, polymers, or polyesters (GRP). Degradation in aqueous environments generally occurs by fiber/matrix debonding. Since glass fibers are attacked by moisture, which drastically reduces their strength, glass fibers are given a protective coating.

Graphite/carbon fiber/epoxy composites (CFRP) have seen some recent use in marine environments. In many cases, they are generally used in contact with metals. In a seawater environment, the graphite fibers act as the cathode for accelerated galvanic corrosion of the metals.

Electrochemical impedance spectroscopy was used by Wall et al. [7.87] to monitor the damage in graphite fiber/bismaleimide composites in contact with aluminum, steel, copper, and titanium immersed into aerated 3.5 wt.% NaCl solution. Decomposition of the bismaleimide polymer was thought to occur by the action of hydroxyl ions, which break imide linkages. The production of hydroxyl ions occurred through the following reaction:

$$O_2 + 2H_2O + 4e^- \rightarrow 4OH^- \tag{7.12}$$

at the surface of the graphite fibers. They concluded that the corrosion concentrated at the fiber/matrix interface was caused by cathodic polarization and was dependent upon the overpotential and the cathodic reaction rate. Oxidation of the matrix and fibers was thought to be the cause of ablation of the composite.

Aylor [7.88] reported increased galvanic action (i.e., initial current level) with increased amounts of fiber exposure for a graphite fiber/epoxy composite in contact with either HY80 steel or nickel aluminum bronze subjected to seawater at ambient temperature for 180 days. Even when no fibers were exposed to the environment galvanic corrosion occurred. This phenomenon was attributed by Aylor to the absorption of moisture through the epoxy to the fibers. The galvanic current determined during the tests was found to display several distinct regions. These have been identified by Aylor as:

Region I—activation of surface
Region II—film formation
Region III—reduction of active surface areas
Region IV—buildup of calcareous deposit on composite

These regions were attributed to localized differences in active anodic and cathodic areas, which could also be affected by the stability of the films formed on the surfaces of the metal and composite. The calcareous deposit on the surfaces of the graphite fibers was reported as the result of formation of hydroxyl ions at the cathode with an associated increase in pH and precipitation of $CaCO_3$ and $Mg(OH)_2$. Actual seawater galvanic corrosion rates would be significantly affected by the stability of the films formed in Region II and most likely would be much greater than the rates found in the laboratory tests.

A mica flake-filled polyester when used as a lining material for outlet duct of coal-fired power plant formed the compound jarosite, $KFe_3(SO_4)_2(OH)_6$, at the mica/polyester interface. Subsequent wedging* of these materials resulted in failure of the lining [7.89].

Leonor et al. [7.90] developed a composite composed of a biodegradable starch thermoplastic matrix and the bioactive hydroxyapatite for implantation into the human body. The degradation of the composite implant must be controlled to allow the gradual transfer of load to the healing bone. Thirty weight percent hydroxyapatite is required to cause the formation of calcium phosphate on the surface of the composite for adhesion to the bone. Samples immersed into a simulated body fluid at pH = 7.35 showed no change after 8 hr. With increased immersion time, calcium phosphate nuclei formed, grew in number and size, and coalesced fully covering the surface of the composite within 24 hr. A dense uniform calcium phosphate layer was formed after 126 hr.

7.6 ADDITIONAL RELATED READINGS

Delmonte J. History of Composites. *Reference Book for Composites Technology*; Lee S., Ed.; Technomics Publ. Co.; Lancaster, PA, 1989.

Lewis, D. III. Continuous fiber-reinforced ceramic matrix composites: A historical overview. In *Handbook on Continuous Fiber-Reinforced*

*Wedging is a procedure where ceramic bodies are prepared by hand kneading. This is done to uniformly disperse water and remove air pockets and laminations.

Ceramic Matrix Composites; Lehman, R.L., El-Rahalby, S.K., Wachtman, J.B., Jr., Eds.; CIAC Purdue Univ, IN and Am. Ceram. Soc. Westerville, OH, 1995; 1–34.

Advanced Synthesis and Processing of Composites and Advanced Ceramics; Logan K.V. Ed.; *Ceramic Transactions*. Am. Ceram. Soc. Westerville, OH, 1995; Vol. 56.

Evans, A.G.; He, M.Y.; Hutchinson, J.W. Interface Debonding and Fiber Cracking in Brittle Matrix Composites. J. Am. Ceram. Soc. 1989, *72*, 2300–2303.

Lowden, R.A. Fiber Coatings and the Mechanical Properties of Fiber-Reinforced Ceramic Composites. Ceram. Trans. 1991, *19*, 619–630.

Taya, M.; Arsenault, R.J. *Metal Matrix Composite Thermomechanical Behavior*; Pergamon Press: New York, 1989; 264 pp.

7.7 EXERCISES, QUESTIONS, AND PROBLEMS

1. Develop a definition for a composite material by listing the various characteristics and explain the reason for each. What is the advantage of using a composite over that of a single component material?
2. Discuss why the adhesion of matrix to reinforcement is the region of greatest importance during corrosion.
3. Discuss how a difference in thermal expansion between the matrix and the reinforcement is related to corrosion.
4. Why is the corrosion process of oxidation a problem for so many composites?
5. How does the thermal expansion mismatch between surface layers formed by corrosion and the underlying substrate materials affect corrosion?
6. Discuss how the manufacturing process of a particular reinforcement fiber may affect the corrosion of a composite?
7. What does the term "embrittlement" mean when related to the corrosion of composites?

8. Discuss the difference that occurs during the oxidation of a composite having a SiC matrix and a SiC fiber with either a BN or carbon interphase.
9. Is it possible for a mixed oxide to demix along an oxygen partial pressure gradient? If so, give an example.
10. Discuss why the oxidation of SiC is much greater in moist environments compared to dry ones.

REFERENCES

7.1. Holmes, M.; Just, D.J. *GRP in Structural Engineering*; Applied Science Publishers: New York, 1983; 10, 282 pp.

7.2. Aveston, J.; Kelly, A. Theory of multiple fracture of fibrous composites. J. Mater. Sci. 1973, *8*, 352–362.

7.3. Curtin, W.A. Theory of mechanical properties of ceramic–matrix composites. J. Am. Ceram. Soc. 1991, *74* (11), 2837–2845.

7.4. Curtin, W.A. In situ fiber strengths in ceramic–matrix composites from fracture mirrors. J. Am. Ceram. Soc. 1994, *77* (4), 1075–1078.

7.5. Marshall, D.B.; Evans, A.G. Failure mechanisms in ceramic-fiber/ceramic–matrix composites. J. Am. Ceram. Soc. 1985, *68* (5), 225–231.

7.6. Davidge, R.W.; Briggs, A. The tensile failure of brittle matrix composites reinforced with unidirectional continuous fibres. J. Mater. Sci. 1989, *24*, 2815–2819.

7.7. Evans, A.G.; Marshall, D.B. *Fiber Reinforced Ceramic Composites*; Mazdiyasni, K.S., Ed.; Noyes Pub.: Park Ridge, NJ, 1990.

7.8. Courtright, E.L. Engineering limitations of ceramic composites for high performance and high temperature applications. In *Proc. 1993 Conf. on Processing, Fabrication and Applications of Advanced Composites*; *Long Beach, CA*; Upadhya, K., Ed.; ASM: Ohio, Aug 9–11,1993; 21–32.

7.9. Munson, K.L.; Jenkins, M.G. Retained tensile properties and performance of an oxide–matrix continuous-fiber ceramic composite after elevated-temperature exposure in ambient air. In *Thermal and Mechanical Test Methods and Behavior of Continuous-Fiber Ceramic Composites,* ASTM STP 1309; Jenkins, M.G., Gonczy, S.T., Lara-Curzio, E., Asbaugh, N.E., Zawada L.P., Eds.; ASTM: West Conshohocken, PA, 1997; 176–189.

7.10. Wu, X.; Holmes, J.W.; Hilmas, G.E. Environmental properties of ceramic matrix composites. In *Handbook on Continuous Fiber-Reinforced Ceramic Matrix Composites*; Lehman, R.L., El-Rahalby, S.K., Wachtman, JB., Jr., Eds.; CIAC Purdue Univ. IN and Am. Ceram. Soc.: Westerville, OH, 1995; 431–471.

7.11. Galasso, F.S. *Advanced Fibers and Composites*; Gordon and Breach Science Publishers: New York, 1989; 178 pp.

7.12. Metcalfe, A.G.; Schmitz, G.K. Mechanism of stress corrosion in E glass filaments. Glass Technol. 1972, *13* (1), 5–16.

7.13. Clark, T.J.; Arons, R.M.; Stamatoff, J.B. Thermal degradation of Nicalon™ SiC fibers. In *Ceramic Engineering and Science Proceedings*; Smothers, W.J., Ed.; Am. Ceram. Soc.: Westerville, OH, 1985, *6* (7–8), 576–588.

7.14. Clark, T.J.; Jaffe, M.; Rabe, J.; Langley, N.R. Thermal stability characterization of SiC ceramic fibers: I, Mechanical property and chemical structure effects. Ceram. Eng. Sci. Proc. 1986, *7* (7–8), 901–913.

7.15. Sawyer, L.C.; Chen, R.T.; Haimbach, F. IV; Harget, P.J.; Prack, E.R.; Jaffe, M. Thermal stability characterization of SiC ceramic fibers: II, Fractography and structure. Ceram. Eng. Sci. Proc. 1986, *7* (7–8), 914–930.

7.16. Filipuzzi, L.; Camus, G.; Naslain, R.; Thebault, J. Oxidation mechanisms and kinetics of 1D-SiC/C/SiC composite materials: I, An experimental approach. J. Am. Ceram. Soc. 1994, *77* (2), 459–466.

7.17. Nolan, T.A.; Allard, L.F.; Coffey, D.W.; Hubbard, C.R.; Padgett, R.A. Microstructure and crystallography of titanium nitride whiskers grown by a vapor–liquid–solid process. J. Am. Ceram. Soc. 1991, *74* (11), 2769–2775.

7.18. Caputo, J.; Lackey, W.J.; Stinton, D.P. Development of a new, faster process for the fabrication of ceramic fiber-reinforced ceramic composites by chemical vapor infiltration. Ceram. Eng. Sci. Proc. 1985, *6* (7–8), 694–706.

7.19. Fareed, A.S.; Schiroky, G.H.; Kennedy, C.R. Development of BN/SiC duplex fiber coatings for fiber-reinforced alumina matrix composites fabricated by directed metal oxidation. Ceram. Eng. Sci. Proc. 1993, *14* (9–10), 794–801.

7.20. Studt, T. Breaking down the barriers for ceramic matrix composites. R & D Mag., Aug 1991; 36–42.

7.21. Bender, B.; Shadwell, D.; Bulik, C.; Incorvati, L.; Lewis, D. III. Effect of fiber coatings and composite processing on properties of zirconia-based matrix SiC fiber composites. Ceram. Bull. 1986, *65* (2), 363–369.

7.22. Singh, R.N.; Brun, M.K. Effect of boron nitride coating on fiber–matrix interactions. Ceram. Eng. Sci. Proc. 1987, *8* (7–8), 636–643.

7.23. French, J.E. Ceramic matrix composite fabrication and processing: Polymer pyrolysis. In *Handbook on Continuous Fiber-Reinforced Ceramic Matrix Composites*; Lehman, R.L. El-Rahalby, S.K., Wachtman, J.B. Jr., Eds.; CIAC Purdue Univ, IN and Am. Ceram. Soc.: Westerville, OH, 1995; 269–299.

7.24. Fareed, A.S.; Schiroky, G.H.; Kennedy, C.R. Development of BN/SiC duplex fiber coatings for fiber-reinforced alumina matrix composites fabricated by direct metal oxidation. Ceram. Eng. Sci. Proc. 1993, *14* (9–10), 794–801.

7.25. Ogbuji, L.U.J.T. Pest-resistance in SiC/BN/SiC composites. J. Eur. Ceram. Soc. 2003, *23*, 613–617.

7.26. Cooper, R.F.; Hall, P.C. Reactions between synthetic mica and simple oxide compounds with application to oxidation-resistant ceramic composites. J. Am. Ceram. Soc. 1993, *76* (5), 1265–1273.

7.27. Rice, R.W. Toughening in ceramic particulate and whisker composites. Ceram. Eng. Sci. Proc. 1990, *11* (7–8), 667–694.

7.28. Cawley, J.D.; Ünal, Ö.; Eckel, A.J. Oxidation of carbon in continuous fiber reinforced ceramic matrix composites. In *Ceramic Transactions: Advances in Ceramic–Matrix Composites*; Bansal, P., Ed; Am. Ceram. Soc.: Westerville, OH, 1993; Vol. 38, 541–552.

7.29. Heredia, F.E.; McNulty, J.C.; Zok, F.W.; Evans, A.G. Oxidation embrittlement probe for ceramic matrix composites. J. Am. Ceram. Soc. 1995, *78* (8), 2097–2100.

7.30. Borom, M.P.; Bolon, R.B.; Brun, M.K. Oxidation mechanism of $MoSi_2$ particles in mullite. Adv. Ceram. Mater. 1988, *3* (6), 607–611.

7.31. Borom, M.P.; Brun, M.K.; Szala, L.E. Kinetics of oxidation of carbide and silicide dispersed phases in oxide matrices. Adv. Ceram. Mater. 1988, *3* (5), 491–497.

7.32. Luthra, K.L. Oxidation of SiC-containing composites. Ceram. Eng. Sci. Proc. 1987, *8* (7–8), 649–653.

7.33. Mukerji, J.; Biswas, S.K. Synthesis, properties, and oxidation of alumina–titanium nitride composites. J. Am. Ceram. Soc. 1990, *73* (1), 142–145.

7.34. Tampieri, A.; Bellosi, A. Oxidation resistance of alumina-titanium nitride and alumina-titanium carbide composites. J. Am. Ceram. Soc. 1992, *75* (6), 1688–1690.

7.35. Revankar, V.; Hexemer, R.; Mroz, C.; Bothwell, D.; Goel, A.; Bray, D.; Blakely, K. Novel process for titanium nitride whisker synthesis and their use in alumina composites. In *Advanced Synthesis and Processing of Composites and Advanced Ceramics, Ceramic Transactions*; Logan, K.V., Ed.; American Ceramic Society: Westerville, OH, 1995; Vol. 56, 135–146.

7.36. Wang, T.C.; Chen, R.Z.; Tuan, W.H. Oxidation resistance of Ni-toughened Al_2O_3. J. Eur. Ceram. Soc. 2003, *23*, 927–934.

7.37. Nelson, H.G. A challenge to materials: Advanced hypersonic flight hydrogen and high temperature materials. In *Proc. 1993 Conf. on Processing, Fabrication and Applications of Advanced Composites, Long Beach, CA*; Upadhya, K., Ed.; ASM: Ohio, Aug 9–11, 1993; 11–20.

7.38. Sambell, R.A.; Bowen, D.; Phillips, D.C. Carbon fiber composites with ceramic and glass matrices, Part 1, Discontinuous fibers. J. Mater. Sci. 1972, *7*, 663–675.

7.39. Brennan, J.J. Interfacial characterization of glass and glass-ceramic matrix/Nicalon SiC composites. In *Tailoring Multiphase and Composite Ceramics*; Mater. Res. Soc. Symp. Proc.; Plenum Press: New York, 1986; Vol. 20, 549–560.

7.40. Cooper, R.F.; Chyung, K. Structure and chemistry of fiber–matrix interfaces in silicon carbide fiber-reinforced glass-ceramic composites: An electron microscopy study. J. Mater. Sci. 1987, *22* (9), 3148–3160.

7.41. Bonney, L.A.; Cooper, R.R. Reaction layer interfaces in SiC-fiber-reinforced glass-ceramics: A high-resolution scanning transmission electron microscopy analysis. J. Am. Ceram. Soc. 1990, *73* (10), 2916–2921.

7.42. Chaim, R.; Heuer, A.H. Carbon interfacial layers formed by oxidation of SiC in SiC/Ba-stuffed cordierite glass-ceramic reaction couples. J. Am. Ceram. Soc. 1991, *76* (7), 1666–1667.

7.43. Shin, H.H.; Berta, Y.; Speyer, R.F. Fiber–matrix interaction in SiC fiber reinforced ceramic and intermetallic composites. In *Ceramic Transactions, Advances in Ceramic–Matrix Composites*; Bansal, N.P., Ed.; Am. Ceram. Soc.: Westerville, OH, 1993; Vol. 38, 235–248.

7.44. Pantano, C.G.; Spear, K.E.; Qi, G.; Beall, D.M. Thermochemical modeling of interface reactions in glass matrix composites. In *Ceramic Transactions, Advances in Ceramic–Matrix Composites*; Am. Ceram. Soc.: Westerville, OH, 1993; Vol. 38, 173–198.

7.45. McCracken, W.J.; Clark, D.E.; Hench, L.L. Surface characterization of ceramed composites and environmental sensitivity. In *Ceramic Engineering and Science Proceedings*; Am. Ceram. Soc.: Westerville, OH, 1980, Vol. 1 (7–8A), 311–317.

7.46. Prewo, K.M.; Brennan, J.J.; Layden, G.K. Fiber reinforced glasses and glass-ceramics for high performance applications. Ceram. Bull. 1986, *65* (2), 305–313.

7.47. Mendelson, M.I. SiC/glass composite interphases. In *Ceramic Engineering and Science Proceedings*; Smothers, W.J., Ed.; 1985; Vol. 6 (7–8), 612–621.

7.48. Herron, M.A.; Risbud, S.H. Characterization of oxynitride glass-ceramic matrix SiC fiber composites. In *Ceramic Engineering and Science Proceedings*; Smothers, W.J., Ed.; Am. Ceram. Soc.: Westerville, OH, 1985, Vol. 6 (7–8), 622–631.

7.49. Herron, M.A.; Risbud, S.H. Characterization of SiC-fiber-reinforced Ba–Si–Al–O–N glass-ceramic composites. Ceram. Bull. 1986, *65* (2), 342–346.

7.50. Wetherhold, R.C; Zawada, L.P. Heat treatments as a method of protection for a ceramic fiber-glass matrix composite. J. Am. Ceram. Soc. 1991, *74* (8), 1997–2000.

7.51. Bischoff, E.; Ruhle, M.; Sbaizero, O.; Evans, A.G. Microstructural studies of the interfacial zone of a SiC-fiber-reinforced lithium aluminum silicate glass-ceramic. J. Am. Ceram. Soc. 1989, *72* (5), 741–745.

7.52. Pannhorst, W.; Spallek, M.; Brückner, R.; Hegeler, H.; Reich, C.; Grathwohl, G.; Meier, B.; Spelmann, D. Fiber-reinforced glasses and glass ceramics fabricated by a novel process. Ceram. Eng. Sci. Proc. 1990, *11* (7–8), 947–963.

7.53. Luthra, K.L.; Park, H.D. Oxidation of silicon carbide-reinforced oxide–matrix composites at 1375 and 1575°C. J. Am. Ceram. Soc. 1990, *73* (4), 1014–1023.

7.54. Hermes, E.E.; Kerans, R.J. Degradation of non-oxide reinforcement and oxide matrix composites. In *Materials Research Society Symposium Proceedings, Vol. 125: Materials Stability and Environmental Degradation*; Barkatt, A., Vernik, E.D., Jr., Smith, L.R., Eds.; Mat. Res. Soc.: Pittsburgh, PA, 1988; 73–78.

7.55. Baudin, C.; Moya, J.S. Oxidation of mullite–zirconia–alumina–silicon carbide composites. J. Am. Ceram. Soc. 1990, *73* (5), 1417–1420.

7.56. Panda, P.C.; Seydel, E.R. Near-net-shape forming of magnesia–alumina spinel/silicon carbide fiber composites. Ceram. Bull. 1986, *65* (2), 338–341.

7.57. Fitzer, E. Oxidation of molybdenum disilicide. In *Ceramic Transactions Vol 10: Corrosion and Corrosive Degradation of Ceramics*; Tressler, R.E., McNallan, M., Eds.; Am. Ceram. Soc.: Westerville, OH, 1990; 19–41.

7.58. Bentur, A.; Ben-Bassat, M.; Schneider, D. Durability of glass-fiber-reinforced cements with different alkali-resistant glass fibers. J. Am. Ceram. Soc. 1985, *68* (4), 203–208.

7.59. Ready, D.W. High temperature gas corrosion of ceramic composites. Ceram. Eng. Sci. Proc. 1992, *13* (7–8), 301–318.

7.60. Arun, R.; Subramanian, M.; Mehrotra, G.M. Oxidation behavior of TiC, ZrC, HfC dispersed in oxide matrices. In *Ceramic Transactions Vol 10: Corrosion and Corrosive Degradation of Ceramics*; Tressler, R.E., McNallan, M., Eds.; Am. Ceram. Soc.: Westerville, OH, 1990; 211–223.

7.61. Falk, L.K.L.; Rundgren, K. Microstructure and short-term oxidation of hot-pressed $Si_3N_4/ZrO_2(+Y_2O_3)$ ceramics. J. Am. Ceram. Soc. 1992, *75* (1), 28–35.

7.62. Lin, H.-T.; Becher, P.F. Stress-temperature-lifetime response of Nicalon fiber-reinforced silicon carbide composites in air. In *Thermal and Mechanical Test Methods and Behavior of Continuous-Fiber Ceramic Composites, ASTM STP 1309*; Jenkins, M.G., Gonczy S.T., Lara-Curzio, E., Asbaugh, N.E., Zawada, L.P., Eds.; ASTM: West Conshohocken, PA, 1997; 128–141.

7.63. Verrilli, M.J.; Calomino, A.M.; Brewer, D.N. Creep-rupture behavior of a Nicalon/SiC composite. In *Thermal and Mechanical Test Methods and Behavior of Continuous-Fiber Ceramic Composites, ASTM STP 1309*; Jenkins, M.G., Gonczy, S.T., Lara-Curzio, E. Ashbaugh, N.E., Zawada, L.P., Eds.; ASTM: West Conshohocken, PA, 1997; 158–175.

7.64. Kim, H.-E.; Moorhead, A.J. Corrosion and strength of SiC-whisker-reinforced alumina exposed at high temperatures to H_2–H_2O atmospheres. J. Am. Ceram. Soc. 1991, *74* (6), 1354–1359.

7.65. Williams, E.L. Diffusion of oxygen in fused silica. J. Am. Ceram. Soc. 1965, *48* (4), 190–194.

7.66. Singhal, S.C. Effect of water vapor on the oxidation of hot-pressed silicon nitride. J. Am. Ceram. Soc. 1976, *59* (1–2), 81–82.

7.67. Narushima, T.; Goto, T.; Iguchi, Y.; Hirai, T. High-temperature oxidation of chemically vapor-deposited silicon carbide in wet oxygen at 1823 to 1923 K. J. Am. Ceram. Soc. 1990, *73* (12), 3580–3584.

7.68. Hallum, G.W; Herbell, T.P. Effects of high temperature hydrogen exposure on sintered α-SiC. Adv. Ceram. Mater. 1988, *3* (2), 171–175.

7.69. Strife, J.R. Fundamentals of protective coating strategies for carbon–carbon composites. In *Damage and Oxidation Protection in High Temperature Composites*; Haritos, G.K., Ochoa, O.O., Eds.; ASME: New York, 1991; Vol. 1, 121–127.

7.70. Labruquere, S.; Blanchard, H.; Pailler, R.; Naslain, R. Enhancement of the oxidation resistance of interfacial area in C/C composites. Part II: oxidation resistance of B–C, Si–B–C and Si–C coated carbon preforms densified with carbon. J. Eur. Ceram. Soc. 2002, *22*, 1011–1021.

7.71. Labruquere, S.; Gueguen, J.S.; Pailler, R.; Naslain, R. Enhancement of the oxidation resistance of interfacial area in C/C composites. Part III: the effect of oxidation in dry or wet air on mechanical properties of C/C composites with internal protections. J. Eur. Ceram. Soc. 2002, *22*, 1023–1030.

7.72. Thomas, J.M. In *Chemistry and Physics of Carbon*; Walker, P.L., Jr., Ed.; Marcel Dekker: New York, 1965; Vol. 1, 135–168.

7.73. Li, S.-B.; Xie, J.-X.; Zhang, L.-T.; Cheng, L.-F. Mechanical properties and oxidation resistance of Ti_3SiC_2/SiC composite synthesized by in situ displacement reaction of Si and TiC. Materials Letters 2003, *57,* 3048–3056.

7.74. Lowden, R.A.; Stinton, D.P.; Besmann, T.M. Ceramic matrix composite fabrication and processing: chemical vapor infiltration. In *Handbook on Continuous Fiber-Reinforced Ceramic Matrix Composites*; Lehman, R.L., El-Rahalby, S.K., Wachtman, J.B., Jr., Eds.; CIAC Purdue Univ, IN and Am. Ceram. Soc.: Westerville, OH, 1995; 205–268.

7.75. Accountius, O.; Sisler, H.; Sheblin, S.; Bole, G. Oxidation resistances of ternary mixtures of the carbides of titanium, silicon, and boron. J. Am. Ceram. Soc. 1954, *37* (4), 173–177.

7.76. Hu, J.; Chen, C.S.; Xu, L.X.; Yao, C.K.; Zhao, L.C. Effect of whisker orientation on the stress corrosion cracking behavior of alumina borate whisker reinforced pure Al composite. Materials Letters 2002, *56,* 642–646.

7.77. Cornie, J.A.; Chiang, Y.-M.; Uhlmann, D.R.; Mortensen, A.; Collins, J.M. Processing of metal and ceramic matrix composites. Ceram. Bull. 1986, *65* (2), 293–304.

7.78. Taya, M.; Arsenault, R.J. *Metal Matrix Composites Thermomechanical Behavior*; Pergamon Press: New York, 1989; 264 pp.

7.79. Trzaskoma, P.P. Corrosion behavior of a graphite fiber/magnesium metal matrix composite in aqueous chloride solution. Corrosion 1986, *42* (10), 609–613.

7.80. Aylor, D.M.; Kain, R.M. In *Recent Advances in Composites in the United States and Japan, ASTM STP 864*; Vinson, J.R., Taya, M., Eds.; ASTM: Philadelphia, PA, 1986; 718–729.

7.81. Nolan, T.A.; Allard, L.F.; Coffey, D.W.; Hubbard, C.R.; Padgett, R.A. Microstructure and crystallography of titanium nitride whiskers grown by a vapor–liquid–solid process. J. Am. Ceram. Soc. 1991, *74* (11), 2769–2775.

7.82. Cook, J.; Mahapatra, R.; Lee, E.W.; Khan, A.; Waldman, J. Oxidation behavior of $MoSi_2$ composites. In *Ceramic Engineering and Science Proceedings*; Wachtman, J.B., Ed.; Am. Ceram. Soc.: Westerville, OH, 1991; Vol. 12 (9–10), 1656–1670.

7.83. Coblenz, W.S. Fibrous Monolithic Ceramic and Method for Production. US Patent No. 4,772,524, September 20, 1988.

7.84. Baskaran, S.; Nunn, S.D.; Halloran, J.W. Fibrous monolithic ceramics: IV, Mechanical properties and oxidation behavior of the alumina/nickel system. J. Am. Ceram. Soc. 1994, *77* (5), 1256–1262.

7.85. *High Temperature and Environmental Effects on Polymeric Composites, STP 1174*; Harrais, C.E., Gates, T.S., Eds.; ASTM: Philadelphia, 1993.

7.86. *High Temperature and Environmental Effects on Polymeric Composites,* 2nd Vol., STP 1302; Gates, T.S., Zureick, A.-H., Eds.; ASTM: Philadelphia, 1997.

7.87. Wall, F.D.; Taylor, S.R.; Cahen, G.L. The simulation and detection of electrochemical damage in BMI/graphite fiber composites using electrochemical impedance spectroscopy. In *High Temperature and Environmental Effects on Polymeric Composites, STP 1174*; Harris C.E., Gates, T.S., Eds.; ASTM: Philadelphia, PA, 1993; 95–113.

7.88. Aylor, D.M. The effect of a seawater environment on the galvanic corrosion behavior of graphite/epoxy composites coupled to metals. In *High Temperature and Environmental Effects on*

Polymeric Composites, STP 1174; Harris, C.E., Gates, T.S., Eds.; ASTM: Philadelphia, PA, 1993; 81–94.

7.89. Koch, H.; Syrett, B.C. Progress in EPRI research on materials for flue gas desulphurization systems. In *Dewpoint Corrosion*; Holmes, D.R., Ed.; Ellis Horwood Ltd.: Chichester, UK, 1985; 105–124.

7.90. Leonor, I.B.; Ito, A.; Onuma, K.; Kanzaki, N.; Reis, R.L. In vitro bioactivity of starch thermoplastic/hydroxyapatite composite biomaterials: an in situ study using atomic force microscopy. Biomaterials 2003, *24*, 579–585.

8

Properties and Corrosion

> Homogeneous bodies of materials—I was told—do not exist, homogeneous states of stress are not encountered.
>
> OTTO MOHR

8.1 INTRODUCTION

Probably the most important property that is affected by corrosion is that of mechanical strength. Other properties are also affected by corrosion; however, they generally do not lead to failure, as is often the case with changes in strength. Strength loss is not the only mechanical effect of corrosion, because there are also many cases where the effects of corrosion lead to increased strength. Increases in strength due to corrosion are the result of healing of cracks and flaws in the surface layers of a specimen due, quite often, to the diffusion of impurities from the bulk to the surface. This change in chemistry at the surface may lead to the formation of a compressive layer on the surface because of differential thermal expansion between the surface layer and the bulk. Compressive surface layers may also form as a result of surface alteration layers having a larger specific volume than the bulk.

Environmentally enhanced strength loss may arise through the following phenomena:

1. Cracking of the surface alteration layers due to excessive mismatch in thermal expansion between the surface and the bulk
2. Melting of secondary phases at high temperature
3. Lowering of the viscosity of a glassy grain boundary phase at high temperature
4. Surface cracking caused by polymorphic transitions in the crystalline phases at the surface
5. Alteration that forms low strength phases
6. Formation of voids and pits, especially true for corrosion by oxidation
7. Crack growth

The term used to describe these phenomena is called *stress corrosion* or *stress corrosion cracking* (SCC), which occurs when a material is subjected to a corrosive environment while being under the influence of an external mechanical load. Stress corrosion cracking implies that the pair of parameters, applied stress and corrosive environment, must both be active. Removal of either the applied stress or the corrosive environment will prevent cracking.

Oxidation often leads to compositional and structural alteration, especially of surface layers and grain boundary phases, of a ceramic that subsequently leads to considerable changes in the physical properties. Such alterations can lead to changes in density, thermal expansion, and thermal and electrical conductivity. The influence that these changes exert upon mechanical properties can be deduced only through a thorough investigation of the mechanisms and kinetics of corrosion. For example, the oxidation of silicon-based ceramics has been shown to be either active or passive depending upon the partial pressure of oxygen present during exposure (see Chapter 5, Section 5.2.2 for a discussion of the oxidation of SiC and Si_3N_4). When the pO_2 is low, gaseous SiO is formed, leading to rapid material loss and generally to a loss in strength. When the pO_2 is high, SiO_2 is formed leading to strength increases depending upon the actual temperature and time of exposure, and whether or not the strength test is conducted at room or an elevated temperature. The investigator should be well aware that conducting mechanical property tests in air (which may also include moisture) at elevated temperatures constitutes exposure to a corrosive environment for many materials.

The failure of ceramics after long exposure to a constant applied load, well below the critical stress, is called *static fatigue* or *delayed failure*. If the load is applied under constant stress rate conditions, it is called *dynamic fatigue*. If the load is applied, removed, and then reapplied, the failure after long-time cycling is called *cyclic fatigue*. It is now well known that brittle fracture is quite often preceded by subcritical crack growth that leads to a time dependence of strength. It is the effect of the environment upon the subcritical crack growth that leads to the phenomenon termed stress corrosion cracking. Thus fatigue (or delayed failure) and stress corrosion cracking relate to the same phenomenon. In glassy materials, this delayed failure has been related to glass composition, temperature, and the environment (e.g., pH). Failure is a result of the chemical reaction that takes place preferentially at strained bonds at the crack tip with the rate being stress sensitive. Some crystalline materials exhibit a delayed failure similar to that in glasses.

The experimental relationship between crack velocity and the applied stress (i.e., stress intensity factor K_I) is therefore of utmost importance. Attempts to fit various mathematical relationships to the experimental data have led to an assortment of equations of either the commonly used power law type or of some exponential form. The power law:

$$v = A(K_I/K_{IC})^n \qquad (8.1)$$

where A is a material constant (strong dependency upon environment, temperature, etc.), n is the stress corrosion susceptibility parameter (weak dependency upon environment), and K_I is the applied stress intensity. K_{IC}, which denotes the critical stress intensity factor, has been used most often because of its simplicity. It is the value of n (and also A) that determines a material's susceptibility to subcritical crack growth. Final lifetime predictions are very sensitive to the value of n. The power law, however, does not always lead to the best representation. Jakus et al. [8.1] evaluated the prediction of static fatigue lifetimes from experimental dynamic fatigue data for four different materials and environments. These were hot-pressed silicon nitride at 1200°C in air, alumina in moist air, optical glass fiber in air, and soda–lime glass in water. They found that the exponential forms of the crack velocity equations allowed better predictions of lifetimes for the silicon nitride and optical glass fiber, but the power law form of the crack velocity equation allowed better predictions for alumina and soda–lime glass. Thus they concluded that one should collect data for several different loading conditions, and then select the crack velocity equation that best represents all the data for making lifetime predictions. Matthewson [8.2] has reported that one particular optical fiber material gave a best fit to the exponential form when tested in ambient air but gave a best fit to the power law when tested at 25°C in a pH=7 buffer solution. Matthewson suggested that one kinetics model unique to all environments probably does not exist, and that since the power law yields the most optimistic lifetimes, it is unsatisfactory for design purposes.

Crack velocity can be evaluated by direct and indirect methods. In the direct methods, crack velocity is determined as a function of the applied stress. These involve testing by techniques such as the double cantilever beam method, the double torsion method, and the edge or center cracked specimen method. Indirect methods, which are normally performed on opaque samples, infer crack velocity data from strength measurements. A common indirect method is to determine the time-to-failure as a function of the applied load. In addition to the constant load technique, the constant strain technique has also been used. Other methods that have been used to evaluate the effects of corrosion upon the mechanical properties of ceramics include:

1. The percent loss in fracture strength after exposure to a corrosive environment (strength test conducted at room temperature).
2. The fracture strength at some elevated temperature during exposure to a corrosive environment.
3. The evaluation of creep resistance during exposure to a corrosive environment.
4. The determination of the strength distribution (at room temperature) after exposure to a corrosive environment and a static load. Generally, this type of evaluation indicates the dynamic nature of the flaw population.

Because silicate glasses are isotropic and homogeneous, most of the investigations into mechanisms have been carried out on these materials.

8.2 MECHANISMS

8.2.1 Crystalline Materials

There have been several mechanisms described in the literature, some of which are attributable to variations in the environment. Probably the most important area where questions still arise is what actually is occurring at the crack tip. Although the mechanism described by Evans and coworkers [8.3–8.5] involves the effects of an intrinsic, small quantity of a secondary amorphous phase, the overall effect should be very similar to the case when a solid is in contact with a corrosive environment that either directly supplies the amorphous phase to the crack tips or forms an amorphous phase at the crack tips through alteration. In essentially single-phase polycrystalline alumina, Johnson et al. [8.3] attributed cracking to the penetration into the grain boundaries of amorphous phase that was contained at the crack tip of intrinsic cracks, which subsequently caused localized creep embrittlement. Crack blunting will occur if the amorphous phase becomes depleted at the crack tip.

Strength degradation at high temperatures according to Lange [8.6,8.7] was a result of crack growth at stress levels below the critical applied stress required for fracture. This type of crack growth is called *subcritical crack growth* and is caused by cavitation of the glassy grain boundary phase located at grain junctions. The stress field surrounding the crack tip causes the glassy phase to cavitate facilitating grain boundary sliding, thus allowing cracks to propagate at stress levels less than critical. Surface and grain boundary self-diffusion were reported by Chuang [8.8] to be the accepted controlling factors in cavity growth at high temperatures, although other factors such as grain sliding and dislocation slip may also be present.

A mechanism for stress corrosion cracking at high temperatures was believed by Cao et al. [8.5] to be a result of stress-enhanced diffusion through the corrosive amorphous phase from crack surfaces, causing accelerated crack propagation along grain boundaries. They made the following assumptions:

1. Flat crack surfaces behind the crack tip.
2. Principal flux toward the crack tip.
3. Equilibrium concentration of the solid in the liquid.
4. Reduced solid in liquid at the crack tip caused by crack surface curvature.
5. Sufficiently slow crack tip velocity to allow viscous flow of liquid into the tip.
6. Chemical potential gradient normal to the crack plane was ignored.

Cao et al. pointed out that this mechanism was most likely to occur in materials where the amorphous phase was discontinuous. Systems that contained a small dihedral angle (see

Chapter 2, Section 2.5.3, "Surface Energy" for a discussion of dihedral angles) at the grain boundary and contained low-viscosity amorphous phases were the ones that were the most susceptible to rapid crack propagation. Thus the wetting of the solid by the amorphous phase was of primary importance, because phases that wet well formed small dihedral angles that induced sharp crack tips.

8.2.2 Glassy Materials

It is a well-known fact that silicate glasses can be strengthened by etching in hydrofluoric acid. This phenomenon has been explained by Hillig and Charles [8.9] to be one that involved the increase in the radius of curvature of the tips of surface cracks caused by the uniform rate of attack, which depended on the curvature, by the corrosive medium. This increase in the radius of curvature or rounding of the crack tips increased the critical stress required for failure. Bando et al. [8.10] gave direct transmission electron microscopy (TEM) evidence of crack tip blunting in thin foils of silica glass, supporting the dissolution/precipitation theory of crack tip blunting suggested by Ito and Tomozawa [8.11], although it is not clear that the precipitated material caused any significant strength increase. However, under the influence of an applied stress, Charles [8.12] concluded that the corrosion reaction rate was stress-sensitive, leading to an increased rate of attack at the crack tip and thus a decrease in the radius of curvature (i.e., a sharper crack tip) and a lower strength.

The fact that glass suffers from static fatigue has been known for many years, and studies over the past few decades have elucidated the reasons for this behavior. It is now believed that the reaction between water vapor and the glass surface is stress-dependent and leads to eventual failure when glass is subjected to static loading. As reported by Wiederhorn [8.13], three regions of behavior are exhibited when crack velocity is plotted vs. applied force (depicted in Fig 8.1). In the first region, the crack velocity (as low as 10^{-10} m/sec) is dependent on the

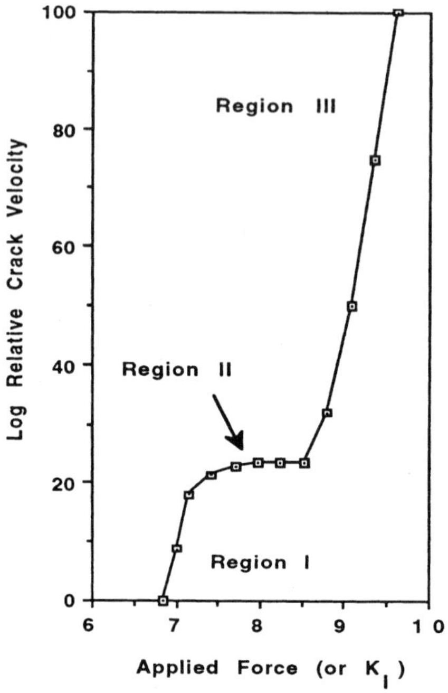

FIGURE 8.1 Crack velocity vs. applied force (or K_I). (From Ref. 8.13.)

applied force, with the exact position of the curve and its slope being dependent on humidity. At higher humidities, the crack growth occurs more rapidly and at a lower force. In addition, Wiederhorn and Bolz [8.14] have shown that the slope and position of the curve in this first region is dependent upon the glass composition (stress corrosion resistance was in the order fused silica > aluminosilicate>borosilicate>soda–lime silicate>lead silicate), and Wiederhorn and Johnson [8.15] have shown that it is dependent upon the pH. In the second region, the crack velocity is essentially independent of the applied force. At higher humidities, this portion of the curve shifts to higher velocities. In the third region, the crack velocity is again dependent upon the applied force, but the slope is much steeper, indicative of a different mechanism for crack propagation. This third region is also independent of the humidity.

Wiederhorn [8.16] has shown that the crack growth in glasses is dependent upon the pH of the environment at the crack tip and is controlled by the glass composition. Wiederhorn and Johnson [8.15] clarified that even further by reporting that at high crack velocities, the glass composition (for silica, borosilicate, and soda–lime glasses) controls the pH at the crack tip, and that at low crack velocities the electrolyte controls the pH at the crack tip. They studied the crack velocity as a function of the applied stress intensity, which they determined by the following equation for a double cantilever beam specimen:

$$K_I = \frac{PL(3.467 + 2.315\ t/L)}{(wa)^{1/2} t^{3/2}} \tag{8.2}$$

where:

P = applied load
L = crack length
w = total thickness
a = web thickness
t = half-width

The actual shape of the velocity vs. K_I curves is determined by a balance between the corrosion process, which tends to increase the crack tip radius, and the stress–corrosion process, which tends to decrease the crack tip radius [8.17].

Wiederhorn et al. [8.18] gave an equation of the following type for determining the crack velocity in aqueous media:

$$v = v_o a_{H_2O} \exp(-DG^*/RT) \tag{8.3}$$

where:

v = crack velocity
v_o = empirical constant
a_{H_2O} = activity of water
G^* = free energy of activation
R = gas constant
T = temperature

derived from reaction rate theory, assuming that crack velocity was directly proportional to the reaction rate. In addition, they assumed that the reaction order was equal to 1 with respect to water in solution. This, it was pointed out, was reasonable at the high relative humidities of their work, but was most likely incorrect at low relative humidities, where it is probably one-half based on the work of Freiman [8.19] in alcohols. The activity of water vapor over a solution is equal to the ratio of the actual vapor pressure to that of pure water. For water dissolved in a nonaqueous liquid, this ratio is equivalent to the relative humidity over the solution. This is why the crack velocity is dependent upon the relative humidity and not the concentration of the water [8.19]. Thus it is important not to assume that a liquid is inert just because it has a low solubility for water. In the region of high crack velocities (i.e., region III), it is the chain length of the alcohol for N between 6 and 8 that determines crack velocity.

The pH at the crack tip was dependent upon the reaction of the solution at the crack tip with the glass composition and diffusion between the bulk solution and the solution at the crack tip. Ion exchange at the crack tip between protons from the solution and alkalies from the glass produced $(OH)^-$ ions, and thus a basic pH at the crack tip. Ionization of the silicic acid and silanol groups at the glass surface produced an acid pH at the crack tip. Estimated crack tip pH ranged from about 4.5 for silica glass to about 12 for soda–lime glass. At high crack velocities, reaction rates at the crack tip are fast and the glass composition controls the solution pH. At low velocities, diffusion depletes the solution at the crack tip, which is then similar to the bulk solution. Wiederhorn and Johnson [8.15] concluded that silica exhibited the greatest resistance to static fatigue in neutral and basic solutions, whereas borosilicate glass exhibited the greatest resistance in acid solutions.

Michalske and Bunker [8.20] gave an equation that related the crack velocity of a silica glass to the applied stress intensity (K_I) for environments of ammonia, formamide, hydrazine, methanol, N-methylformamide, and water. This equation is given below:

$$V = V_o \exp(nK_I) \tag{8.4}$$

where:

V = crack velocity
V_o = empirical constant
K_I = applied stress intensity
n = slope of the exponential plot

Crack velocity vs. applied stress intensity plots (same as Fig. 8.1) yielded region I behavior for ammonia, formamide, hydrazine, methanol, *N*-methylformamide, and water. A small amount of residual water contained in aniline, *n*-propylamine, and *tert*-butylamine yielded a behavior representative of regions I and II. Moist N_2 exhibited a behavior represented by all three regions. Michalske and Bunker interpreted the mechanism for each region based upon the representations given in Table 8.1.

All the chemicals that exhibited region I behavior only have at least one lone pair electron orbital close to a labile proton. Using the shift in the vibrational frequencies of the OH groups on silica surfaces, Michalske and Bunker concluded that all nine of the chemicals tested acted as effective bases toward the silica surface silanol groups and thus one would expect a similar behavior based solely upon chemical activity.

TABLE 8.1 Representation of Crack Growth for Each Region of Fig. 8.1

Region	Represented by
I	Velocity ($<10^{-5}$ m/sec) exponentially dependent upon the applied K_I. Crack growth controlled by corrosive action of water (ion exchange) that ruptures Si–O–Si bonds as shown in Fig.8.2.
II	Velocity ($10^{-5} < v < 10^{-3}$ m/sec) independent of applied K_I. Crack growth controlled by transport rate of water to the crack tip.
III	Velocity ($>10^{-3}$ m/sec) exponentially dependent upon the applied K_I. Crack growth controlled purely by mechanical bond rupture.

Michalske and Bunker, therefore, developed a steric hindrance model to explain why anniline, *n*-propylamine, and *tert*-butylamine exhibited a bimodal behavior. These molecules were the largest of all those examined and a critical diameter of <0.5 nm for molecular diffusion to the crack tip opening was suggested. They also noted that, as the size of the corrosive environment molecule exhibiting region I behavior only increased, its effectiveness decreased. This whole area of the effects of steric hindrance and chemical activity upon stress corrosion fracture kinetics appears to be one of some importance, not only to glass, but also to crystalline materials.

For environments to enhance the crack growth, they must be both electron and proton donors [8.21]. In soda–lime glass, the modifier ions do not participate directly in the fracture process, but may change the reactivity of the Si—O bridging bonds and affect the elastic properties of the network bonds [8.22]. Thus, static fatigue is controlled by the stress-enhanced reaction rate between the Si—O bond and the environment at the crack tip.

Michalske and Freiman [8.21] described a three-step mechanism for the reaction of water with strained Si—O bonds. These were:

1. Water molecule aligns its oxygen lone electron pair orbitals toward the Si with hydrogen bonding to the oxygen of the silica (a strained Si—O bond enhances reaction at this site).
2. Electron transfer from oxygen of water to Si along with proton transfer to oxygen of silica.
3. Rupture of hydrogen bond to oxygen of water and the transferred hydrogen yielding Si—OH bonds on each fracture surface.

This mechanism is depicted in Fig. 8.2. This mechanism appears to be a general one, at least for cations that are attracted to the oxygen's (of water) lone electron pair. Michalske et al. [8.23] have shown that this dissociative chemisorption mechanism is the same for alumina, although the details differ. In alumina, it is not necessary for the bonds to be strained for adsorption to occur.

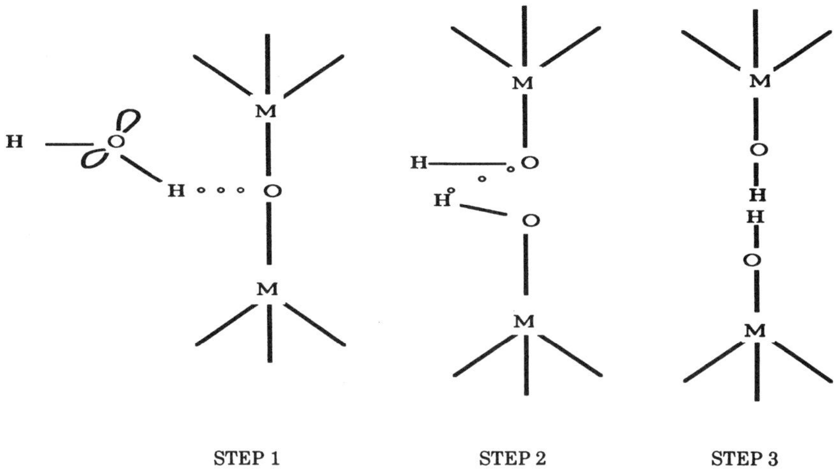

FIGURE 8.2 Mechanism of bond rupture. (From Ref. 8.21.)

White et al. [8.24] reported that Li^+ ions in solution negated the enhanced rates noted in high pH solutions, where OH^- controls the rate of bond breaking by readily associating with the OH^-, and not allowing OH^- to react with the Si–O–Si bond at the crack tip. This type of reaction is not exhibited by other alkalies, because they do not readily react with OH^- ions.

A more recent review of this topic of stress corrosion in silicate glasses was given by Gy [8.25].

8.3 DEGRADATION OF SPECIFIC MATERIALS

8.3.1 Degradation by Oxidation

Carbides and Nitrides

When evaluating the effects of corrosion, one must be alert to the changes that occur if samples are corroded, then cooled to room temperature for mechanical testing. McCullum et al. [8.26] found that the room-temperature flexural strength of SiC increased with exposure time to an air environment at 1300°C, whereas it decreased for Si_3N_4. They attributed this increased strength for SiC to the formation of a thin silica surface layer that healed surface flaws. The decreased strength for Si_3N_4 was attributed to the formation of a much thicker silica surface layer that cracked upon cooling. This cracking of the surface oxide layer was caused by stresses arising from the volume difference between the nitride and the oxide, and to the polymorphic transformation of either cristobalite or tridymite. Exposure times beyond 100 hr did not yield continued lower strengths for the Si_3N_4, because the layer thickness remained essentially constant for exposure times greater than 100 hr. Hench et al. [8.27] reported that if Si_3N_4 contained 8 wt.% ZrO_2 as a hot-pressing additive, an amorphous silicate film formed ($\leq$18 μm thick) that provided an effective diffusion barrier to oxygen when heated to about 1430°C in air for 100hr. The room-temperature MOR of this material decreased about 40% after oxidation, whereas the MOR of a material containing 3 wt.% MgO decreased by more than 60%. The difference was attributed to the oxidized layer formed on the zirconia-containing material being essentially amorphous, and that of the magnesia-containing material being totally crystalline and much thicker (~3×).

When tested at temperature, McCullum et al. [8.26] found the flexural strength for SiC remained constant with increasing exposure times to oxidation, with values less than when tested at room temperature. This lower strength obtained when tested at temperature when compared to room temperature strength, was attributed to the formation of a compressive layer on the surface when cooled. In contrast, the Si_3N_4 exhibited a slight increase in strength with exposure time when tested at temperature with values greater than the room temperature values. This was attributed to the integrity of the surface layer at temperature. These data are generalized in Fig. 8.3.

McCullum et al. [8.26] gave the following equation to evaluate the dynamic fatigue of several different SiC and Si_3N_4 samples:

$$S_f = AVs^{(1/n+1)} \tag{8.5}$$

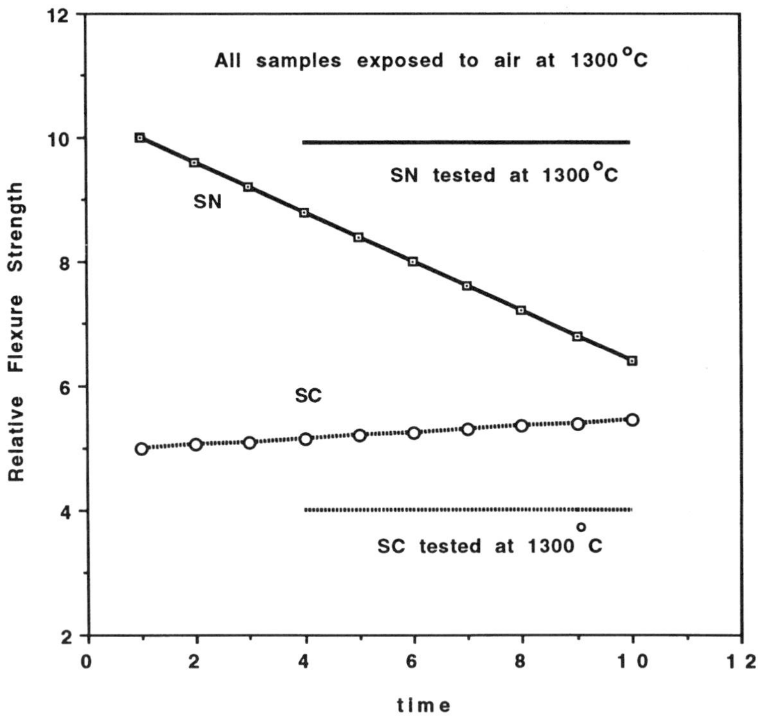

FIGURE 8.3 Fracture strength vs. time for Si_3N_4 and SiC. (From Ref. 8.26.)

where:

S_f = fracture strength
A′ = material constant
s = loading rate
n = stress corrosion susceptibility constant

Although they showed no strength vs. loading rate plots, one can obtain the value of n from the slopes of these plots. The values for the stress corrosion susceptibility constant, obtained in this manner for several different materials, ranged from infinity to about 8, over temperatures ranging from 20 to 1400°C, respectively.

Because of a larger quantity of sintering aids for the Si_3N_4 as compared to the SiC samples, the nitride samples were more susceptible to strength loss due to increased temperatures and decreased loading rates than the carbide. Variations in the amount and chemistry of the sintering aids in Si_3N_4 caused a variation in the mechanical behavior. Tensile test results followed the same general trends as flexural test results. In general, the tensile strength values were lower than the flexural values.

In the evaluation of a sintered and a hot-pressed Si_3N_4 under pure oxidation and oxidation under a static load, Easler et al. [8.28] found that, when exposed to air at 1370°C for times ranging from 0.5 to 50 hr, fracture data indicated that the range in flaw sizes decreased, whereas it increased after exposure for 1 hr under a static load. For the sintered material, strengths increased for oxidation under a static load of 23 or 45 MPa; however, the higher load condition resulted in a wider range of flaw sizes. In contrast, the hot-pressed material exhibited lower strengths for static loads of 45 and 160 MPa during oxidation. Under pure oxidation, the strengths of both materials increased for short (0.5 hr) oxidation times, and then decreased at longer times. The increased strengths were attributed to flaw tip blunting. The reduced strengths under static loading conditions were attributed to subcritical crack growth. Easler et al. concluded that the strength-controlling mechanisms, at least for silicon nitride, were dynamic in nature and very material-specific.

Rapid oxidation in air of Y-doped, sintered reaction bonded Si_3N_4 at 1000°C was reported by Govila et al. [8.29] to lower the strength and cause early failure. The fracture origins were determined to be β-Si_3N_4 needles. An excessive weight gain was reported to occur at 1000°C that was attributed to oxygen and nitrogen absorption of the matrix and secondary phases, one of which was reported to be $YSiO_2N$. The oxidation of $YSiO_2N$ to $Y_2Si_2O_7$ is accompanied by a 12% molar volume change. This anomalously high weight gain was accompanied by a 15% loss in the room-temperature strength. Stress rupture tests indicated the presence of stress-enhanced oxidation at 1000°C, with failure times ranging from 19 to 93 hr at an applied load of 138 MPa, and from 14 to 31 hr at an applied load of 276 MPa. Losses in strength at temperatures greater than 1200°C were attributed to the softening of the glassy grain boundary phase, which leads to creep by grain boundary sliding. Samples exposed to oxidation at 1200°C at an applied load of 344 MPa, did not fail, even after 260 hr, although some slight deformation had occurred.

In an effort to determine the effects of oxidation upon the flexural strength of Si_3N_4, Kim and Moorhead [8.30] evaluated the room-temperature four-point bend strength of HIP-SN (with 6 wt.% Y_2O_3 and 1.5 wt.% Al_2O_3) after exposure in either H_2/H_2O or Ar/O_2 at 1400°C for 10 hr. In both atmospheres, the strength was dependent on the amount of oxidant present. However, the actual variation in strength was different, depending upon the alteration of the surface layers formed and their characteristics. In the H_2/H_2O atmosphere at low pH_2O, a nonprotective and not well-attached glass-like layer containing crystalline $Y_2Si_2O_7$ formed. Because this layer was relatively uniform with no new strength-limiting flaws being formed (although some large bubbles were found at the surface/substrate interface), the maximum reduction in strength was limited to about 20% at a pH_2O of 2×10^{-5} MPa. A significant strength increase occurred as the pH_2O was increased, which the authors attributed to blunting of preexisting cracks by the interfacial silicate phase. This silicate phase was a continuous dense layer of $Y_2Si_2O_7$ containing small isolated bubbles believed to be formed by nitrogen generation during oxidation of the Si_3N_4. In the Ar/O_2 atmosphere, a similar reduction and subsequent increase in strength was not found. Instead, at low pO_2, an increase in strength occurred with increasing pO_2. The maximum strength occurred at pO_2 (10^{-5} MPa) that yielded the greatest weight loss. Even at low pO_2, a surface reaction product of $Y_2Si_2O_7$ formed in isolated pockets at grain junctions, presumably by the reaction of Y_2O_3 solid with SiO gas. Kim and Moorhead attributed the increased strengths observed to the formation of more $Y_2Si_2O_7$ as the pO_2 increased. At approximately a pO_2 of 10^{-5} MPa, where the maximum strength was observed, the $Y_2Si_2O_7$ layer became interconnected and, although not continuous, blunted strength limiting flaws. At higher pO_2, where weight gains were observed and a continuous layer containing $Y_2Si_2O_7$ and cristobalite formed, the increase in strength was not as significant. In this region, competition between crack blunting and formation of new flaws (cracks and bubbles) was suggested as the reason for the slightly lower strengths. This particular study by Kim and Moorhead pointed out very well the effects that the surface layer characteristics have upon the mechanical properties. Similar strength increases were found by Wang et al. [8.31] for two silicon nitride materials, one containing 13.9% Y_2O_3 plus 4.5% Al_2O_3 and the other containing 15% Y_2O_3 plus 5% Al_2O_3, when exposed to air at 1200°C for 1000 hr prior to strength testing at 1300°C. Strength increases as high as 87% were reported when compared to the unoxidized 1300°C strength, although the preoxidized 1300°C strength was slightly less than the unoxidized room temperature strength. Wang et al. attributed these strength increases to healing of surface flaws and crack blunting during oxidation, along with purification of the grain boundaries that raised the viscosity of the glassy boundary phase. These beneficial effects were not present when oxidation was conducted at 900°C.

Lange and Davis [8.32] have suggested that oxidation can lead to surface compressive stresses that, if optimum, may lead to increased apparent strengths. If the compressive stresses become too severe, then spalling may occur leading to lowered strengths. They demonstrated this concept with Si_3N_4 doped with 15% and 20% CeO_2 exposed to oxidation in air, at temperatures ranging from 400 to 900°C. The apparent critical stress intensity factor (K_a) increased for short exposure times at 400, 500, and 600°C. This increase in K_a was attributed to oxidation of the Ce-apatite secondary phase and subsequent development of a surface compressive layer. At longer times (~ 8 hr) and the two higher temperatures, surface spalling caused a decrease in

K_a. At higher temperatures (i.e., 1000°C), the compressive stresses that may cause spalling were relieved by extrusion of the oxide product from the interior of the material. Thus, prolonged oxidation at 1000°C did not degrade this material.

Oxynitrides

In a study of β′ and O′ SiAlON solutions, O'Brien et al. [8.33] found that the oxygen (or nitrogen) content significantly affected the performance of these materials. The grain boundary glassy phase viscosity increased as the nitrogen content increased, which subsequently slowed the healing of flaws (see Chapter 2, Section 2.2.3 on Glasses and Chapter 6, Section 6.2 upon Silicate Glasses for a discussion of the effects of nitrogen upon durability). The higher viscosity glassy phase also trapped evolving gases more easily, creating additional flaws. In general, the mean retained flexural strengths after oxidation at 1273 K for 24 hr of the SiAlON solutions was higher than that of several silicon nitrides, with the strengths being generally proportional to the oxidation resistance. O'Brien et al. concluded that the retained strengths after oxidation were dependent upon the characteristics of the surface oxide layer that formed. At higher temperatures, the potential for flaw healing was dependent upon the amount and composition of the glassy phase formed.

A zirconium oxynitride with the stoichiometry $ZrO_{2-2x}N_{4x/3}$ was reported by Claussen et al. [8.34] to form as a secondary phase in hot-pressed ZrO_2–Si_3N_4. This phase readily oxidized to monoclinic ZrO_2 at temperatures greater than 500°C. Lange [8.35] used the volume change (about 4–5%) associated with this oxidation to evaluate the formation of a surface compression layer on silicon nitride compositions containing 5–30 vol.% zirconia. To develop the correct stress distribution for formation of the surface compressive layer, the secondary phase that oxidizes must be uniformly distributed throughout the matrix. When oxidized at 700°C for 5 hr, a material containing 20 vol.% ZrO_2 exhibited an increase in strength from 683 to 862 MPa. Lange attributed this increase in strength to the oxidation-induced phase change of the zirconium oxynitride to monoclinic zirconia.

8.3.2 Degradation by Moisture

Lifetimes that are predicted from different fatigue tests will vary. Slow crack growth has been reported by Kawakubo and Komeya [8.36] to accelerate under cyclic conditions, especially of the tension–compression type cycle at room temperature for sintered silicon nitride. They also reported a plateau at about 70–90% of the stress intensity factor, when crack velocity was plotted vs. K_I. Three regions in the data were observed, very similar to that reported for glasses as shown in Fig. 8.1. As the materials studied had a glassy grain boundary phase, the fatigue mechanism was assumed to be the same as that reported for glassy materials [8.13] (i.e., stress corrosion cracking due to moisture in the air). Fett et al. [8.37] reported that at 1200°C, the lifetimes for cyclic loads were higher than for static loads. Tajima et al. [8.38] reported that a gas pressure sintered silicon nitride was resistant to slow crack growth up to 900°C, but then was susceptible to slow crack growth at 1000°C because of the softening of the glassy grain boundary phase. A higher fatigue resistance was reported for higher frequencies of the load cycle due to the viscoelastic nature of the glassy grain boundary phase.

8.3.3 Degradation by Other Atmospheres

Carbides and Nitrides

Clark [8.39] reported that Nicalon™ SiC fibers when aged in nitrogen or humid air at 1200°C for 2 hr, lost about one-half of their tensile strength. A more gradual strength decrease was observed for fibers that were exposed to hot argon. Although the time dependence of strength loss for the different aging environments was similar, the mechanisms causing strength loss were quite different. For exposure to nitrogen, Clark attributed the strength loss to crack propagation from existing flaws; for exposure to argon, he attributed the loss to grain growth and porosity; and for exposure to humid air, he attributed the strength loss to fiber coalescence at the silica surface, to poor adherence of the surface silica layer, to a cracked crystalline silica surface layer, and to bubbles at the silica/fiber interface. Clark also pointed out that thermal stability should not be based solely upon weight change data, because for this fiber, the weight gain produced by oxidation to silica was offset by weight loss due to CO evolution.

Siliconized, boron-doped, and aluminum-doped SiC samples were exposed to gaseous environments containing mixtures of predominantly N_2, H_2, and CO, representative of metallurgical heat-treatment atmospheres at 1300°C for up to 1000 hr by Butt et al. [8.40]. They reported significant strength losses for all three materials for times less than 100 hr when exposed to a gas mixture containing about 40% nitrogen. At longer exposure times, no additional strength loss occurred. The aluminum-doped SiC, unlike the other two, exhibited a slight strength increase after 1000 hr when exposed to a gas mixture containing 98.2% nitrogen. The strength losses were attributed primarily to pitting that was related to the presence of transition metal impurities.

It has been shown by Li and Langley [8.41] that ceramic fibers composed of Si–C–N–O experienced various degrees of strength degradation when aged in atmospheres of various hot gases. The rate of strength loss experienced by fibers aged in these hot gases was related to the rate of diffusion of the gases formed by decomposition. The gases of decomposition (N_2, CO, and SiO) diffused through the fiber porosity and any surface boundary layers present. The diffusion of these product gases can be controlled by aging the fibers in atmospheres of these gases. Thus, greater strength loss was exhibited when fibers were aged in argon compared to aging in nitrogen. This effect can be seen by examining the data of Table 8.2.

Zirconia-Containing Materials

Brinkman et al. [8.42] studied the effects of a diesel engine environment upon the strength of two commercial zirconias

TABLE 8.2 Effects of Aging in Various Atmospheres upon Strength of Si–C–N–O Fibers [8.41]

Unaged fiber (MPa)	Aged at 1400°C for 0.5 hr in (MPa)	
	N_2	Ar
1517	717	276

Source: Ref. 8.41.

partially stabilized with magnesia. The combustion environment at temperatures between 500 and 900°C contained Fe, Zn, Ca, Mg, and P contaminants from the fuel along with water vapor. The average flexural strength decreased by about 32% after exposure for 100 hr for the material rated as thermally shock resistant, and decreased by only 9% for the one rated as maximum strength. When the surface reaction products were removed from the thermally shock resistant material before strength testing, the strength decreased 22%. Both materials when exposed to air for 100 hr at 700 and 750°C exhibited decreases in strength of 6–8%, indicating a much more significant effect of the actual diesel engine environment. They found that the strength decreased as the amount of monoclinic zirconia increased. Thus the primary mechanism of degradation was attributed to localized increases in the monoclinic content.

8.3.4 Degradation by Molten Salts

Carbides and Nitrides

The strength loss of α-SiC and siliconized-SiC tubes exposed to a combustion flame into which a sodium silicate/water solution was injected was evaluated by Butt and Mecholsky [8.43]. The corrosive exposure was for times up to 373 hr, at temperatures 900–1050°C, with an oxygen partial pressure of about 4 kPa. Strength losses exceeded 50% for the α-SiC, and were 25–45% for the siliconized-SiC. Strength tests were conducted on C-ring samples after most of the reaction products were removed. Those samples for which the reaction products were not removed prior to strength testing exhibited no significant loss of strength, although an increase in scatter of the data was reported. Surface or corrosion pits were identified as the fracture origin for both types of SiC. In addition, the α-SiC exhibited grain boundary attack, whereas the siliconized-SiC exhibited oxidation of the silicon matrix and attack of the large SiC grains.

In a study of the effects of molten salt upon the mechanical properties of silicon nitride, Bourne and Tressler [8.44] reported that hot-pressed silicon nitride exhibited a more severe degradation in flexural fracture strength than did reaction sintered silicon nitride, although the weight loss of the hot-pressed material was less than that of the sintered one as reported by Tressler et al. [8.45] in a previous study. Their strength data are shown in Fig. 8.4. The exposure to a eutectic mixture of NaCl and Na_2SO_4 was more severe than to molten NaCl alone for the hot-pressed material, whereas for the reaction sintered material the effect was about the same. The differences between these two materials were attributed to the diffusion of contaminants along grain boundaries in the hot-pressed material and penetration of contaminants into pores of the reaction sintered material. This was based upon the observation that the grain boundaries of the hot-pressed material were more severely affected than those of the reaction sintered material, which did not contain an oxide grain boundary phase. The lowered fracture strengths resulted from an increase in the critical flaw size and a decrease in the critical stress intensity factor. The slight increase in fracture strengths at 1200°C was a result of a slight increase in the critical stress intensity factor. The NaCl/Na_2SO_4 eutectic mixture, being more oxidizing than the NaCl melt, caused a greater increase in the critical flaw size.

In the application of ceramics to turbine engines, the static fatigue life is of prime importance. Compared to the other types of mechanical testing in corrosive environments, little work has been reported on the long time exposure effects to static fatigue

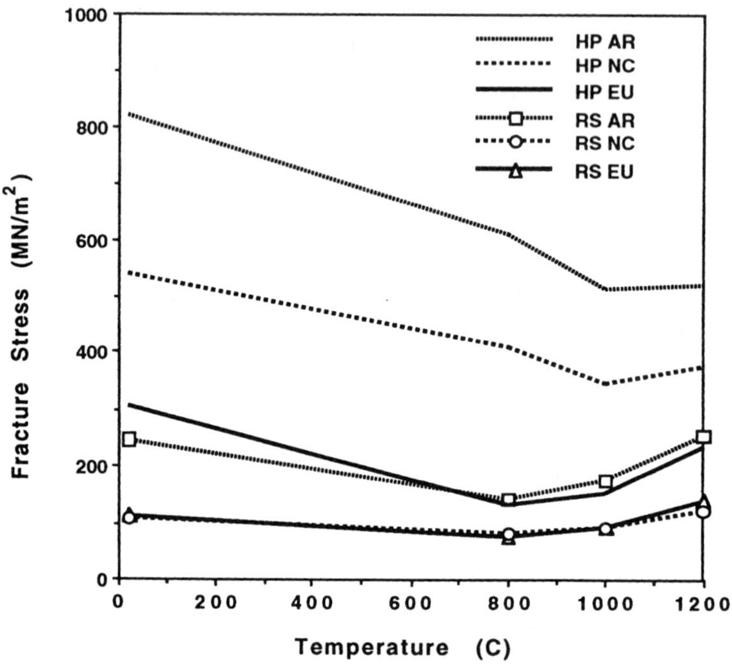

FIGURE 8.4 Fracture strength vs. molten salt composition. HP = hot pressed, RS = reaction sintered, AR = as-received, NC = NaCl, and EU = eutectic mixture of NaCl + Na_2SO_4. (From Ref. 8.44.)

life. Swab and Leatherman [8.46] reported that, at stresses between 300 and 500 MPa, there was a significant decrease in the time-to-failure for Si_3N_4 containing magnesia exposed to Na_2SO_4 at 1000°C. At stresses above 500 MPa and below 300 MPa, little change in the time-to-failure was noted. Because molten salt was not replenished during the test, corrosion pits were unable to grow to a size sufficient to decrease the time-to-failure at stress levels between 300 and 500 MPa. Although the decrease in room temperature strength for a yttria-containing silicon nitride after exposure to sodium sulfate was about 35%, it retained a greater strength than the magnesia-containing material (549 MPa for the Y-containing material vs. 300 MPa for the Mg-containing material) [8.47]. Fox and Smialek [8.48] tested sintered silicon nitride in a simulated gas turbine rig, where the corrosive environment was continued throughout the 1000°C/40 hr of the test. Room-temperature MOR fracture origins were located at pits in 17 of 22 samples. Pit formation was attributed to gas evolution during the oxidation of the silicon nitride and subsequent reaction of the silica with sodium sulfate-forming a low viscosity sodium silicate liquid. Fracture stresses were on the order of 300 MPa after exposure.

Boron- and carbon-doped injected molded sintered α-SiC sprayed with thin films of Na_2SO_4 and Na_2CO_3 were exposed to several gas mixtures at 1000°C for 48 hr by Smialek and Jacobson [8.49]. The gas mixtures used were 0.1%SO_2 in oxygen and 0.1%CO_2 in oxygen in combination with the sulfate or carbonate thin films, respectively. The sulfate-covered sample was also exposed to pure air. Strength degradation was most severe in the sulfate/SO_2 exposure (49% loss in strength), intermediate in the sulfate/air exposure (38% loss in strength), and least severe in the carbonate/CO_2 exposure. The latter exposure caused a statistically insignificant decrease in strength when analyzed by Student's *t*-test.* The primary mode of degradation was the formation of pits that varied in size and frequency depending upon the corrosion conditions. The size of the pits correlated quite well with the strength degradation (i.e., larger pits caused greater strength loss). Jacobson and Smialek [8.50] attributed this pit formation to the disruption of the silica scale by the evolution of gases and bubble formation.

Zirconia-Containing Materials

Although a considerable amount of scatter existed in the data of Swab and Leatherman [8.46], they concluded that Ce-TZP survived 500 hr at 1000°C in contact with Na_2SO_4 at stress levels below 200 MPa. At stress levels greater than 250 MPa,

*The application of the Student's *t*-test can be found in any elementary statistics book.

failure occurred upon loading the samples. Swab and Leatherman also reported a 30% decrease in the room-temperature strength of Y-TZP after 500 hr at 1000°C in the presence of Na_2SO_4. This lowered strength for Y-TZP was probably a result of leaching of the yttria from the surface, which caused the transformation of the tetragonal phase to the monoclinic phase.

8.3.5 Degradation by Molten Metals

The strength degradation of sintered α-silicon carbide was evaluated in both an as-received and as-ground (600 grit) condition after exposure to molten lithium by Cree and Amateau [8.51]. Transgranular fracture was exhibited for all samples when treated at temperatures below 600°C. At temperatures above 600°C, both transgranular and intergranular fracture occurred. The transgranular fracture strengths were generally greater than 200 MPa, whereas the intergranular strengths were less than 200 MPa. The low-strength intergranular failure was attributed to lithium penetration along grain boundaries beyond the depth of the uniform surface layer that formed on all samples. Grain boundary degradation was caused by the formation of Li_2SiO_3, from the reaction of oxidized lithium and silica. The formation of lithium silicate was accompanied by an increase in volume by as much as 25%, depending upon the temperature of exposure. The localized stresses caused by this expansion promoted intergranular crack propagation.

8.3.6 Degradation by Aqueous Solutions

Bioactive Materials

Bioactive ceramics include those materials that rapidly react with human tissue to form direct chemical bonds across the interface. Poor bonding across this interface and a sensitivity to stress corrosion cracking has limited the use of some materials. Alumina is one material that has received a reasonable amount of study. Porous alumina has been shown to lose 35%

of its strength in vivo after 12 weeks [8.52]. Seidelmann et al. [8.53] have shown that alumina loses about 15% of its strength after exposure to deionized water or blood when subjected to a constant stress. They also concluded that the service life of a hip endoprosthesis was dependent upon the density of the alumina. Ritter et al. [8.54] studied the effects of coating alumina with a bioactive glass that retarded the fatigue process.

Bioactive glasses, although bonding well to bone and soft tissue, generally lack good mechanical properties. Bioactive glasses are especially sensitive to stress corrosion cracking. Barry and Nicholson [8.55] reported that a soda–lime phosphosilicate bioactive glass was unsuitable for prosthetic use at stresses above 15 MPa, thus limiting its use to tooth prostheses. This glass sustained a tensile stress of 17 MPa for only 10 years in a pH=7.4 environment. Troczynski and Nicholson [8.56] then studied the fatigue behavior of particulate and fiber-reinforced bioactive glass of the same composition. The reinforcement materials were either −325 mesh silver powder or silicon carbide whiskers. These materials were mixed with powdered glass and hot-pressed at 700°C and 30 MPa for 30 min. The composite containing the silver particulates exhibited a decreased sensitivity to stress corrosion cracking, while the composite containing the silicon carbide whiskers exhibited a sensitivity similar to that of the pure glass. Comparison of the 10-year lifetimes of the two composites indicated that the particulate-containing material survived a static stress of 22 MPa, and the whisker-containing material survived a static stress of 34 MPa. Fractography results indicated agglomerate-initiated failure for the composites as opposed to surface machining defects for the pure bioactive glass.

Nitrides

In the evaluation of several hot isostatically pressed silicon nitrides, Sato et al. [8.57] found that the dissolution in HCl of the sintering aids (Y_2O_3 and Al_2O_3) from the grain boundaries decreased the three-point flexural strength. Their test variables included acid concentration, temperature, duration of dissolution, and crystallinity of the grain boundary phase. In general, the flexural strength decreased with increasing dissolution of Y^{3+} and Al^{3+} cations. Strengths were decreased by at least 50% after being exposed to 1 M HCl solution for 240 hr at 70°C. As expected, the grain boundary phase, having the highest degree of crystallinity, exhibited the highest strength (i.e., it is easier to leach cations from a glass than from a crystal). A control composition containing no sintering aids exhibited little, if any, strength degradation after the HCl treatment, although the strengths were considerably below those materials containing sintering aids (initially 240 vs. 600 MPa).

Glassy Materials

In their investigation of silica optical fibers, Dabbs and Lawn [8.58] presented data that questioned the acceptance of the Griffith flaw concept, which assumed that the flaws were exclusively cracklike and were free of preexisting influences. The real problem lies in predicting fatigue parameters for ultrasmall flaws from macroscopic crack velocity data. Abrupt changes in lifetime characteristics can occur as a result of evolution of flaws long after their inception. To conduct experiments with well-defined flaws, many investigators are now using microindentation techniques. It has been reported by Lawn and Evans [8.59] that the formation of radial cracks from indentations is dependent upon the applied load. There exists a threshold load below which no radial cracks are generated; however, radial cracks may spontaneously form at the corners of subthreshold indentations long after the initial indent has been implanted if the surface is exposed to water [8.60]. Dabbs and Lawn reported data for silica optical fibers showing an abrupt increase in strength under low load conditions below the threshold for formation of radial cracks. They attributed this behavior to a transition from crack propagation-controlled failure to one of crack initiation-controlled failure. Although the subthreshold indents had no well-developed radial cracks, they were still the preferred site for fracture origin and, therefore, must overcome crack initiation first. This crack initiation step, being close to the sample's free surface, was thus sensitive to environmental interactions. This low load region exhibited three general features when compared to the high load region where failure was controlled by crack propagation: an increase in strength, an increase in fatigue susceptibility, and an increase in scatter of the data.

Matthewson and Kurkjian [8.61], however, have suggested that dissolution of high-strength silica fibers, with the subsequent formation of surface pits, was the cause of enhanced fatigue at low stress levels, and not the spontaneous crack "pop-in" as suggested by Dabbs and Lawn. "Pop-in" does occur for weaker fibers. Their dissolution theory of enhanced fatigue was supported by the data of Krause [8.62], who reported a two- to threefold reduction in strengths after exposure to water under zero stress. Because the time-to-failure was essentially linear with pH over the entire pH range, Matthewson and Kurkjian stated that the link between fatigue and dissolution was unclear. Matthewson et al. [8.63] showed that by incorporating colloidal silica into a polymer coating, substantial improvements in static fatigue and zero stress aging behavior could be obtained. This essentially delayed the onset of the fatigue knee (discussed below), leading to greater times-to-failure. The abrupt change of slope (or change in the fatigue parameter, n) in plots of applied stress vs. time-to-failure has been called the *fatigue knee* (see Fig. 8.5). If one were to extrapolate short-term data to longer times, a very much shorter fatigue life would be predicted. This fatigue knee, which has been well established for liquid environments, has also been recently established for vapor environments [8.64]. Matthewson et al. [8.63] have shown that the reduction in strength of silica fiber exposed to water under zero stress occurred at a time similar to that of the fatigue knee, and thus attributed both phenomena to the formation of surface pits by dissolution.

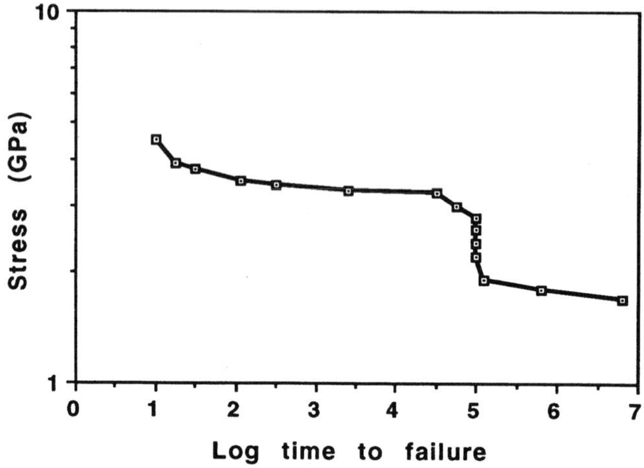

FIGURE 8.5 Applied stress vs. time-to-failure; the so-called fatigue knee.

These data all strongly suggested that enhanced fatigue at low stress levels was caused by the initiation of new surface flaws from dissolution pits, and not by the propagation of cracks from preexisting defects. Thus it is best not to base lifetimes upon extrapolated data, but to study the behavior in the strength range of interest.

Ito and Tomozawa [8.11] investigated the effects of exposure to water and to $Si(OH)_4$ aqueous solution at room temperature and 88°C upon the strength of high-silica glass rods. Mechanical strengths were determined at a constant stressing rate at room temperature in the two aqueous solutions and in liquid nitrogen. The room-temperature strength after exposure to $Si(OH)_4$ at 88°C increased more rapidly during the first 250 hr than that of rods exposed to water. Strengths leveled off after 240 hr for the $Si(OH)_4$ exposure, whereas strengths for rods exposed to water increased gradually throughout the entire range of exposure times approaching those of the $Si(OH)_4$ exposed rods after 360 hr. Maximum obtained strengths were about 30% higher than for unexposed samples. The strengths of samples exposed to room temperature solutions were essentially unchanged. The weight loss at 88°C in water was much higher than in $Si(OH)_4$ by a factor of about 10. As the strength increase was observed only when an observable weight loss was recorded, Ito and Tomozawa attributed the strength increase to a mechanism involving glass dissolution that increased the crack tip radii (i.e., crack blunting). If dissolution were the only phenomenon involved, strengths for water-exposed samples should be higher than those for $Si(OH)_4$ exposed samples, because the dissolution was greater for samples exposed to water. Because solubility is a function of surface curvature, and if solubility and dissolution were proportional, the dissolution rate would decrease with decreasing crack tip radius. This leads to a variation in dissolution rate around the crack tip leading to diffusion of dissolved glass and the combined effect of dissolution and precipitation [8.65]. Ito and Tomozawa, therefore, attributed the strength increasing mechanism to one of crack tip blunting caused by dissolution and precipitation.

Crack tip blunting by a different mechanism was suggested by Hirao and Tomozawa [8.66] for soda–lime, borosilicate, and high-silica glasses that had been annealed at or near their transition temperatures for 1 hr in air or a vacuum. Diffusion of water vapor into the glasses as they were being annealed in air was confirmed by infrared spectroscopy. The more rapid strength increases for glasses annealed in air compared to those annealed in a vacuum were attributed to the faster rate of viscous flow (causing more rapid crack tip blunting) in the less-viscous water-containing glasses, indicating that the release of residual stresses by annealing was not the cause for the strength increase as suggested by Marshall and Lawn [8.67]. Hirao and Tomozawa thus suggested that the conventional idea of glass fatigue caused by crack propagation alone is not sufficient, and must include a crack-sharpening step.

Environmentally enhanced crack growth was shown to be dependent upon composition in zirconia and barium fluoride glasses by Freiman and Baker [8.68]. They observed extended crack growth after 15 min in several different liquids, and found them to increase in the order dry oil, heptane, acetonitrile, and water. The fact that crack growth in acetonotrile was greater than in heptane suggested that it was not the presence of dissolved water in the liquids but the acetonitrile molecule that led to the enhanced crack growth.

It should be obvious that stress corrosion cracking is a rather complex phenomenon, and that its evaluation is not as straightforward as it might first appear. Exactly how crack tip blunting increases strength is still unclear. Decreases in strength are generally attributed to bond rupture at the crack tip caused by the presence of water molecules; however, it has been shown that other molecules (i.e., acetonitrile) act in a similar manner. Life-time predictions are based upon the selection of the proper crack velocity equation, and it has been shown that it is best to use an equation that represents the data of several loading conditions. In addition, the equation selected most likely will not be unique to all environments.

8.4 ADDITIONAL RELATED READING

Advances in Ceramics, Fractography of Glasses and Ceramics, Varner, J.R., Frechette, V.D., Eds.; Am. Ceram. Soc., Westerville, OH, 1988; Vol. 22, 442 pp.

Ceramic Transactions, Fractography of Glasses and Ceramics II, Frechette, V.D., Varner, J.R., Eds.; Am. Ceram. Soc., Westerville, OH, 1991; Vol. 17, 548 pp.

Fracture in Ceramic Materials, Evans A.G., Ed.; Noyes Publications, Park Ridge, NJ, 1984, 420 pp.

8.5 EXERCISES, QUESTIONS, AND PROBLEMS

1. Describe stress corrosion cracking and the consequences that relate to engineering materials.
2. Describe the differences among static, dynamic, delayed, and cyclic fatigue.

3. How does stress corrosion cracking relate to the type of fatigue listed in question #2?
4. How does one determine whether to use the power law or the exponential form to represent best the static fatigue lifetimes?
5. Discuss how cracks may propagate at a stress level less than that of the critical one for crack growth?
6. Discuss the three regions of behavior related to crack velocity and applied force for glassy materials. What role does relative humidity play?
7. Explain how Si_3N_4 may decrease in strength (room temperature) and SiC increase in strength (room temperature) after being exposed to air art 1300°C.
8. Discuss the differences that one may find when determining strengths at temperature vs. at room temperature and why this difference occurs.
9. Is it possible for an oxidative corrosion reaction to produce zero weight gain or loss? Explain.
10. Discuss the problems that one may encounter when extrapolating data to extended lifetimes.

REFERENCES

8.1. Jakus, K.; Ritter, J.E. Jr; Sullivan, J.M. Dependency of fatigue predictions on the form of the crack velocity equation. J. Am. Ceram. Soc. 1981, *64* (6), 372–374.

8.2. Matthewson, M.J. Models for fiber reliability. Proc. Int. Symp. Fiber Optic Networks & Video Communications: Berlin, Germany, Apr., 1993.

8.3. Johnson, S.M.; Dalgleish, B.J.; Evans, A.G. High temperature failure of polycrystalline alumina: III. Failure times. J. Am. Ceram. Soc. 1984, 67 (11), 759–763.

8.4. Evans, A.G.; Blumenthal, W. High temperature failure mechanisms in ceramic polycrystals. In *Deformation of Ceramics II;* Tressler, R.E., Bradt, R.C. Eds.; (1984). Plenum Publishing Co.: New York, 1984; 487–505.

8.5. Cao, H.C.; Dalgleish, B.J.; Hsueh, C-H.; Evans, A.G. High-temperature stress corrosion cracking in ceramics. J. Am. Ceram. Soc. 1987, *70* (4), 257–264.

8.6. Lange, F.F. High-temperature strength behavior of hot-pressed Si_3N_4: evidence for subcritical crack growth. J. Am. Ceram. Soc. 1974, *57* (2), 84–87.

8.7. Lange, F.F. Evidence for cavitation crack growth in Si_3N_4. J. Am. Ceram. Soc. 1979, *62* (3–4), 222–223.

8.8. Chuang, T-J. A diffusive crack-growth model for creep fracture. J. Am. Ceram. Soc. 1982, *65* (2), 93–103.

8.9. Hillig, W.B.; Charles, R.J. Surfaces, stress-dependent surface reactions, and strength. In *High Strength Materials;* Zackey, V.F., Ed.; Wiley Sons: New York, 1965; 682–705. Chapter 17.

8.10. Bando, Y.; Ito, S.; Tomozawa, M. Direct observation of crack tip geometry of SiO_2 glass by high-resolution electron microscopy. J. Am. Ceram. Soc. 1984, *67* (3), C36–C37.

8.11. Ito, S.; Tomozawa, M. Crack blunting of high-silica glass. J. Am. Ceram. Soc. 1982, *65* (8), 368–371.

8.12. Charles, R.J. J. Appl. Phys. 1958, *29*, 1549, 1554.

8.13. Wiederhorn, S.M. Influence of water vapor on crack propagation in soda–lime glass. J. Am. Ceram. Soc. 1967, *50* (8), 407–414.

8.14. Wiederhorn, S.M.; Bolz, L.H. Stress corrosion and static fatigue of glass. J. Am. Ceram. Soc. 1970, *53* (10), 543–548.

8.15. Wiederhorn, S.M.; Johnson, H. Effect of electrolyte pH on crack propagation in glass. J. Am. Ceram. Soc. 1973, *56* (4), 192–197.

8.16. Wiederhorn, S.M. A chemical interpretation of static fatigue. J. Am. Ceram. Soc. 1972, *55* (2), 81–85.

8.17. Simmons, C.J.; Freiman, S.W. Effect of corrosion processes on subcritical crack growth in glass. J. Am. Ceram. Soc. 1981, *64* (11), 683–686.

8.18. Wiederhorn, S.M.; Freiman, S.W.; Fuller, E.R. Jr.; Simmons, C.J. Effect of water and other dielectrics on crack growth. J. Mater. Sci. 1982, *17* (12), 3460–3478.

8.19. Freiman, S.W. Effect of alcohols on crack propagation in glass. J. Am. Ceram. Soc. 1974, *57* (8), 350–353.

8.20. Michalske, T.A.; Bunker, B.C. Steric effects in stress corrosion fracture of glass. J. Am. Ceram. Soc. 1987, *70* (10), 780–784.

8.21. Michalske, T.A.; Freiman, S.W. A molecular mechanism for stress corrosion in vitreous silica. J. Am. Ceram. Soc. 1983, *66* (4), 284–288.

8.22. Freiman, S.W.; White, G.S.; Fuller, E.R., Jr. Environmentally enhanced crack growth in soda–lime glass. J. Am. Ceram. Soc. 1985, *68* (3), 108–112.

8.23. Michalske, T.A.; Bunker, B.C.; Freiman, S.W. Stress corrosion of ionic and mixed ionic/covalent solids. J. Am. Ceram. Soc. 1986, *69* (10), 721–724.

8.24. White, G.S.; Freiman, S.W.; Wiederhorn, S.M.; Coyle, T.D. Effects of counterions on crack growth in vitreous silica. J. Am. Ceram. Soc. 1987, *70* (12), 891–895.

8.25. Gy, René Stress corrosion of silicate glass: a review. J. Non-Cryst. Solids 2003, *316*, 1–11.

8.26. McCullum, D.E.; Hecht, N.L.; Chuck, L.; Goodrich, S.M. Summary of results of the effects of environments on mechanical behavior of high-performance ceramics. Ceram. Eng. Sci. Proc. 1991, *12* (9,10), 1886–1913.

8.27. Hench, L.L.; Ohuchi, F.; Freiman, S.W.; Wu, C.Cm.; McKinney, K.R. Infrared reflection analysis of Si_3N_4 oxidation. Ceramic Engineering and Science Proceedings, 1980 ; Vol. 1 (7–8A), 318–330.

8.28. Easler, T.E.; Bradt, R.C.; Tressler, R.E. Effects of oxidation and oxidation under load on strength distributions of Si_3N_4. J. Am. Ceram. Soc. 1982, *65* (6), 317–320.

8.29. Govila, R.K.; Mangels, J.A.; Baer, J.R. Fracture of yttria-doped, sintered reaction-bonded silicon nitride. J. Am. Ceram. Soc. 1985, *68* (7), 413–418.

8.30. Kim, H-E.; Moorhead, A.J. High-temperature gaseous corrosion of Si_3N_4 in H_2–H_2O and Ar–O_2 environments. J. Am. Ceram. Soc. 1990, *73* (10), 3007–3014.

8.31. Wang, L.; He, C.; Wu, J.G. Oxidation of sintered silicon nitride materials. In *Proceedings of the 3rd International Symposium on Ceramic Materials and Components for Engines;* Tennery, V.J., Ed.; Am. Ceram. Soc.: Westerville, OH, 1989; 604–611.

8.32. Lange, F.F.; Davis, B.I. Development of surface stresses during the oxidation of several Si_3N_4/CeO_2 materials. J. Am. Ceram. Soc. 1979, *62* (11–12), 629–630 .

8.33. O'Brien, M.H.; Huang, C.M.; Coon, D.N. Oxidation and retained strength of in-situ O′–β′ SiAlON composites. Ceram. Eng. Sci. Proc. 1993, *14* (7–8), 350–357. Part 1.

8.34. Clausen, N.; Wagner, R.; Gauckler, L.J.; Petzow, G. Nitride-stabilized cubic ZrO_2. Presented at the 79th Annual Mtg. Am. Ceram. Soc., Chicago, Apr 23, 1977; Am. Ceram. Soc. Bull [abstr]. 1977; *56* (3), 301. Paper no. 136-B-77.

8.35. Lange, F.F. Compressive surface stresses developed in ceramics by an oxidation-induced phase change. J. Am. Ceram. Soc. 1980, *63* (1–2), 38–40.

8.36. Kawakubo, T.; Komeya, K. Static and cyclic fatigue behavior of a sintered silicon nitride at room temperature. J. Am. Ceram. Soc. 1987, *70* (6), 400–405.

8.37. Fett, T.; Himsolt, G.; Munz, D. Cyclic fatigue of hot-pressed Si_3N_4 at high temperatures. Adv. Ceram. Mater. 1986, *1* (2), 179–184.

8.38. Tajima, Y.; Urashima, K.; Watanabe, M.; Matsuo, Y. Static, cyclic and dynamic fatigue behavior of silicon nitride. In *Proceedings of the 3rd International Symposium on Ceramic Materials and Components for Engines;* Tennery, V.J., Ed.; Am. Ceram. Soc.: Westerville, OH, 1989; 719–728.

8.39. Clark, T.J. Fracture properties of thermally aged ceramic fiber produced by polymer pyrolysis. In *Advances in Ceramics, Fractography of Glasses and Ceramics;* Varner, J.R., Frechette, V.D., Eds.; Am. Ceram. Soc.: Westerville, OH, 1988; Vol. 22, 279–293.

8.40. Butt, D.P.; Tressler, R.E.; Spear, K.E. Silicon carbide materials in metallurgical heat-treatment environments. Am. Ceram. Soc. Bull. 1992, *71* (11), 1683–1690.

8.41. Li, C-T.; Langley, N.R. Development of a fractographic method for the study of high-temperature failure of ceramic fibers. In *Advances in Ceramics, Fractography of Glasses and Ceramics;* Varner, J.R. Frechette, V.D., Eds.; Am. Ceram. Soc.: Westerville, OH, 1988; Vol. 22, 177–184.

8.42. Brinkman, C.R.; Begun, G.M.; Cavin, O.B.; Foster, B.E.; Graves, R.L.; Kahl, W.K.; Liu, K.C.; Simpson, W.A. Influence of diesel engine combustion on the rupture strength of partially stabilized zirconia. In *Proceedings of the 3rd International Symposium on Ceramic Materials and Components for Engines;* Tennery, V.J Ed.; Am. Ceram. Soc.: Westerville, OH, 1989; 549–558.

8.43. Butt, D.P.; Mecholsky, J.J. Effects of sodium silicate exposure at high temperature on sintered α-silicon carbide and siliconized silicon carbide. J. Am. Ceram. Soc. 1989, *72* (9), 1628–1635.

8.44. Bourne, W.C.; Tressler, R.E. Molten salt degradation of Si_3N_4 ceramics. Ceram. Bull. 1980, *59* (4), 443–446, 452.

8.45. Tressler, R.E.; Meiser, M.D.; Yonushonis, T. Molten salt corrosion of SiC and Si_3N_4 ceramics. J. Am. Ceram. Soc. 1976, *59* (5–6), 278–279.

8.46. Swab, J.J.; Leatherman, G.L. Static-fatigue life of Ce-TZP and Si_3N_4 in a corrosive environment. J. Am. Ceram. Soc. 1992, *75* (3), 719–721.

8.47. Leatherman, G.L.; Katz, R.N.; Bartowski, G.; Chadwick, T.; King, D. The effect of sodium sulfate on the room temperature strength of a yttria containing silicon nitride. Ceram. Eng. Sci. Proc. 1993, *14* (7–8), 341–349.

8.48. Fox, D.S.; Smialek, J.L. Burner rig hot corrosion of silicon carbide and silicon nitride. J. Am. Ceram. Soc. 1990, *73* (2), 303–311.

8.49. Smialek, J.L.; Jacobson, N.S. Mechanism of strength degradation for hot corrosion of α-SiC. J. Am. Ceram. Soc. 1986, *69* (10), 741–752.

8.50. Jacobson, N.S.; Smialek, J.L. Corrosion pitting of SiC by molten salts. J. Electrochem. Soc. 1986, *133* (12), 2615–2621.

8.51. Cree, J.W.; Amateau, M.F. Degradation of silicon carbide by molten lithium. J. Am. Ceram. Soc. 1987, *70* (11), C318–C321.

8.52. Frakes, J.T.; Brown, S.D.; Kenner, G.H. Delayed failure and aging of porous alumina in water and physiological media. Am. Ceram. Soc. Bull. 1974, *53* (2), 183–187.

8.53. Seidelmann, U.; Richter, H.; Sotesz, U. Failure of ceramic hip endoprostheses by slow crack growth-lifetime prediction. J. Biomed. Mater. Res. 1982, *16*, 705–713.

8.54. Ritter, J.E. Jr; Greenspan, D.C.; Palmer, R.A.; Hench, L.L. Use of fracture mechanics theory in lifetime predictions for alumina and bioglass-coated alumina. J. Biomed. Mater. Res. 1979, *13*, 251–263.

8.55. Barry, C.; Nicholson, P.S. Stress corrosion cracking of a bioactive glass. Ad. Ceram. Mater. 1988, *3* (2), 127–130.

8.56. Troczynski, T.B.; Nicholson, P.S. Stress corrosion cracking of bioactive glass composites. J. Am. Ceram. Soc. 1990, *73* (1), 164–166.

8.57. Sato, T.; Tokunaga, Y.; Endo, T.; Shimada, M.; Komeya, K.; Komatsu, M.; Kameda, T. Corrosion of silicon nitride ceramics in aqueous hydrogen chloride solutions. J. Am. Ceram. Soc. 1988, *71* (12), 1074–1079.

8.58. Dabbs, T.P.; Lawn, B.R. Strength and fatigue properties of optical glass fibers containing microindentation flaws. J. Am. Ceram. Soc. 1985, *68* (11), 563–569.

8.59. Lawn, B.R.; Evans, A.G. A model for crack initiation in elastic/plastic indentation fields. J. Mater. Sci. 1977, *12* (11), 2195–2199.

8.60. Lawn, B.R.; Dabbs, T.P.; Fairbanks, C.J. Kinetics of shear-activated indentation crack initiation in soda–lime glass. J. Mater. Sci. 1983, *18* (9), 2785–2797.

8.61. Matthewson, M.J.; Kurkjian, C.R. Environmental effects on the static fatigue of silica optical fiber. J. Am. Ceram. Soc. 1988, *71* (3), 177–183.

8.62. Krause, J.T. Zero stress strength reduction and transitions in static fatigue of fused silica fiber lightguides. J. Non-Cryst. Solids 1980, *38–39*, 497–502.

8.63. Matthewson, M.J.; Rondinella, V.V.; Kurkjian, C.R. The influence of solubility on the reliability of optical fiber. *Optical Materials Reliability and Testing*, SPIE 1992, *1791*, 52–60.

8.64. Matthewson, M.J.; Yuce, H.H. Kinetics of degradation during aging and fatigue of fused silica optical fiber. In *Fiber Optics Materials and Components*; Proc. SPIE 1994, *2290,* 204–210.

8.65. Iler, R.K. *The Chemistry of Silica*; Wiley & Sons: New York, 1979; 3–115.

8.66. Hirao, K.; Tomozawa, M. Kinetics of crack tip blunting of glasses. J. Am. Ceram. Soc. 1987, *70* (1), 43–48.

8.67. Marshall, D.B.; Lawn, B.R. Residual stresses in dynamic fatigue of abraded glass. J. Am. Ceram. Soc. 1981, *64* (1), C6–C7.

8.68. Freiman, S.W.; Baker, T.L. Effects of composition and environment on the fracture of fluoride glasses. J. Am. Ceram. Soc. 1988, *71* (4), C214–C216.

9

Methods to Minimize Corrosion

> Failure is only the opportunity to begin again more intelligently.
>
> HENRY FORD

9.1 INTRODUCTION

The control of the chemical reactivity of ceramics with their environment is one of the most important problems facing the ceramics industry today. Through the study of corrosion phenomena, one can learn best how to provide the control of the chemical reactivity that will provide a maximum service life expectancy at a minimum cost. Most methods used to minimize corrosion have generally been methods that slow the overall reaction rates. However, once a complete understanding is available, one can attempt possibly to change the reaction mechanism to something less harmful, in addition to slowing the rate. Corrosion reactivity is affected by the following items (not necessarily listed in the order of importance):

1. Heat transfer
2. Mass transfer
3. Diffusion-limited processes
4. Contact area
5. Mechanism
6. Surface-to-volume ratio
7. Temperature
8. Time

The following discussion will address some of these items and how they may be used to minimize the effects of corrosion by discussing various examples.

9.2 CRYSTALLINE MATERIALS—OXIDES

The most obvious method of providing better corrosion resistance is to change materials; however, this can be done only to a certain extent. There will be ultimately only one material that does the job best. Once this material has been found, additional corrosion resistance can be obtained only by property improvement or, in some cases, by altering the environment. Different parts of an industrial furnace generally involve variations in the corrosive environment, necessitating the use of different materials with the best properties for a particular location within the furnace. Furnace designers have thus for a long time used a technique called *zoning* to maximize overall service life by using different materials in different parts of the furnace.

9.2.1 Property Optimization

Since exposed surface area is a prime concern in corrosion, an obvious property to improve is the porosity. Much work has been done in finding ways to make polycrystalline materials less porous or denser. The most obvious is to fire the material during manufacture to a higher temperature. Other methods of densification have also been used. These involve various sintering or densification techniques: liquid-phase sintering, hot pressing, and others. If additives are used to cause liquid phase sintering, care must be exercised that not too much secondary phase forms, which might lower corrosion resistance, although porosity may be reduced.

Alterations in major component chemistry may aid in increasing corrosion resistance, but this is actually a form of finding a new or different material, especially if major changes are made.

The history of glass-contact refractories is a good example of corrosion resistance improvement in a polycrystalline material. Porous clay refractories were used originally. Changes in chemistry by adding more alumina were made first to provide a material less soluble in the glass. The first major improvement was the use of fusion-cast aluminosilicate refractories. These provided a material of essentially zero porosity. The next step was the incorporation of zirconia into the chemistry. Zirconia is less soluble than alumina or silica in most glasses. Because of the destructive polymorphic transformation of zirconia, a glassy phase had to be incorporated into these refractories. This glassy phase added a less corrosion-resistant secondary phase to the refractory. Thus the higher resistance of the zirconia was somewhat compromised by the lower resistance of the glassy phase. The final product, however, still had a corrosion resistance greater than the old product without any zirconia. Today, several grades of ZrO_2–Al_2O_3–SiO_2 fusion-cast refractories are available. Those with the highest amount of zirconia and the lowest amount of glassy phase have the greatest corrosion resistance.

As discussed in Chap. 2, Sec. 2.5.2 (Porosity and Surface Area), *thermal transpiration* is the migration of a gas along a thermal gradient. As long as the pore size distribution is optimized, the transpiring gas will flow toward the hot face.

This transpiring gas must be selected so that it will alter the reaction at the hot face in a beneficial way. One obvious way is to dilute the effects of a corroding gas. Although the author knows of no examples of the use of thermal transpiration of gases to minimize or eliminate corrosion, there is no reason why it should not work.

Another example from the glass industry is the development of furnace regenerator refractories through the optimization of materials made of fireclay by using higher purity raw materials and then increased firing temperatures. Changes in chemistry were then made by switching from the fireclay products to magnesia-based products. Again, improvements were made by using higher purity raw materials and then increased firing temperatures. Minor changes in chemistry were also made during the process of property improvement. Changes in processing involving prereaction of raw materials have also been done. The evolution of regenerator refractories for the flat glass industry up to the mid-1970s has been described by McCauley [9.1]. The latest development in regenerator refractories has been the use of fusion cast alumina–zirconia–silica cruciform products. These are in the shape of a cross and are stacked in interlocking columns. This represents not only a change in chemistry, but also a change in the shape of the product, both of which lead to better overall performance.

A part of the concept of improvement through chemistry changes is that of improving resistance to corrosion of the bonding phases. Bonding phases normally have a lower melting point and lower corrosion resistance than does the bulk of the material. The development of high alumina refractories is a good example of improvement based on the bonding phase. The best conventional high alumina refractories are bonded by mullite or by alumina itself. To change this bond to a more corrosion-resistant material compatible with alumina, knowledge of phase equilibria played an important role. Alumina forms a complete series of crystalline solutions with chromia, with the intermediate compositions having melting points between the two end members. Thus a bonding phase formed by adding chromia to alumina would be a solution of chromia in alumina with a higher melting point than the bulk alumina and thus a higher corrosion resistance. In addition to the more resistant bonding phase, these materials exhibit a much higher hot modulus of rupture (more than twice mullite or alumina-bonded alumina). Nothing is ever gained, however, without the expense of some other property. In this case, the crystalline-solution-bonded alumina has a slightly lower thermal shock resistance than does the mullite-bonded alumina. Owing to the excellent resistance of these materials to iron oxide and acid slags, they have found applications in the steel industry.

The development of tar-bonded and tar-impregnated basic refractories to withstand the environment of the basic oxygen process of making steel is yet another example of a way to improve the corrosion resistance of a material. Tar-bonded products are manufactured by adding tar to the refractory grain before pressing into shape. In this way, each and every grain is coated with tar. When the material is heated during service, the volatiles burn off, leaving carbon behind to fill the pores. An impregnated product is manufactured by impregnating a finished brick with hot tar. This product, once in service, will similarly end up with carbon in the pores. Impregnated products do not have as uniform a carbon distribution as do the bonded types. Newer products incorporate graphite into the raw material mix. The carbon that remains within the refractory increases the corrosion resistance to molten iron and slags by physically filling the pores, by providing a nonwetting surface, and by aiding in keeping iron in the reduced state, which then does not react with the oxides of the refractory. Any oxygen that diffuses into the interior of the refractory causes carbon oxidation that slightly increases the pore pressure and thus minimizes slag and metal penetration. A thin layer on the hot face (1–2 mm) does lose its carbon to oxidation and various slag components penetrate and react within this layer. This corrosion, however, is much slower than with a product that contains no carbon.

An additional improvement upon the carbon-containing magnesia refractories has been the incorporation of magnesium metal, as reported by Brezny and Semler [9.2]. Upon magnesium volatilization and diffusion toward the hot face, oxidation and precipitation enhance the formation of the dense magnesia-rich zone that forms behind the hot face and thus minimizes slag and metal penetration in addition to oxygen diffusion to the interior of the refractory (see Chap. 5, Sec. 5.3.2 for a discussion on the formation of this magnesia-rich dense zone).

Others [9.3] have tried to take advantage of the nonwetting enhancement caused by antiwetting additives; however, their success was questionable.

The automotive industry in their efforts to develop a gas turbine engine has conducted a considerable amount of research on low expansion lithium aluminosilicates (LAS) and magnesium aluminosilicates (MAS) for a rotary wheel heat exchanger. The LAS materials are based upon solid solutions of the high-temperature polymorphs of two different compounds—eucryptite ($Li_2O{\cdot}Al_2O_3{\cdot}SiO_2$) and spodumene ($Li_2O{\cdot}Al_2O_3{\cdot}4SiO_2$). Both of these materials have an upper use temperature of about 1200°C. Both have a very low thermal expansion (eucryptite being slightly negative) which gives them excellent thermal shock resistance. These materials, however, suffer from corrosion problems when used in dirty environments. To overcome these corrosion problems, an aluminosilicate (AS) material was developed by the acid leaching of lithium from LAS prior to application. This material had acceptable thermal expansion, although not as low as LAS, but did not distort or crack as much.

The development of direct bonded basic refractories is an example where the chemistry was changed to take advantage of the formation of multiple phases and the effects of surface energy upon penetrating liquids. See Chap. 2, Sec. 2.5.3 for a discussion of the surface energy effects of multiphase systems. The direct bonded basic refractory contains magnesia and spinel crystalline phases along with a grain boundary phase that is partially amorphous and partially small spinel crystals. At service temperatures, the silicate amorphous phase becomes molten and will penetrate between like grains more easily than between unlike grains. Thus the presence of multiple phases keeps the liquid from penetrating between all the grains and causing failure of the refractory. Therefore the possibility exists of creating more corrosion-resistant materials by incorporating a small amount of a second phase that would form along grain boundaries and act to retard the penetration of a corroding liquid.

Since corrosion of ceramics quite often involves the diffusion of various cations and anions through an interfacial reaction layer, changes to the chemistry that would either provide a layer through which diffusion is more difficult or provide species that would form a reaction layer immune to continued corrosion should be investigated. This would undoubtedly involve considerable research into the diffusion of various cations and anions through various materials. Only then will it be possible to tailor a composition to provide minimum corrosion.

9.2.2 External Methods of Improvement

In Chap. 2 on Fundamentals, the importance of temperature was stressed several times. Various techniques have been used to lower the temperature of the interface or hot face of the material (lower hot face temperatures mean less corrosion). Many applications of a ceramic material subject the material to a thermal gradient. By altering the material or providing a means to increase the heat flow through the material, the hot-face temperature can be lowered significantly, or more accurately, the slope of the thermal gradient is increased as shown in Fig. 2.3. One means of doing this is by forcibly cooling the cold face. This provides faster heat removal and thus lowers the hot-face temperature. Most industrial furnaces use some means of forced cooling on the cold face by cooling-air systems or water-cooled piping. In a few cases, water has actually been sprayed onto the cold face of the refractory using the heat of vaporization of the water to extract heat from the refractory. If the thermal gradient through the material becomes too steep, failure may occur (this depends upon the thermal expansion characteristics of the material).

Another method that has been used to lower the hot-face temperature is to place metal plates either within individual bricks or between them. A large portion of the heat is thus conducted through the metal plate. A similar technique has been used by manufacturing a product containing oriented graphite particles. The steel industry has used many of these techniques in their blast furnaces. The most common technique today is the use of water-cooled internal metal plates (or boxes).

Another way to take advantage of increased cooling is initially to use a thinner material. This will automatically cause a thinner reaction layer to form on the surface. In general, glass furnace basin wall linings should not be greater than 10–12 in. thick. Anything greater than about 12 in. does not normally increase overall life but adds an economic penalty in refractory cost per campaign. The thickness at the flux-line generally is 9 in. so that effective air cooling can be used. In fact, most linings could probably be less than 10 in.; however, the thermal–mechanical environment will determine the ultimate thickness that should be used.

If a refractory lining is insulated, a greater portion of the refractory will be at a higher temperature and corrosion will proceed at a faster rate. In these cases, a balance must be obtained between service life and energy conservation. Because of the potential for increased corrosion of insulated linings, the properties of the lining material must be carefully evaluated before insulation is installed. In many cases, the engineer may want to upgrade the lining material if it is to be insulated.

Historically, it has been recommended that flue gas temperatures be 20–30°C higher than the dew point [9.4]. However, because of thermal efficiencies and the related cost, this has been lowered to 5–10°C higher than the dew point. Once condensation has occurred, reevaporation of the water or other volatile can concentrate corrosive species causing a more severe corrosion problem; thus condensates should be removed as rapidly as possible.

The addition of redox couples in photoelectrochemical corrosion of electronically conductive materials in acids acts on the environment to minimize corrosion. An example is the addition of cobalt as the redox couple to scavenge SO_4^- that is formed by the reaction of a positive hole with the sulfate ion [9.5]. The positive hole is photogenerated in the valence band of an illuminated titania semiconductor. The reactions listed below act to minimize corrosion:

$$SO_4 = + p^+ \rightarrow SO_4^- \tag{9.1}$$

$$CO^{2+} + SO_4^- \rightarrow CO^{3+} + SO_4^{2-} \tag{9.2}$$

In the preservation of national monuments, engineers have tried various materials to fill the porous limestones, marbles, etc. to slow corrosion. Not only are the fillers used to eliminate open porosity, but also they are used to consolidate or strengthen friable portions of the structure. The most important parameter of the fillers is that they too must be corrosion-resistant. Many materials have been used to provide this filler/strengthening characteristic from waxes to acrylic polymers to silicic ester-type products [9.6]. According to Amoroso and Fassina [9.6], these materials must have the following basic characteristics:

1. Control the diffusion of water
2. Protect against atmospheric pollutants
3. Possess a low coefficient of thermal expansion
4. Be inert toward the parent structure
5. Not modify the original appearance

Although the preservation of national monuments around the world has been in progress for over 100 years, it has been only recently that true advances have been made in their preservation. At first, it would seem that the preservation of monuments is not very much like the slowing or elimination of corrosion of ceramics; however, the two areas are very much alike. The major difference is that for the monuments, one is concerned with their final appearance, whereas with ceramics, in general, that is not the case.

9.3 CRYSTALLINE MATERIALS—NONOXIDES

9.3.1 Property Improvement

Most of the items discussed earlier can also be applied to these materials. The one property improvement that should be discussed a little further is that of porosity. For example, Si_3N_4 is predominantly covalent and does not densify on heating as do conventional ionic ceramics. In applications such as turbine blades, a theoretically dense material is desired. Only through special densification procedures can theoretically dense materials be obtained. In the past, this could be accomplished for Si_3N_4 only through hot pressing with large amounts (up to 10 wt.%) of additives at very high temperatures and pressures. SiC, in contrast, could be prepared in the fully dense state with only a few percent of additives. Newer techniques have recently been developed using gas pressure sintering and much lower amounts of additives that allow the production of materials that are fully dense. The additives in these processes cause a liquid phase to form at high temperatures, and therefore densification can proceed through liquid-phase sintering. This liquid either crystallizes or forms a glass phase upon cooling. Much work has been done in attempting to obtain either crystalline phases with higher melting points or glassy compositions with higher viscosities to improve the high-temperature properties. The densification processes using lower amounts of additives (generally <2 wt.%) help to maximize the high-temperature properties.

Improved corrosion resistance of porous materials can be obtained by impregnating with either a material of the same composition as the bulk or with a material that, in the case of

SiC or Si_3N_4, is later exposed to a carbiding or nitriding treatment. Other pore-filling materials can also be used, such as nitrates or oxychlorides. Decomposition reactions then produce pore-filling oxides. Impregnation with organosilicon compounds will yield SiC as the pore filler.

Corrosion resistance can sometimes be improved by changing the processing method. Chemical vapor deposition (CVD) is one of the most attractive methods to produce high purity dense materials because the sintering process is not required if a bulk material can be obtained directly from the raw vapors or gases. Microstructures of CVD products are strongly dependent upon the deposition temperature and total gas pressure. Chemical vapor deposition can produce materials with no grain boundary phases but which are highly oriented. It is a well-known fact that CVD materials contain residual internal stresses. At present, the effects of these stresses upon high-temperature strength and corrosion are not well known.

Preoxidation under some conditions can form a protective oxide layer that will minimize or possibly eliminate continued corrosion [9.7]. In addition, impurities present, generally in the form of sintering aids, may migrate toward the surface and become part of the protective oxide layer. This layer can then be removed resulting in a purer material with subsequent improvement in mechanical properties.

The development of nitride-based materials today has progressed to the point of studying materials in $Si_aM_bO_cN_d$ systems, where M has been confined mostly to trivalent cations. Most work has been in systems where M = Al, Y, and/or Be. These materials form secondary grain boundary phases which are highly oxidation-resistant and thus provide a better material than conventional Si_3N_4 materials.

Cemented carbide cutting tools made from WC wear rapidly due to local welding of the tool to the steel piece being cut. To overcome this welding, additions of TiC were made to the WC to form a TiO_2 surface layer that protected the tool from rapid wear. WO_3 also formed, but it was volatile and produced no protective layer. In addition, small amounts of TaC and NbC were added to increase the overall oxidation resistance by increasing the melting temperature of the carbide solution formed.

9.3.2 External Methods of Improvement

One method of minimizing corrosion not widely practiced is that of coating the ceramic with a layer of more resistant material. Probably the best method to coat a ceramic is by a layer of CVD [9.8] or plasma-sprayed material of the same composition as the substrate [9.9]. Chemical vapor deposition, in general, provides a better coating than plasma-sprayed coatings since it is difficult to form pore-free coatings with uniform thickness using plasma spraying. This provides a well-attached, pure, nonporous layer that has a good thermal expansion match with the substrate. Coating conditions can be varied to produce layers of amorphous material covered by crystalline material of the same composition. This sometimes provides a more complex diffusion path that minimizes oxidation.

Although plasma or flame spraying can be used to deposit most materials, control of the spraying parameters confines the coating to mainly oxides. Other methods investigated have been cathode sputtering [9.8,9.10], glow-discharge cathode sputtering, electron beam evaporation, and detonation deposition. These methods are not necessarily confined to the coating of nonoxides; oxides can also be coated.

Wittmer and Temuri [9.11] in their work on oxidation of carbon–carbon composites have described a method of protection by coating first with a well-adhering solid oxygen barrier and then coating with a glass-forming material to seal any cracks that may develop from thermal expansion mismatch. The carbon–carbon composite coating system used for the Space Shuttle nose cap is composed of an inner SiC layer covered by a silicate glaze. This is probably the most successful example of the use of oxygen barrier coatings to protect carbon–carbon composites [9.12].

9.4 GLASSY MATERIALS

9.4.1 Property Optimization

The development of more resistant glasses has been predominantly through optimization of compositions. Historically, small amounts of alumina have been added to the basic soda–lime–silicate composition to improve durability. In general, lowering the alkali content increases the durability. This, however, has practical limits based upon melting temperatures, viscosities, softening points, and working ranges. Borosilicate glasses are, in general, more resistant than soda–lime silicate glasses. In general, silicate glasses are less resistant to alkali solutions than they are to acid solutions. Table 6.1 of Chap. 6 lists the corrosion resistance of many glasses of varying compositions.

One technique of composition variation to improve durability that has not received much attention is that of incorporation of nitrogen into the glass structure. Frischat and Sebastian [9.13] have shown that soda–lime–silica glasses containing 1.1 wt.% nitrogen exhibited considerable improvement toward leaching by water at 60°C over compositions containing no nitrogen. This improvement was attributed to a denser structure for the nitrogen-containing glass.

Small changes in the chemistry of the glass can cause a significant change in the dissolution mechanism as shown by Lehman and Greenhut [9.14]. They reported that 1 mol% P_2O_5 addition to a lead silicate glass caused the formation of lead phosphosilicate crystals on the glass surface when exposed to 1% acetic acid at 22°C. They attributed the reduction in dissolution to the reduction of the apparent average interdiffusion coefficient of lead by a factor of 11.3. This is an example of changing the material chemistry to form an interface reaction product that reduces the diffusion rate of the species being leached.

9.4.2 External Methods of Improvement

The development of coating technology has provided a means to improve corrosion resistance, abrasion resistance, and strength. Combinations of coatings applied while the glass is hot and after it has cooled have been developed that form a permanent bond to the glass. These coatings are not removed by cooking or washing.

The most commonly used metallic hot-end coatings are tin and titanium. As the piece goes through the annealing lehr, the metal oxidizes, forming a highly protective ceramic coating. Tin is easier to work with since a thicker coating can be applied before problems of iridescence occur. These hot-end metallic coatings give the glass a high glass-to-glass sliding friction and thus a cold-end coating must be applied over these metallic coatings. The cold-end coatings usually have a polyethylene or fatty acid base.

Another type of coating is one that reacts with the surface of the glass to form a surface layer that is more corrosion-resistant than the bulk composition. Chemically inert containers are needed to contain various beverages and pharmaceuticals. To provide increased corrosion resistance, these containers are coated internally to tie up the leachable components. Internal treatment with a fluoride gas provides a new surface that is more corrosion-resistant than the original and is more economical than the older sulfur treatment.

Although not a true coating technique, the manufacturers of flat glass have, for many years, treated the surface of their glass with SO_2 gas just prior to the glass being annealed to increase the weatherability of their products. This surface treatment allows the sodium in the surface layers to react with the SO_2 forming sodium sulfate. The sulfate deposit that forms on the surface due to this reaction is then washed off prior to inspection and packing. The first step in weathering is then diminished due to the low alkali content of the surface.

It has been shown by Harvey and Litke [9.15] that matrix dissolution of an aluminosilicate glass apparently does not occur if the leaching solution is saturated first with solution products of the same glass composition. This technique is an example of how dissolution can be minimized by decreasing the driving force for corrosion by lowering the concentration gradient between the material and leachant, thus minimizing or eliminating the diffusion of cations and anions across the interfacial boundary. Using a different approach to minimize dissolution of a predominantly soda-borosilicate glass, Buckwalter and Pederson [9.16] have shown that the sorption of metal ions onto the glass surface and/or the buffering of the leachate solution caused by the corrosion of metal containers significantly lowered the rate of aqueous corrosion.

REFERENCES

9.1. McCauley, R.A. Evolution of flat glass furnace regenerators. Glass Ind. 1978, *59* (10), 26–28, 34.

9.2. Brezny, R.; Semler, C.E. Oxidation and diffusion in selected pitch-bonded magnesia refractories. J. Am. Ceram. Soc. 1984, *67* (7), 480–483.

9.3. Siljan, O.-J.; Rian, G.; Pettersen, D.T.; Solheim, A.; Schøning, C. Refractories for molten aluminum contact Part I: thermodynamics and kinetics. Refract. Appl. News 2002, *7* (6), 17–25.

9.4. Meadowcroft, D.B.; Cox, W.M. Dewpoint corrosion: mechanisms and solutions. In *Dewpoint Corrosion*; Holmes, D.R., Ed.; Ellis Horwood Ltd.: Chichester, UK, 1985. Chapter 2.

9.5. Harris, L.A.; Cross, D.R.; Gerstner, M.E. Corrosion suppression on rutile anodes by high energy redox reactions. J. Electrochem. Soc. 1977, *124* (6), 839–844.

9.6. Amoroso, G.G.; Fassina, V. *Stone Decay and Conservation*; Elsevier: Amsterdam, 1983; 453 pp.

9.7. Lange, F.F.; Davis, B.I.; Metcalf, M.G. Strengthening of polyphase Si_3N_4 materials through oxidation. J. Mater. Sci. 1983, *18* (5), 1497–1505.

9.8. Davies, G.B.; Holmes, T.M.; Gregory, O.J. Hot corrosion behavior of coated covalent ceramics. Adv. Ceram. Mater. 1988, *3* (6), 542–547.

9.9. Gogotsi, Yu.G.; Lavrenko, V.A. Corrosion protection and development of corrosion-resistant ceramics. *Corrosion of High-Performance Ceramics*; Springer-Verlag: Berlin, 1992; 151–162. Chp. 7.

9.10. Gregory, O.J.; Richman, M.H. Thermal oxidation of sputter-coated reaction-bonded silicon nitride. J. Am. Ceram. Soc. 1984, *67* (5), 335–340.

9.11. Wittmer, D.E.; Temuri, M.Z. Thermochemical studies in selected metal–carbon–oxygen systems. J. Am. Ceram. Soc. 1991, *74* (5), 973–982.

9.12. Strife, J.R. Fundamentals of protective coating strategies for carbon–carbon composites. In *Damage and Oxidation Protection in High Temperature Composites*; Haritos, G.K., Ochoa, O.O., Eds.; ASME: New York, 1991; Vol. 1, 121–127.

9.13. Frischat, G.H.; Sebastian, K. Leach resistance of nitrogen-containing $Na_2O–CaO–SiO_2$ glasses. J. Am. Ceram. Soc. 1985, *68* (11), C305–C307.

9.14. Lehman, R.L.; Greenhut, V.A. Surface crystal formation during acid corrosion of phosphate-doped lead silicate glass. J. Am. Ceram. Soc. 1982, *65* (9), 410–414.

9.15. Harvey, K.B.; Litke, C.D. Model for leaching behavior of aluminosilicate glasses developed as matrices for immobilizing high-level wastes. J. Am. Ceram. Soc. 1984, *67* (8), 553–556.

9.16. Buckwalter, C.Q.; Pederson, L.R. Inhibition of nuclear waste glass leaching by chemisorption. J. Am. Ceram. Soc. 1982, *65* (9), 431–436.

Glossary

Cor·rode, v.t. to eat into or wear away gradually, as by rusting or by the action of chemicals.

WEBSTER'S NEW WORLD DICTIONARY

Alteration The change or modification of a material through interaction with its environment, generally by the formation of a new phase. This reaction need not be deleterious.

Atmospheric Corrosion The degradation of materials by natural atmospheric environments. *Atmospheric corrosion* is a term often used by the metallurgist, whereas an equivalent term used by ceramists is *weathering*.

Biodurability The ability of a ceramic to withstand the action of a biological environment.

Condensation Corrosion Equivalent to *dew point corrosion.*

Corrosion The chemical interaction of a ceramic with its environment, generally producing a deleterious effect. This chemical reaction can, in some cases, be put to beneficial use.

Dealkalization The corrosion of a ceramic through the selective solution of the alkalies into the corroding medium. Generally used to describe the removal of alkalies from glasses.

Degradation A general decrease or lowering of the quality of a ceramic; often through corrosive action.

Dewpoint Corrosion The deterioration of a solid ceramic material caused by the condensation of a corrosive liquid from a saturated gas when the temperature is lowered below the point (the dew point) where the liquid will condense. A form of *atmospheric corrosion* and equivalent to *condensation corrosion*.

Dissolution Corrosion The corrosion of a ceramic through the solution of its various components into the corroding medium (generally a liquid).

Dissolution: Congruent, direct, or homogeneous When the total ceramic chemistry dissolves simultaneously into the environment.

Dissolution: Incongruent, Indirect, or Heterogeneous When the ceramic dissolves in such a way as to leave behind a material chemically different than the original ceramic as an interface between the ceramic and the corroding medium. These terms generally imply that the dissolution is selective.

Dissolution: Selective The corrosion of a ceramic through the selective solution of one or more (but not all) species into the corroding medium.

Dissolution: Anisotropic When the ceramic material dissolves at different rates parallel to different crystallographic or geometric directions.

Dissolution: Isotropic When the ceramic material dissolves at the same rate for all crystallographic or geometric directions.

Durability The ability of a ceramic to withstand the action of its environment.

Electrochemical Corrosion The corrosion that takes place when the reaction occurring involves electronic charge transfer. Generally this type occurs when ceramics are in contact with aqueous media, but may also occur in other media.

Galvanic Corrosion The corrosion that takes place when two chemically dissimilar ceramics are in contact with one another, both of which are in contact with the same electrolyte. Reaction occurs only when current flows in an external circuit. A type of electrochemical corrosion.

Hot corrosion Normally used to designate high temperature oxidation of ceramics in contact with molten salt deposits. This definition should probably not be used, since the term *Hot Corrosion* is nonspecific and could apply to any type of corrosion at an elevated temperature.

Intergranular or Grain Boundary Corrosion The corrosion through any mechanism that takes place preferentially along grain boundaries or between grains.

Leaching To remove through dissolution a portion of a ceramic material.

Leaching: Selective Removes one species in preference to another. The use of the word *selective* in this case is superfluous.

Oxidation Embrittlement The embrittlement of a composite caused by the oxidation of the interface between the reinforcement material and the matrix resulting in a strong bond between the two.

Pesting The formation of a powder-like deposit on the exposed surface of metallic silicides (i.e., $MoSi_2$) during oxidation.

Photoelectrochemical Corrosion Electrochemical corrosion that takes place when the charge transfer involves the positive holes formed by photon illumination. Also called *photodissolution*.

Stress Corrosion Corrosion by any mechanism that is enhanced by the presence of either a residual or applied stress.

Thermo-Oxidative Stability The resistance to oxidation at elevated temperatures. Generally used in the discussion of composite materials.

Weathering This term describes the atmospheric effects upon materials of the construction industry, mostly structural clay products, sandstone, limestone, marble, and glass; and is essentially the attack by water vapor, CO_2, and SO_2.

Epilog

The literature and data available on the corrosion of ceramics indicate that corrosion occurs by either one of several possible mechanisms or a combination of these mechanisms. Many similarities exist between the corrosion of crystalline and glassy ceramics, although in general glass corrodes more rapidly under identical environmental conditions.

Corrosion in either crystalline or glassy ceramics can occur by a direct process where the ceramic congruently dissolves into the corroding medium. Reaction rates are generally linear, being proportional to the duration of the test. One way to minimize this type is to saturate the corroding medium with the same chemical species that are dissolving from the ceramic. Another way to minimize this type of corrosion is to add something to the ceramic that will diffuse to the surface and react with the corroding medium forming a protective interface layer.

In another type, apparently the more common type of corrosion process, indirect, in either crystalline or glassy ceramics, species from both the corroding medium and the ceramic counterdiffuse and react at the interface forming a glassy, a crystalline, or a gaseous interface reaction product. If the interface reaction product is solid, continued corrosion can occur only by continued diffusion through the interface. In some cases, the interface reaction product may be multilayered. The reaction layer thickness may vary from a few nanometers to several hundred micrometers. Reaction rates are generally parabolic, being proportional to the square root of time. One way to minimize this type of corrosion is to prereact the ceramic to form an initial interface reaction layer that, if protective, will slow continued reaction. Another way is to add something to the ceramic that will form a layer through which diffusion will be more difficult.

In the first case discussed above, the corroding medium can be either a liquid or a gas; however, in gaseous corrosion one may not consider the *dissolution* to be congruent if the products are two different gases, as in the active oxidation of SiC to SiO and CO_2. In the second case above, the medium can again be either liquid or gaseous with either all or part of the ceramic forming the layer. In most cases, only part of the ceramic forms the layer (i.e., selective dissolution). In corrosion by liquids, the mechanisms are different if the corroding medium is a glass/slag versus water. In water, the first step is usually ion exchange, whereas in glass/slag attack, the first step is counter-diffusion, not quite the same as ion exchange, although ion exchange may take place in glass/slag attack.

Multicomponent ceramics generally corrode by a mixed mechanism with each step exhibiting a different and unique reaction rate. In these cases, the overall reaction rate will exhibit a mixed rate law, being neither linear nor parabolic.

Extended duration tests have indicated that the mechanism of corrosion may change after some extended time. This is especially true for oxide layers formed on nonoxide ceramics during gaseous corrosion. This change in mechanism is due to one or more of the following changes: crystallization of amorphous layers, alteration of crystalline phases as diffusion continues, cracking due to crystallization and alteration, and spalling. The few studies that have shown these changes indicate that one must be careful in making life-time predictions based upon data from short-time laboratory tests.

In all cases, an increase in temperature increases the rate of corrosion. The mechanism of corrosion, however, may change as temperature is increased due to crystallization of amorphous reaction layers, polymorphic transitions, melting of crystalline layers, vaporization of various species in the layer, cracking, etc.

One method of minimizing corrosion that requires more emphasis appears to be the various coating methods. These could be used to advantage in composites where the initial step is, for example, oxidation of SiC fibers. By coating the fibers before incorporation into the matrix, oxidation may be slowed or even eliminated. The object is to find a material through which the diffusion of oxygen is a minimum and then use this material to coat the fibers. The technique of electrostatic attraction in an aqueous dispersion appears attractive as a coating method for materials such as fibers.

Although the above discussion may be an oversimplification of the corrosion processes that occur in ceramics, it is a step in the direction of simplifying and unifying the whole area. All of the data and discussion about corrosion point towards the need for more in-depth diffusion and solubility studies of the various species in the different corroding media encountered in practice.

Corrosion, being an interfacial process, requires a thorough understanding of the surface structure of the materials being corroded. Thus the study of single crystals is the best method to determine the fundamentals of corrosion mechanisms. Although the crystal surface characteristics determine short-term corrosion behavior, they may not be as important for long-term corrosion. Single crystals do lend themselves to the evaluation of the effects that various dopants have upon leaching kinetics. In addition, various types of defects (e.g., vacancies, dislocations, etc.) could be incorporated into the lattice during production of the single crystals.

A large amount of published data on the corrosion of crystalline and glassy ceramics points toward the fact that more compact structures are more durable. In the study of glasses, references are made to corrosion being a function of glass structures, which are related to parameters such as composition, the number of nonbridging oxygens, the amount of cross-linking of the network structure, the degree of network packing, the density, the strength of the bonding, and the amount of covalent bonding. References have also been made to compact, strongly bonded glass structures being those with low thermal expansion and high softening points. Thus a technique that would determine the structural tightness may be sufficient to rank the durability of various materials, at least in the various compositional classes and to a specific environment. In addition to thermal expansion and softening point determinations, the determination of hardness may also yield information related to durability. Hardness is a measurement, however, that must be performed with some care, since hardness varies with the applied load and cracking and friction may interfere with the measurements. There has been no systematic study reported in the literature of the corrosion of ceramics related to properties such as expansion, hardness, or softening point.

Only through a thorough understanding of all the parameters involved can the engineer make an intelligent selection of the material that will best resist corrosion for a particular application. Only through intelligent materials selection can the cost of corrosion be minimized. Since the application of ceramics requires the optimization of properties other than corrosion resistance, a compromise among corrosion resistance, properties, and cost is generally needed.

Index

Acronyms
  of composites, 295
  of standards organizations, 150
Ab initio calculations, 31
Acid/base effects, 77, 78
  Bronsted & Lowry theory, 77
  Lewis theory, 77
Activation energy, 93, 98
  barrier, 93, 94
Activated process, 94
Activity
  hydrogen ion, 50
  oxygen ion, 26, 52
  silica, 84
  sodium oxide, 23, 194
  water, 33, 342
Al_2O_3 (*see* Alumina)
Al_2O_3-Cr_2O_3, 166
AlN, 201
Albite, 32
Alteration, 389
Alumina, 15, 23, 24, 166–170, 178, 191, 193, 201–206, 234, 336, 338
Amphibole, 31
Amphoteric, 77
Anatase, 187
Anorthite, 15
Atmospheric corrosion, 389
AZS refractory, 21, 166–170, 375

Beta-alumina, 199
BeO, 234
Be_2SiO_4, 193
Bioactive materials, 37, 277, 278, 358, 359
Bioceramics, 36–38

Biodurability, 390
Bioglass™, 54
Borides
  HfB_2, 231
  NbB_2, 231
  SiB_6, 233
  TaB_2, 231
  TiB_2, 191, 231, 232
  ZrB_2, 231
Boundary layer
  caused by liquid attack, 166
  caused by vapor attack, 201
  double layer on Si_3N_4, 214
  thickness, 15, 57
  types, of glass, 50
Brookite, 187

CaO, 166
$Ca(OH)_2$, 45
Carbon, 153
Carbides
  B_4C, 201, 231
  HfC, 231
  SiC (*see* Silicon carbide)
  TiC, 201, 231
  VC, 189
  WC, 230
  ZrC, 230, 231
Catalyst
  autocatalyst, 95
  heterogeneous, 95
  homogeneous, 95
Cavitation, 338
Cement, 149, 153, 182–185
Characterization methods
  chemical analysis, 134, 135
  data reduction, 139
  density, 136, 137
  gravimetry, 136
  microstructure, 130–134
  MOR, 138
  optical microscopy, 131, 132
  porosity, 137, 138
  sample preparation, 126–129
[Characterization methods]
  sample selection, 126–129
  SEM/EDS, 133, 134
  surface analysis, 135
  surface area, 137, 138
  TEM, 134
  test conditions, 129, 130
  visual observation, 130, 131
  XRD, 132, 133
Chrome refractory, 205
Chrysotile, 71, 181
Clay, 182
Composites
  acronyms, 295
  Al_2O_3/Ni, 318
  $Al_{18}B_4O_{33}$/Al, 316
  carbon/carbon, 313–315
  cermet, 293
  C/glass, 303
  debonding, 293
  desirable properties, 292
  Dimox™, 294
  graphite/Al alloy, 316
  graphite/BMI, 319
  graphite/epoxy, 318
  graphite/Mg alloy, 317
  $HfB_2/MoSi_2$, 318
  HfC/hafnia, 309
  hydroxyapatite/thermoplastic, 320
  interphase, 293, 298, 299
  matrix
    ceramic, 300–316
    metal, 316–318
    polymer, 318–320
  mica/polyester, 320
  $MoSi_2$/alumina, 302
  $MoSi_2$/mullite, 308
  nanocomposite, 293
  Ni/alumina, 303
  $NiTiO_3$/titania, 309
  Reinforcement
    fibers
      boron, 296
      carbon, 296

[Composites]
glass, 296
graphite, 296, 303
Nicalon™, 297, 298, 301
SiC, 297
Sylramic™, 311, 312
TiN, 298, 302
particulates
Al_2O_3, 300
BN, 300
Co, 300
SiC, 300
TiB_2, 300
TiC, 300
ZrO_2, 300
SiC/Al, 317
SiC/alumina, 301, 312
SiC/glass, 303
SiC/MgAlSilicate, 305
$SiC/MoSi_2$, 317, 318
SiC/mullite, 301, 306
SiC/SiC, 310, 311
SiC/spinel, 306
SiC/Ti, 317
SiC/TiB_2, 315
SiC/Ti_3SiC_2, 315
$TiB_2/MoSi_2$, 318
SiC/zirconia, 306
TiC/alumina, 302
TiC/zirconia, 309
TiN/alumina, 301
$ZrB_2/MoSi_2$, 318
ZrC/zirconia, 309
ZrO_2/Si_3N_4, 309
Concrete, 182
Contact angle, 73
Cordierite, 34
Corrosion (*see also* Dissolution)
acid/base effects of, 77, 78
as activated process, 19
analysis of, 123–139
atmospheric, 43, 44, 389
beneficial, 2, 24
condensation, 23, 41, 44, 64, 390
[Corrosion (*see also* Dissolution)]
cost of, 4
definitions, 389–392
dew point, 44–47, 390
electrochemical, 25–28, 391
by gas (*see* Environments)
flux-line, 18, 19
galvanic, 6, 25, 44, 317, 319, 320, 391
general theory of, 10
glass-line, 18
grain boundary, 6, 391
hot corrosion, 23, 391
intergranular, 391
leaching, selective, 32, 55, 187, 273, 274, 391
by liquids
aqueous solutions, 30–47, 82, 177–191, 257–281, 312, 313, 352, 358–364
molten glass, 15–23, 155, 156, 165–177
molten metals
Al, 29, 199, 200
Al/Mg, 29, 200
Fe, 198
Li, 358
molten salts (*see* Environments)
mechanisms, 9, 10, 12, 18–67, 267–280, 337–345
metal-line, 18
pH dependence of, 33, 35, 42, 48, 51, 180, 265, 275, 340
and polymorphism, 47
and properties, 333–364
rate of
by gases, 59, 61, 62
by liquids, 15–17, 23, 30
aqueous, 264
free convection of glass, 16, 17
forced convection of glass, 17
of spheres, 62
by sublimation, 63
by vaporization, 63
by weathering, 38–44, 64

[Corrosion (*see also* Dissolution)]
reactivity, 373
by solids, 65–67
stress corrosion, 6, 334–346, 338, 340, 344, 347, 352, 358, 359, 364
study of, 1, 2
surface charge, effects of, 67, 77
surface energy, effects of, 71–76
temperature dependence of, 19
thermal gradient, effects of, 19, 20
upward drilling, 20, 21
Crack growth, 334–346, 338, 340–344, 348, 352, 363
Crack growth, environmentally enhanced, 363
Crack initiation, 361
Crack propagation (*see* Crack growth)
Crack tip blunting, 338, 339, 348–350, 363, 364
Crack velocity (*see* Crack growth)
Cristobalite, 179
CuO, 192

Dealkalization, 48, 262, 390
Delayed failure, 335
Dew points
of industrial gases, 46
of Na_2SO_4, 23
of sulfuric acid, 46
Differential scanning calorimetry, DSC, 99
Differential thermal analysis, DTA, 99
Diffusion
bulk, 66
chemical, 65
coefficients of, 65, 101–103
Fick's 1st law, 100
Fick's 2nd law, 101
grain boundary, 66
inter-, 65
mechanism of, 104
rate of, 101
self, 65
[Diffusion]
stress assisted, 105
stress enhanced, 338
surface, 66
thermal, 67
DLCM, digital library of ceramic microstructures, 3
Dihedral angle
and composition, 75, 76
critical, 74
Dissolution (*see also* Corrosion)
anisotropic, 13, 391
chemical, 36
congruent, 12, 390
direct, 12, 390
electrochemical, 36
heterogeneous, 12, 390
homogeneous, 12, 390
incongruent, 12, 390
indirect, 12, 390
isotropic, 13, 391
kinetics of, 54
matrix, 48, 50, 258, 266
pH, dependence of, 33, 262, 265
selective, 12, 390
Dolomite, 21, 22, 154
Downward drilling, 30

Erosion, 6
Electrical double-layer, 27, 35
Ellingham diagram, 28, 85–88
Environments
acetonitrile, 267, 364
acids
acetic, 158–160, 179, 273, 276
aqua-regia, 190
boric, 14
citric, 152, 157, 273, 275
colas, 275
fulvic, 33
humic, 33
hydrochloric, 156, 157, 179, 185, 187, 190, 274, 276
hydrofluoric, 187, 190, 260

[Environments]
nitric, 179, 273, 267
oxalic, 179, 273
phosphoric, 185, 260, 273
sulfuric, 179, 185, 188, 189
tartaric, 273
alcohol, 267, 342
alkaline solution, 161, 179
ammonia, 261, 342
aniline, 343, 344
aqueous solutions, 30–47, 177–191, 267, 277, 312, 352, 358–364
with CO_2, 183
lime, 149
magnesium sulfate, 160
milk, 275
physiological, 274
potassium carbonate, 153, 179
sodium borate, 179
sodium carbonate, 154, 179, 261
sodium hydroxide, 156, 179, 260
sodium pyrophosphate, 157
sodium sulfate, 189, 193, 195
sodium sulfide, 194
trisodium phosphate, 156
bases
K_2CO_3, 185
KOH, 179, 185, 260
LiOH, 185, 189
Na_2CO_3, 179
NaOH, 179, 185, 190, 260
NH_4OH, 260
carbon (*see also* Graphite), 234, 235
formamide, 342, 267
gases
alkali vapors, 57, 64, 160, 201, 207, 228
atmospheric, 13, 42, 47, 64, 183–185, 206, 209, 234
Cl vapors, 230
CO, 152, 232
CO_2, 64, 232, 357
CO_2/O_2, 228

[Environments]
H_2, 202, 218, 227
H_2/H_2O, 219
$H_2/H_2O/Ar$, 228
$H_2/H_2O/CH_4$, 218
H_2/H_2S, 219, 227
$H_2/H_2S/H_2O$, 218
K vapors, 204, 228
mercury, 64
N_2/SO_2, 227
$N_2/SO_2/CH_4$, 228
nitrogen, 227
oxidation, 58–60, 209–218, 223–233
reducing, 29, 59, 60, 64, 195, 199, 237
SO_2, 261, 357
SO_3, 192, 229
steam, 153–155
UF_6, 205, 209
vacuum, 63
vanadium vapors, 207
water vapor, 61, 153–155, 219, 228, 312, 339–345
glasses, molten, 15–23, 155, 165–177
borosilicate, 173
silicate
Al-Ca, 166
Al-Ca-Mg, 166
Ca-Fe, 166
Ca-Mg, 166
Ca-Na, 166
coal slag, 166
Li-Al, 173
K-Pb, 166, 170
oxynitride, 175
silica, 166
glycine, 273
graphite (*see also* Carbon), 234
halogens, 260
heptane, 364
hydrazinc, 267, 342
metals, 236, 237
metals, molten, 197–201

[Environments]
Al, 29, 199
Al-Mg, 29, 200
Fe, 199
Li, 274
NiTaC, 192, 200
methanol, 342
n-propylamine, 343, 344
oils, 261
salts, molten
carbonates, 192, 194
chlorides, 197
cryolite, 191, 193
fluorides, 193, 196
K_2CO_3, 192
Li_3AlF_6, 24
molybdates, 193
NaCl, 197, 355
Na_2SO_4, 23, 191–195, 355
vanadates, 193
seawater, 260
slag, 158, 159, 166
tert-butylamine, 343, 344
water, 30–47, 260, 267, 268, 277, 312, 344, 361
xylene, 153
Embrittlement, 298, 305, 308, 311, 338
Equations
Arrhenius, 19
Avrami-Erofe'ev, 97
Berthoud, 15
Cooper-Kingery, 16
crack velocity, 336, 341, 342
Fick's, 100, 101
Hertz-Langmuir, 61
Hogensen & Healy, 51
Hrma, 17
Jander, 97
kinetic, 97
Nerst, 82
Noyes-Nerst, 15
Washburn, 68
Eucryptite, 378
Fatigue
cyclic, 335
dynamic, 335, 346
knee, 361, 362
static, 335
Field trial, 124
Forsterite, 32
Freudenbergite, 187
Free energy
Gibbs, 79
Helmholtz, 79
of hydration, 49
of reaction, 59, 80–82

Glass
bioactive, 53–56, 277, 359
borosilicate, 259, 271–275
chalcogenide, 280, 281
fiber, 53, 296, 318
fluoride, 278–280
fluorozirconate, 279
halide, 280, 281
lead, 259, 275, 276
nuclear waste, 49, 271
optical, 336, 360, 361
oxynitride, 48, 268
phosphorus, 276–278
silicate
lime, 265
lithium, 267
potassium, 264
potassium-lead, 259
REAS, 54, 269
soda, 265, 267
soda-lime, 64, 265, 269
YAS, 55, 269
ZBL, 279, 280

Heat
capacity, 81
of reaction, 81
of transport, 104
Hollandite, 186, 187

Hydrolysis, 32, 48, 190, 258, 266, 279
Hydrolysis, rate of, 52

Ion-exchange, 32, 49, 258, 263, 265, 268, 270, 274, 277
Isoelectric point, 35, 36

Kinetic
  decomposition, 60
  equations, 97
Kinetics, 91–100

La_2O_3, 24, 193
$LaAlO_3$, 193
$LaCoO_3$, 192
$LaNiO_3$, 192
Landauite, 187
Leaching, selective, 32, 391
Lead release, 275, 276
$LiFeO_2$, 192
Limestone, 39, 41, 43, 182
Loveringite, 187

Mag-chrome refractory, 198
Marble, 39, 41, 43, 183
Marangoni effect, 18
Mechanism
  of bond rupture, 345
  of corrosion, 48, 394
mercury intrusion, 68, 69
$MgAl_2O_4$, 166, 193
MgO, 24, 166, 235, 300
MgO-containing refractory, 198, 207
Minimization of corrosion, 373–387
  by chemical means, 375–379
  by coating, 384–386
  by forced cooling, 379, 380
  by preoxidation, 383
  through processing, 383
  by property optimization, 374–379, 382–385
Molecular dynamics, 30
MoO_2, 188
MoO_3, 188
Mo_4O_{11}, 188
Mo_8O_{23}, 188
Mo_9O_{26}, 188
$MoSi_2$, 233, 302, 308, 317, 318
Mullite, 13, 15, 59, 80, 81, 84, 166, 170–174, 191, 200, 235, 298, 299, 301, 302, 306

NiO, 87, 188, 192
Nitrides
  AlN, 189, 190, 200, 201, 222, 223
  $NiTiO_3$, 309
  Si_3N_4, 175, 177, 193–197, 201, 209–221
  TaN, 223
  TiN, 298, 301, 302

Olivine, 31
Opal, 179
Oxidation
  active, 210, 211, 223, 226
  active-to-passive transition, 210, 211
  embrittlement, 300
  passive, 210, 211, 223, 226
  rate of, 217, 225, 227
  of SiC, 59, 223–230, 345–348
  of Si_3N_4, 59, 209–218, 345–351
Oxynitrides, 351

Perovskites, 187
Pesting, 233, 312
pH, at crack tip, 342
Phase diagram, 85, 86
Phase stability diagram, 86, 90
Photoelectrochemical
  aging, 188
  corrosion, 392
Porosity, effects of, 68–71
Potential
  bias, 26
  oxygen, 26
  redox, 50, 81
Pourbaix diagram, 91, 92
Pressure, of dissociation, 83

Pseudobrookite, 187
PSZ, 47, 353, 354
Pyroxene, 31

Quartz, 32, 39, 48, 179, 180

Reaction
  order, 95, 98
  rate, 93–100
RuO_2, 188
Rutile, 187

Sandstone, 39
SiAlON, 201, 209, 221
SiC (*see* Silicon carbide)
Si-C-N-O, 303
Si_3N_4, (*see* Silicon nitride)
Silica, 15, 23, 29, 32, 49, 50, 84, 103, 178–182, 199, 234
Silicates, 32–36, 48, 178–182, 201, 206
Silicon carbide, 175–177, 189–190, 193–196, 200, 201, 223–230, 297, 300, 303–306, 310–313, 345–348, 352–355
Silicon nitride, 175, 177, 193–197, 201, 209–221, 309, 310, 359, 345–350, 354–357, 382, 383
Sorét effect, 67, 104
Spinel, 170, 198, 201, 205, 306
Spodumene, 378
Stability fields for SiO_2, Si_2N_2O, and Si_3N_4, 213
Stress corrosion, 47, 334, 392
Stress corrosion cracking, 300, 334
Stress corrosion susceptibility
  parameter, 336, 347
Subcritical crack growth, 338
Superconductors, 186, 197, 234–236
Surface area
  exposed to corrosion, 68–71, 137
  to volume ratio, 49, 126, 127
Surface compressive stress, 350
Surface energy, 71–76
Surface tension, 71–76
Surface types of leached glass, 50
SYNROC, 186

Tar-bonded refractories, 377
Tar-impregnated refractories, 377
Test procedures, 143–163
  ASTM test methods, 146–149
  ASTM standard practices, 146–149
  MCC methods, 145
  nonstandard, 162
Thermodynamics, 78–91
  graphical representation, 85–91
  mathematical representation, 79–85
Thermal transpiration, 70
Thermogravimetry, TG, 98
ThO_2, 24, 193, 189
TiC, 201, 230, 231, 300, 302, 309
TiO_2, 186, 187
Titanates, 186–188
Toxicity, 180
TZP, 208

Upward drilling, 21, 22
UO_2, 189

Vaporization
  direct, 63
  selective, 63
VC, 189
Volatility
  diagrams, 86, 89
  maps, 90
Vycor™, 2, 272

Water adsorption, 30
Weathering
  of Cr-containing refractories, 38
  by fog, 42
  of limestone, 39, 41, 43
  of marble, 39, 41, 43
  by rainwater, 41
  of sandstone, 39
  of zirconia, 47
Wetting, 13, 74

WO_2, 188
WO_3, 188

$Y_3Al_5O_{12}$, 166, 193
Y_2BaCuO_5, 186, 197, 234
Y_2O_3, 24, 193
YSZ, 185

Zero point of charge, 35, 77
Zircon, 166, 173, 174, 193, 201
Zirconia, 174, 175, 185, 208, 300, 306, 309, 353, 354, 357
Zirconolite, 187
Zirkelite, 187
Zoning, 374
ZnO, 192
ZrO_2, (*see* Zircon)
  partially stabilized, 47, 353, 354
  yttria stabilized, 185
$ZrSiO_4$, (*see* Zircon)